HANDBOOK OF Decision Analysis

Wiley Handbooks in

OPERATIONS RESEARCH AND MANAGEMENT SCIENCE

The *Wiley Handbooks in Operations Research and Management Science* is a series of books that present both established techniques and cutting-edge developments in the field of operations research and management science. The goal of each handbook is to supply a practical, one-stop reference that treats the theory, formulae, and applications that, together, make up the cornerstones of a particular topic in the field. A self-contained presentation allows each volume to serve as a quick reference on ideas and methods for practitioners, while providing an accessible introduction to key concepts for students. The result is a high-quality, comprehensive collection that serves as a mainstay for novices and professionals alike.

Kouvelis, Dong, Boyalatı, and Li · *Handbook of Integrated Risk Management in Global Supply Chains*

Parnell, Bresnick, Tani, and Johnson · *Handbook of Decision Analysis*

Sokolowski and Banks · *Handbook of Real-World Applications of Modeling and Simulation*

Handbook of Decision Analysis

Gregory S. Parnell
United States Military Academy
West Point, NY

Terry A. Bresnick
Innovative Decisions, Inc.
Vienna, VA

Steven N. Tani
Strategic Decisions Group
Palo Alto, CA

Eric R. Johnson
Bristol-Myers Squibb
Skillman, NJ

WILEY
A John Wiley & Sons, Inc., Publication

Published by John Wiley & Sons, Inc., Hoboken, New Jersey.
Published simultaneously in Canada.

For general information on our other products and services or for technical support, please contact our Customer Care Department within the United States at (800) 762-2974, outside the United States at (317) 572-3993 or fax (317) 572-4002.

Wiley also publishes its books in a variety of electronic formats. Some content that appears in print may not be available in electronic formats. For more information about Wiley products, visit our web site at www.wiley.com.

Library of Congress Cataloging-in-Publication Data:
Handbook of decision analysis / Gregory S. Parnell ... [et al.].
p. cm.
Includes bibliographical references and index.
ISBN 978-1-118-17313-8 (cloth : alk. paper) 1. Decision making. I. Parnell, Gregory S.
HD30.23.H3542 2012
658.4'03–dc23

2012032884

andbooks in

ATIONS RESEARCH AND MANAGEMENT SCIENCE

Handbooks in Operations Research and Management Science is a series that present both established techniques and cutting-edge developments in f operations research and management science. The goal of each is to supply a practical, one-stop reference that treats the theory, and applications that, together, make up the cornerstones of a particular e field. A self-contained presentation allows each volume to serve as a erence on ideas and methods for practitioners, while providing an introduction to key concepts for students. The result is a high-quality, nsive collection that serves as a mainstay for novices and professionals

Dong, Boyalatı, and Li · *Handbook of Integrated Risk Management in Supply Chains*

resnick, Tani, and Johnson · *Handbook of Decision Analysis*

ki and Banks · *Handbook of Real-World Applications of Modeling and tion*

Contents

List of Figures

List of Tables

Foreword

This handbook represents a significant advance for decision professionals. Written for practitioners by practitioners who respect the theoretical foundations of decision analysis, it provides a useful map of the tools and capabilities of effective practitioners. I anticipate that this and future editions will become the primary repository of the body of knowledge for practicing decision professionals.

This Handbook Is Timely

The practice of decision analysis (DA) is at a major inflection point. That high-quality decisions can generate immense value is being demonstrated again and again. Leaders of organizations are increasingly aware of how opportunities are lost by making "satisficing" decisions—that is, decisions that are "good enough." The benefit-to-cost ratio of investing in better decisions is frequently a thousand to one. I know of no better opportunity for value creation anywhere. As Frank Koch,[1] president of the Society of Decision Professionals (SDP), has said, "Benefit to cost ratios . . . are immense simply because the added cost of doing DA is negligible. We would still be paying the analysts and decision makers without DA; they would simply be talking about different things. The incremental cost of having a better, more relevant conversation is zero, so regardless of the benefit, the ratio is infinite! Even if I throw in the cost of training and learning some software, that's measured in thousands and the benefits are clearly measured in millions."

Why is this huge opportunity still a secret from most decision makers? It is because we humans are wired to believe that we are making good decisions even when we leave value on the table. We are wired to be satisfied with good enough. We shape our memories with hindsight and rationalization. The burgeoning set of literature from the behavioral decision sciences documents many of our biases

[1]Frank Koch in a written response to the question: What is the ROI of investing in DA based on your experience at Chevron? Frank Koch retired in 2010 after the Chevron team had been awarded the best practice award for *20 Years of DA at Chevron.*

and draws attention to the gap between true decision quality (DQ) (see Chapter 5) and our natural decision-making tendencies.

Our individual cognitive biases are amplified by social behavior, like groupthink. We assume that advocacy decision processes in use by most organizations produce good decisions, yet they are designed to suppress good alternatives. We assume that agreement is the same as DQ, yet we see a lot of agreement around nonsense. It is not uncommon to hear statements like, "I can't believe it—what were we thinking?"

If DQ can create immense additional value in specific decisions, can we develop DQ as an organizational competence? The answer is yes, and Chevron has shown the way. Over the period in which it has implemented a deep and broad adoption of DQ, Chevron has outperformed its peer group of major oil companies in creating shareholder value. While many organizations have pockets of organizational decision quality (ODQ), to my knowledge, Chevron has the broadest and deepest adoption to date. And by "adoption," I don't just mean better analytics. All the major oil companies have the analytics to deal address uncertainties and risk. The difference is that the whole Chevron organization seems to be in passionate and collaborative pursuit of value creation based on quality decisions linked with effective execution. I believe that Chevron's success is the beginning of a big wave of broad adoption of organizational DQ.[2]

The immense value left behind by our satisficing behaviors represents the biggest opportunity for our business and societal institutions in the coming decades. If we begin to think of these opportunity losses as an available resource, we will want to mine this immense value potential. The courts—led by the Delaware Supreme Court—are raising the bar in their interpretation of a board director's duty of "good faith." In the coming years, board and top management's best defense is their documented practice of DQ.

Decision Professionals: The Practitioner Perspective

The Society of Decision Professionals[3] states that the mission of decision professionals is to:

- Bring DQ to important and complex decisions.
- Provide quality insights and direction through use of practical tools and robust methodologies.
- Promote high professional standards and integrity in all work done by decision professionals.
- Advance the profession to the benefit of mankind through helping decision makers.

[2]See the SDG white paper, *Chevron Overcomes the Biggest Bias of All* (Carl Spetzler, 2011). Available from SDG website, http://www.sdg.com.

[3]See: http://www.decisionprofessionals.com

The role of a decision professional as a practitioner of DA and facilitator of organizational alignment is gaining acceptance. Dozens of organizations have established internal groups of professionals, designed career ladders, and developed specific competency requirements. The recently formed SDP has created a certification process and career ladder that specifies increasing competency levels for practitioners.

While there are important similarities between becoming a successful practitioner and becoming a tenured academic, there are also major differences. The decision professional is motivated by bringing clarity to complex decision situations and creating value potential in support of decision makers. He or she is less interested in the specialization required for peer-reviewed publication. Instead, the practitioner wants to acquire practical tools and relevant skills that include both analytical and facilitation skills (project management, conflict resolution, and other so-called "soft skills").

The ability to address both organizational and analytical complexity (see Figure F.1) are of great importance to the practitioner. As I like to say, "If you can only deal with the analytical complexity, you can get the right answer—but nobody cares. If you can only facilitate in the face of organizational complexity, you can resolve conflicts and gain agreement—but it can be agreement around nonsense." To bring full value, we need to deliver the combination—agreement around the best answer, the answer that generates the greatest value potential.

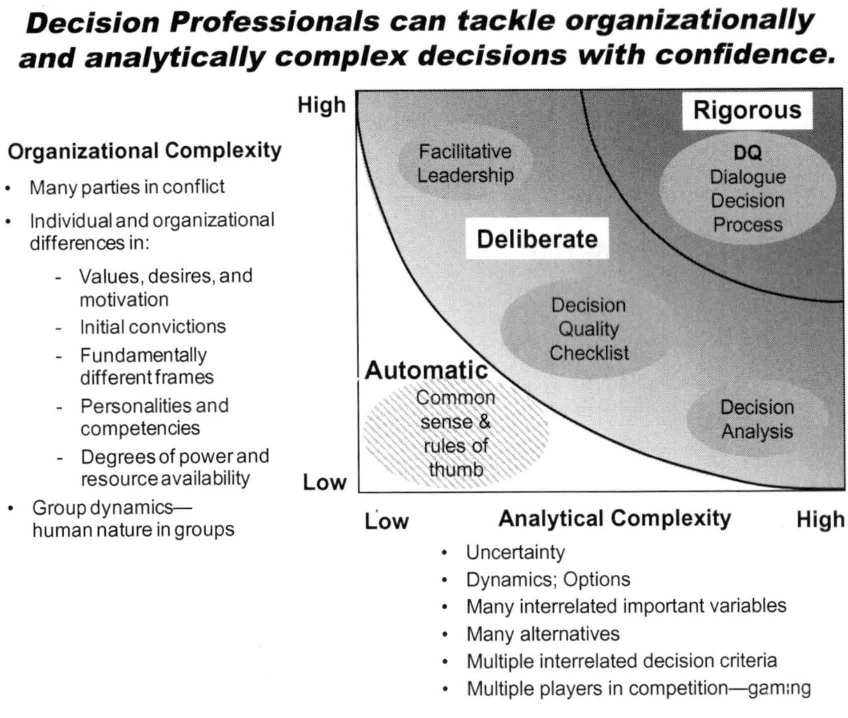

FIGURE F.1 Two dimensions of competence.

Individual decision professionals can deliver this if we have the competency in both areas. However, many practitioners are significantly better in one or the other—either strong analytical capabilities or strong social/emotional intelligence and facilitation skills. Therefore, many practitioners find it best to team up with others to deliver the full value of DQ. To make such teaming effective, there must be mutual respect for the other competency and a recognition that value creation from the combination is the goal. It bears repeating: We need to gain alignment around the best answer—the answer that creates the greatest value potential.

As practitioners we are always approximating and simplifying. We are practical decision engineers and decision facilitators who want robust solutions that are effective in creating a lot of potential value. We are organizational facilitators who are not satisfied unless the best decision is owned by the decision makers and implementers. Incisiveness with tools that produce insight and processes that foster effective engagement are more important to us than another refinement to the axioms of normative decision theory. In my experience, the academic debates at the edges of decision science over the last two decades have contributed surprisingly little to the practice. Seldom is the primary challenge in solving real decision problems a matter of advanced theory.

Our goal should be to make our concepts and methods as simple and accessible as possible. As I am writing this, I am participating in a 2-week program to teach incoming high school freshmen the basics of decision quality and help them apply the concepts to significant school decision projects. I recommend that all decision professionals become engaged with spreading decision skills to youth[4] for the simple reason that it will make one a better decision professional. Senior executives and ninth graders have about the same attention span (albeit for different reasons) and want to get to the essence simply and clearly. Even when we employ advanced tools, our results should always be made transparent.

Our Profession

What does it mean to be in a profession? A profession differs from a trade. In providing a professional service, we recognize that the customer cannot fully judge our service and must trust the integrity of the professional to act in the customer's best interest—even when the customer does not wish to hear it. Our customers are the decision makers—the leaders of organizations. We have the responsibility to speak "truth to power."

We also have the obligation to not "fake it." Decision professionals must be able to recognize which tools are normative (that is consistent with the norms of decision theory) and which are not but may be useful in practice. We also have to recognize destructive or limited practices. A true decision professional avoids making claims that can be proven to violate the basic norms of decision theory.

[4]Check out The Decision Education Foundation at http://www.decisioneducation.org.

As with the medical field, we have to protect our profession from quackery. The profession is beginning to step up to this challenge, taking measures to assure quality and certify competence. This is, of course, a sensitive area in a field that incorporates science, art, and engineering. While I recognize the risks of trying to come to agreement on a definition of decision competence, I support this trend fully and applaud the start that the Society of Decision Professionals has made.

The Biggest Challenge

In this nascent profession, our biggest challenge is to gain greater mindshare among decision makers. The fraction of important and complex decisions being made with the support of decision professionals is still very small. We can make faster progress if we unify our brand and naming conventions. I urge all practitioners to use a common language to make more headway with our audiences.

Here are my suggestions:

- Let's call ourselves "decision professionals" instead of decision analysts, decision consultants, decision advisors, decision facilitators, and so on.
- Let's use the term "decision quality" as the overall name that combines getting to the right answer via DA and gaining organizational alignment via process leadership, decision facilitation, and other soft skills.
- Let's refer to DA as the field that provides decision professionals with the analytical power to find the best alternative in situations of uncertainty, dynamics, complex preferences, and complex systems. The use of the term "DA" also means we will be consistent with the norms of decision theory—usually with a single decision-making body—whose preferences are aligned.
- Multiparty decisions—negotiation, collaboration, competition (game theory)—need to become a part of the decision professional's domain of expertise, whether or not these areas are considered a subset of or adjacent to DA.
- Decision professionals frequently act as mediators and facilitators with "soft skills" to lead decision processes to reach sound conclusions and to gain alignment and commitment for action. While these skills are not usually considered a part of DA, they are as crucial as model building to the decision professional.

On behalf of the profession, I would like to express my gratitude to Greg Parnell, Steve Tani, Eric Johnson, and Terry Bresnick for creating this handbook. This handbook represents a valuable contribution to the practitioner community. I expect that it will be the first edition of many to come.

Carl Spetzler

Preface

Our *Handbook of Decision Analysis* is written for the decision professional. The target audience is the decision analysis practitioner who wants to increase the breadth and depth of his or her technical skills (concepts and mathematics) and soft skills (personal and interpersonal) required for success in our field. We assume the reader has a technical (engineering, science, mathematics, or operations research) or business degree; a course in probability and statistics (Appendix A provides a probability review); and, perhaps, some introduction to single or multiple objective decision analysis in a college course or a professional short course. The book is not designed to introduce new decision analysis mathematics, but rather to make the most common mathematics and best practices available to the practitioner.

The handbook is designed to be supplemental reading for professional decision analysis training courses, a reference for beginning and experienced practitioners, and a supplemental text for an undergraduate or graduate course in decision analysis. Decision analysts work in many industries and government agencies; many work in oil and gas firms, pharmaceutical firms, and military/intelligence agencies. The book should be useful to both domestic and international practitioners.

Our handbook describes the philosophy, technical concepts, mathematics, and art of decision analysis for the decision professional. The handbook includes chapters on the following topics: decision-making challenges; mathematical foundations of decision analysis; decision analysis soft skills; selecting the decision making process for interacting with decision makers and stakeholders; framing the decision; crafting decision objectives; designing creative alternatives to create value; performing deterministic modeling and analysis of alternatives; assessing uncertainty; performing probabilistic modeling and analysis; portfolio decision analysis; communicating with senior decision makers; and implementing decisions.

Figure P.1 provides the organizational structure of the book. Chapter 1 provides an introduction to decision analysis. Chapters 2–4 provide the foundational knowledge required for decision analysis success. Chapters 5–14 provide the decision analysis best practices to create value as sequential, iterative steps. However, the order of the steps should be tailored to the application, and some

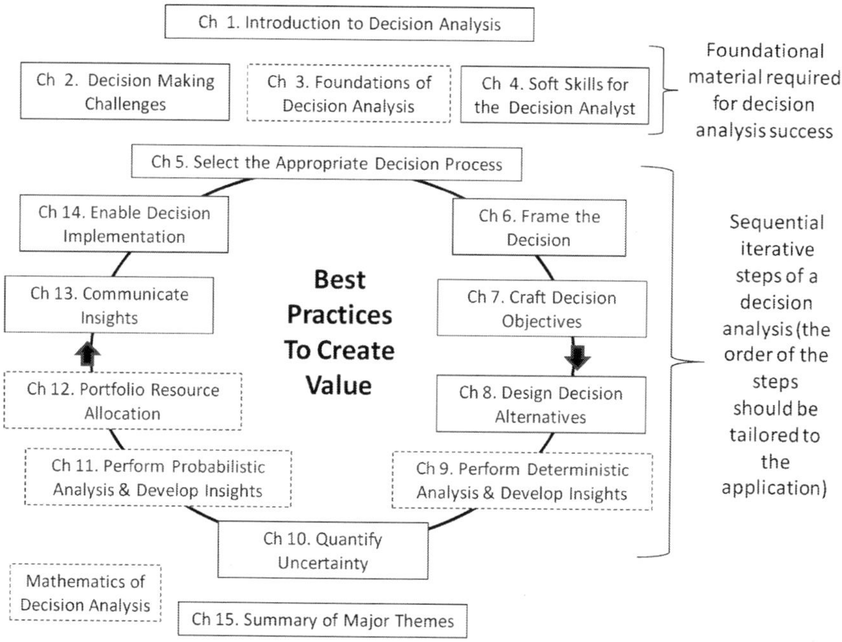

FIGURE P.1. Chapter organization of the *Handbook Decision of Analysis.*

steps may not apply. For example, if the decision is a choice of the best alternative, the portfolio decision analysis chapter would not apply. Also, some steps can be combined. For example, the decision framing and crafting of the decision objectives may be done at the same time. Chapter 15 provides a summary of the major themes of the book. The chapters that provide the mathematics of decision analysis are outlined with dotted lines.

The book also includes key insights from decision analysis applications and behavioral decision analysis research. The handbook references decision analysis textbooks, technical books, and research papers for detailed mathematical proofs, advanced topics, and further professional reading.

The handbook has five unique features:

1. The book provides a balanced presentation of technical skills (decision analysis concepts, mathematics, and modeling) and soft skills (strategic thinking, leading teams, managing teams, researching, interviewing individuals, facilitating groups, and communicating).
2. The book integrates the techniques of single and multiple objective decision analysis instead of presenting them in separate sections of the book. Chapter 3 provides our framework.
3. The book uses three substantive illustrative examples (Roughneck North American Strategy, Geneptin Personalized Medicine, and Data Center) to

illustrate the key decision analysis concepts and techniques, show the diversity of applications, and demonstrate how the techniques are tailored to different decision problems.

4. The book presents multiple qualitative and quantitative techniques for each key decision analysis task as opposed to presenting one technique for all problems. After describing the techniques, we discuss their advantages and disadvantages.
5. Supplementary material for this book can be found by entering ISBN 9781118173138 at booksupport.wiley.com. This website will contain information on the book and the Microsoft® Office Excel® files used for the three illustrative examples.

We, the coauthors, became decision analysts and strive to be decision professionals because we believe in the power of decision analysis to create value for organizations and enterprises. The art and science of decision analysis has changed our professional and personal decision making.

Writing the handbook has been a great opportunity for us to reflect on what we have learned and to describe the best practices that we use. In addition to our mentors and colleagues, we have also learned a lot from each other in the process of writing (and rewriting) this book! We look forward to hearing your comments on the handbook, and we hope that the material helps your development as a decision professional.

Gregory S. Parnell
Terry A. Bresnick
Steven N. Tani
Eric R. Johnson

Acknowledgments

We have benefited greatly from our decision analysis colleagues and mentors. We would like to acknowledge these contributions in four categories: special, individual, chapter, and handbook reviewers.

Special Acknowledgments

The authors of this book wish to express our deep appreciation to Professor Ronald A. Howard for being our teacher, mentor, and colleague. All four of us studied under Ron at Stanford, and he inspired each of us to pursue a career as a decision professional. His formulation of decision analysis is the basis of our understanding of how to address tough decisions, and his ongoing contributions to the field have strengthened our professional practice.

In addition, the authors appreciate the support of the John Wiley & Sons OR/MS series editor Susanne Steitz-Filler, our colleagues on the board of the Society for Decision Professionals, and the anonymous Wiley OR/MS reviewers who all encouraged us to write a decision analysis handbook for decision professionals.

Individual Acknowledgments

Gregory S. Parnell acknowledges colleagues Dennis Buede, Pat Driscoll, Ralph Keeney, Robin Keller, Craig Kirkwood, and Larry Phillips for their valuable professional advice and friendship. In addition, I acknowledge Terry Bresnick for recruiting me to Innovative Decisions, Inc., which has involved me in decision analysis practice as well as education.

Terry A. Bresnick acknowledges Ed Sondik and Rex Brown for providing DA theoretical background; Cameron Peterson, Roy Gulick, and Larry Phillips for making me understand the importance of the "socio" side of decision analysis and molding my perspective on facilitation; Al Grum for being my mentor as I started my career as a decision analyst; and Dennis Buede for being my collaborator and partner for many years in our DA practice.

Eric R. Johnson acknowledges Leonard Bertrand and Jim Matheson for conversations that contributed to my perspective on decision analysis and to the handbook.

Steven N. Tani acknowledges Jim Matheson for providing vital leadership during the early years of the decision analysis profession at Stanford Research Institute and Carl Spetzler for keeping Strategic Decisions Group true to its values through many transitions in its 31-year history. Both of these colleagues have had a large and positive role in shaping my professional career.

Sean Xinghua Hu acknowledges colleague Stanhope Hopkins for refinement of some of the Geneptin case illustrations and key contribution to the development of the case Excel model.

Chapter Acknowledgments

CHAPTER 1 AND 4: SOFT SKILLS

The authors would like to acknowledge the instructors of the Soft Skills Workshops that has been taught at Institute for Operations Research and Management Science (INFORMS) and Military Operations Research Society (MORS). These individuals include Bill Klimack, Freeman Marvin, Jack Kloeber, Don Buckshaw, Dave Leonardi, and Paul Wicker. We have also benefited from discussions with colleagues at DAAG 2012, including Carl Spetzler. They have motivated us to provide explicit identification and description of soft skills that are essential for decision analysts. The identification methodology that includes personal and intrapersonal skills is our contribution to the field.

CHAPTER 2

The authors would like to acknowledge Dennis Buede and Freeman Marvin, our colleagues at Innovative Decisions, Inc. Dennis and Freeman have developed and co-taught courses on decision analysis, systems engineering, and facilitation. Their collaboration and insights have made this chapter more complete and more focused.

Handbook Chapter Reviewers

After drafting the entire book, we sent each chapter to one or two colleagues for review. We acknowledge the following individuals for their timely reviews and excellent suggestions: Ali Abbas, Ritesh Banerjee, Kevin Carpenter, Jim Chinnis, Ellen Coopersmith, Robin Dillon, Jim Felli, Dave Frye, Roy Guilick, Onder Guven, Ralph Keeney, Rob Kleinbaum, Jeff Keisler, Craig Kirkwood, Jack Kloeber, Ken Kuskey, Bill Klimack, Frank Koch, William Leaf-Herrmann, Pat Leach, Freeman Marvin, Dan Maxwell, Jason Merrick, Cam Peterson, Jan Schulze, Carl Spetzler, and Joe Tatman. Of course, any remaining errors or omission are the responsibility of the authors.

About the Authors

The handbook was written by the first four contributors. The primary authors of each chapter are on its title page; however, all four authors contributed to each chapter. The handbook has four illustrative examples; the Roughneck North American Strategy was written by Eric R. Johnson; the Geneptin was written by Sean Xinghua Hu; the Data Center Location written by Gregory S. Parnell; and the Data Center portfolio was written by Terry A. Bresnick.

Dr. Gregory S. Parnell is a professor of systems engineering at the U.S. Military Academy at West Point. His research focuses on decision and risk analysis for defense, intelligence, homeland security, and environmental applications. He has also taught at the U.S. Air Force Academy, Virginia Commonwealth University, and the Air Force Institute of Technology. He has taught over 60 1-week decision analysis professional courses for government and commercial clients. He has been Chairman of the Board and is a senior principal analyst with Innovative Decisions, Inc., a decision analysis consulting firm. Dr. Parnell is a former president of the Decision Analysis Society of the Institute for Operations Research and Management Science (INFORMS) and of the Military Operations Research Society (MORS). He has also served as editor of *Journal of Military Operations Research*. Dr. Parnell has published more than 100 papers and book chapters and has coedited *Decision Making for Systems Engineering and Management*, Wiley Series in Systems Engineering (2nd ed, John Wiley and Sons, 2011). He has received several professional awards, including the U.S. Army Dr. Wilbur B. Payne Memorial Award for Excellence in Analysis, MORS Clayton Thomas Laureate, two INFORMS Koopman Prizes, and the MORS Rist Prize. He is a Fellow of MORS, INFORMS, the International Committee for Systems Engineering, the Society for Decision Professionals, and the Lean Systems Society. He received his BS in Aerospace Engineering from the State University of New York at Buffalo, his ME in Industrial and Systems Engineering from the University of Florida, his MS in Systems Management from the University of Southern California, and his PhD in Engineering-economic Systems from Stanford University. Dr. Parnell is a retired Air Force Colonel and a graduate of the Industrial College of the Armed Forces.

Mr. Terry A. Bresnick is the cofounder and Senior Principal Analyst at Innovative Decisions, Inc. He has had extensive experience in applying decision analysis to complex problems of government and industry. Earlier, as an officer in the U.S. Army, and currently, as a consultant in the private sector, Mr. Bresnick has demonstrated his expertise in the areas of decision analysis, risk analysis, strategic planning, resource allocation and budgetary analysis, evaluation of competing alternatives, cost–benefit analysis, and business area analysis. He has facilitated more than 1,000 decision conferences and/or workshops for government and private sector clients. He has been an Assistant Professor of Systems and Decision Analysis at the U.S. Military Academy, is a certified Financial Planner, a Fellow of the Society of Decision Professionals, and a registered Professional Engineer in the State of Virginia. Mr. Bresnick was awarded the David Rist Prize by the Military Operations Research Society for his work on an innovative military application of decision analysis. He received a BS in Engineering from the U.S. Military Academy, an MBA in Decision Science from George Mason University, and an MS in Statistics and the Degree of Engineer in Engineering-Economic Systems from Stanford University. Mr. Bresnick is a retired Lieutenant Colonel in the U.S. Army.

Dr. Eric R. Johnson has helped clients facing decision challenges throughout his career. This has included work through consultancies, as well as working within the client organization. He has extensive experience in pharmaceuticals and oil, gas, and electric utilities, having worked for Schering-Plough, Portland General Electric, Pharsight, and Decision Strategies, Inc. before taking his current position at Bristol-Myers Squibb. Dr. Johnson is a Fellow and Board member of the Society of Decision Professionals. He is a member of the Decision Analysis Society of INFORMS and won its Decision Analysis Best Practice award in 2002 for a decision analysis of development decisions for a drug code-named Apimoxin. He has a BA in philosophy from Reed College and a PhD in Management Science and Engineering, focusing in decision analysis, from Stanford University.

Dr. Steven N. Tani has been a professional decision analyst since 1975. During his career, he has helped numerous clients in both the private and public sectors make good choices in difficult decision situations. He has also taught many courses in decision analysis and its application. He was manager of the Decision Analysis Executive Seminar Program for SRI International and serves as an instructor in the Strategic Decision and Risk Management program in Stanford University's Center for Professional Development. Dr. Tani is a partner with Strategic Decisions Group (SDG), and in 2004 was named the first SDG Fellow. He is a Fellow in the Society of Decision Professionals and has served as a Board member for that organization. He holds a BS degree in Engineering Science and MS and PhD degrees in Engineering-Economic Systems, all from Stanford University.

Dr. Sean Xinghua Hu is Head of Bionest USA and Managing Partner, North America at Bionest Partners, a global strategy/management consulting firm. He has been a decision analysis practitioner and professional decision analyst for many years. Dr. Hu joined the Life Sciences Division of Strategic Decisions Group (SDG) upon its acquisition by IMS Management Consulting in 2006 and served as its Leader of Personalized Medicine Strategy Consulting. A recognized thought leader in the field of personalized medicine strategy, Dr. Hu has been a pioneer in applying decision analysis framework and analytics to advising pharmaceutical and diagnostic industries in the development and commercialization of personalized medicine and optimization of related R&D and commercial decisions. Dr. Hu was the only representative from the management consulting industry to serve on the multiyear FDA Personalized Medicine Initiative Consortium, responsible for decision analysis/probabilistic modeling. This FDA Consortium effort led to the publication in *Nature Reviews Drug Discovery* (November 2011) a landmark article, of which Dr. Hu is a colead author, describing an analytical approach to evaluate personalized medicine R&D and commercialization strategic alternatives based on decision analysis concept. Among the several academia-oriented extracurricular appointments, Dr. Hu serves on the Editorial Board of the peer-reviewed journal *Personalized Medicine*. Dr. Hu holds a BS degree in Organic Chemistry from Peking University, China, a PhD in Genomics from New York University, and an MBA in Strategic and Entrepreneurial Management from the Wharton School, University of Pennsylvania.

Acronyms

AFT	Alternative-Focused Thinking
BLUF	Bottom Line Up Front
boe	Barrel of oil-equivalent
BRAC	Base realignment and closure
BTU	British thermal units
BU	Business Unit
capex	Capital expenses
CBM	Coalbed Methane
CE	Certain Equivalent
CFO	Chief Financial Officer
COTS	Commercial off-the-shelf software
DA	Decision Analysis
DAAG	Decision Analysis Affinity Group
DEF	Decision Education Foundation
DFT	Decision Focused Transformation
DM	Decision Maker
DoD	Department of Defense
Dx	Diagnostic
E&P	Exploration and Production
EGFR	Epidermal Growth Factor Receptor
EnP	Exploration and Production
ENPV	Expected Net Present Value
EOR	Enhanced Oil Recovery
EV	Expected Value
F&D	Finding and Development
FDA	Food and Drug Administration
FISH	Fluorescent in situ hybridization
Gbps	Gigabytes per second
GRASP	Goals and Outcomes, Room and Logistics, Agenda and Time Available, Support Team, Tools, and Techniques, Participants and Observers

(Continued)

HER2	Human Epidermal Growth Factor Receptor 2
IC	Intelligence Community
ID	Influence Diagram
IDCF	International Decision Conferencing Forum
IHC	Immunohistochemistry
INFORMS	Institute for Operations Research and Management Science
IT	Information Technology
KRAS	v-Ki-ras2 Kirsten rat sarcoma viral oncogene homolog
LAV	Light armored vehicle
LCC	Life Cycle Cost
LDT	Laboratory-Developed Test
mmBtu	Million British thermal units
MODA	Multiple Objective Decision Analysis
MPG	Miles per gallon
MW	Megawatts
NGT	Nominal Group Technique
NPV	Net Present Value
ODQ	Organizational Decision Quality
opex	Operating expenses
P&L	Profit and Loss Statement
P0	0th percentile
P10	10th percentile
P100	100th percentile
P50	50th percentile
P90	90th percentile
PDORA	Purpose, Desired Outcomes, Roles, and Agenda
PI	Profitability Index
PnL	Profit and Loss Statement
PTRS	Probability of Technical and Regulatory Success
PV	Present Value
R&D	Research and development
RAM	Randon access memory
RNAS	Roughneck North American Strategy
ROM	Read-only memory
SDP	Society for Decision Professionals
SH	Stakeholder
SME	Subject Matter Expert
SRI	Stanford Research Institute
SVP	Senior Vice President
TS	Tar Sands
USMC	United States Marine Corps
VBA	Visual Basic for Applications (Microsoft OfficeTM programming language)
VFT	Value-Focused Thinking
VoI	Value of Information
VP	Vice President
WACC	Weighted average cost of capital

CHAPTER ONE

Introduction to Decision Analysis

GREGORY S. PARNELL and TERRY A. BRESNICK

Nothing is more difficult, and therefore more precious, than to be able to decide.
—Napoleon, "Maxims," 1804

(*Continued*)

Handbook of Decision Analysis, First Edition. Gregory S. Parnell, Terry A. Bresnick, Steven N. Tani, and Eric R. Johnson.

1.1 Introduction

The consequences of our decisions directly affect our professional and personal lives. As Napoleon noted in our opening quote, decisions can be difficult, and making good decisions can be very valuable. Our focus is on professional decisions, but the same principles apply to our personal decisions.

We begin by defining a decision. Professor Ronald Howard of Stanford University defines a decision as an irrevocable allocation of resources (Howard, 1988). Consider the contracting process used by many companies and organizations. The company does not make a decision to buy a product or service when they begin thinking about the procurement. They make the decision when they sign a legally binding contract, which obligates them to provide resources (typically dollars) to the supplier of the product or service. Can they change their mind? Absolutely, but they may have to pay contract cancellation fees.

A decision is an irrevocable allocation of resources.

Decisions are made by people vested with the authority and responsibility to make decisions for an organization or enterprise. Many decisions involve stakeholders who are individuals and organizations that could be affected by the future consequences of the decision. Some decisions are easy because few stakeholders are involved, the values are clear, good alternatives are readily identified, and there are few uncertainties. However, some difficult decisions involve many stakeholders with potentially conflicting objectives, complex alternatives, significant uncertainties, and large consequences. The discipline of decision analysis, the focus of this handbook, has been developed to help decision makers with these complex decisions.

There are many definitions of decision analysis. Howard, who coined the term "decision analysis" (Howard, 1966), defines decision analysis as "a body of knowledge and professional practice for the logical illumination of decision problems." In the first book on decision analysis, Howard Raiffa of Harvard University defined decision analysis as an approach that "prescribes how an individual faced with a problem of choice under uncertainty should go about choosing a course of action that is consistent with personal basic judgments and preferences" (Raiffa, 1968). Ralph Keeney of Duke University (Keeney, 1982) provides an intuitive and a technical definition. Keeney's intuitive definition is "a formalization of common sense for decision problems that are too complex for informal use of common sense." His technical definition is "a philosophy, articulated by a set of logical axioms, and a methodology and collection of systematic procedures, based on those axioms, for responsibly analyzing the complexities inherent in decision problems." Professor Larry Phillips of the London School of Economics emphasizes that decision analysis is a socio-technical process to provide insights to decision makers in organizations (Phillips et al., 1990) and (Phillips, 2005). In a popular decision analysis textbook, Clemen and Reilly state that "decision analysis provides effective methods for organizing a problem into a structure that can be analyzed. In particular, elements of a decision's structure include the possible courses of action, the possible outcomes that could result, the likelihood of those outcomes, and eventual consequences (e.g., costs and benefits) to be derived from the different outcomes" (Clemen & Reilly, 2001). We will use the following definition of decision analysis:

> Decision analysis is a philosophy and a social-technical process to create value for decision makers and stakeholders facing difficult decisions involving multiple stakeholders, multiple (possibly conflicting) objectives, complex alternatives, important uncertainties, and significant consequences. Decision analysis is founded on an axiomatic decision theory and uses insights from the study of decision making.

In decision analysis, we distinguish between a good decision and a good outcome. A good decision is one that is logically consistent with our preferences for the potential outcomes, our alternatives, and our assessment of the uncertainties. A good outcome is the occurrence of a favorable event—one that we like. We believe that consistently making good decisions will lead to more good outcomes than otherwise. However, since there is uncertainty, even a good decision process may not always lead to a good outcome. Of course, a bad decision does not always result in a bad outcome—sometimes we can be lucky and obtain a good outcome. Unfortunately, we cannot count on being lucky.

The purpose of our handbook is to describe the best practices that decision analysts have found the most useful in helping decision makers make good decisions when faced with difficult and important choices. Since many individuals and social organizations are involved in complex decisions, to be successful, decision analysis must use a socio-technical process to help those individuals and organizations make decisions. Socially, the purpose of decision analysis is to provide credible, understandable, and timely insights to decision makers and key stakeholders in organizations. Technically, decision analysis is an operations research/management science discipline that uses probability, value, and utility theory (see Chapter 3) to analyze complex alternatives, under significant uncertainty, to provide value for stakeholders with multiple (and possibly conflicting) objectives. Since it relies on the reasonable axioms of choice (Chapter 3), decision analysis identifies decisions that are logically consistent with our preferences, our alternatives, and our assessment of the uncertainties.

This chapter introduces the field of decision analysis and defines some of the key terms that we use in the handbook. The chapter is organized as follows. Section 1.2 further describes decision analysis as a socio-technical process. We introduce the decision analysis process that we use in the handbook and use the process to list the key technical concepts and techniques and the soft skills necessary to help organizations create potential value for themselves and their stakeholders. Section 1.3 emphasizes that decision analysis has many significant applications and compares three important application areas: oil and gas, pharmaceuticals, and defense. We also briefly describe four decision analysis success stories. Section 1.4 defines the decision professional, discusses the education and training of decision professionals, identifies some of their major professional societies, and describes some of their professional service activities. Section 1.5 provides an overview of the handbook and introduces the three substantive illustrative examples used in the handbook. Section 1.6 provides a summary of the key ideas in the chapter.

1.2 Decision Analysis Is a Socio-Technical Process

An effective decision analyst must understand the challenges of decision making in organizations, the mathematical foundations of decision analysis, and the soft skills required to work with decision makers, stakeholders, and experts to perform a decision analysis. In this section, we describe the decision analysis process used in the handbook and use that process (and our experience) to identify the critical soft skills that are essential for the successful use of decision analysis.

There are several decision processes (see Chapter 6) that have been used by decision analysts to integrate the contributions of decision makers (DMs), stakeholders[1] (SH), subject matter experts (SMEs), and decision analysts to reach a good decision. Figure 1.1 shows the decision analysis process that we use to organize the handbook. The decision frame is how we view the decision opportunity. At the center of the figure is a reminder that our purpose is to use best practices to create value for DMs and SH. The steps in the process are shown as 10 boxes around the center. Although sequential arrows are used in the figure, the process is iterative. The order of the steps should be tailored to the application and some steps may not apply. For example, if the decision is a choice of the best alternative, the portfolio resource allocation chapter would not apply. Also, some steps can be combined. For example, the decision framing and crafting of the decision objectives may be done at the same time. In addition, some

[1]For the purpose of this chapter, "stakeholders" refers to all interested and affected individuals besides the DM(s) and SMEs. We will use SH instead of SHs for simplicity.

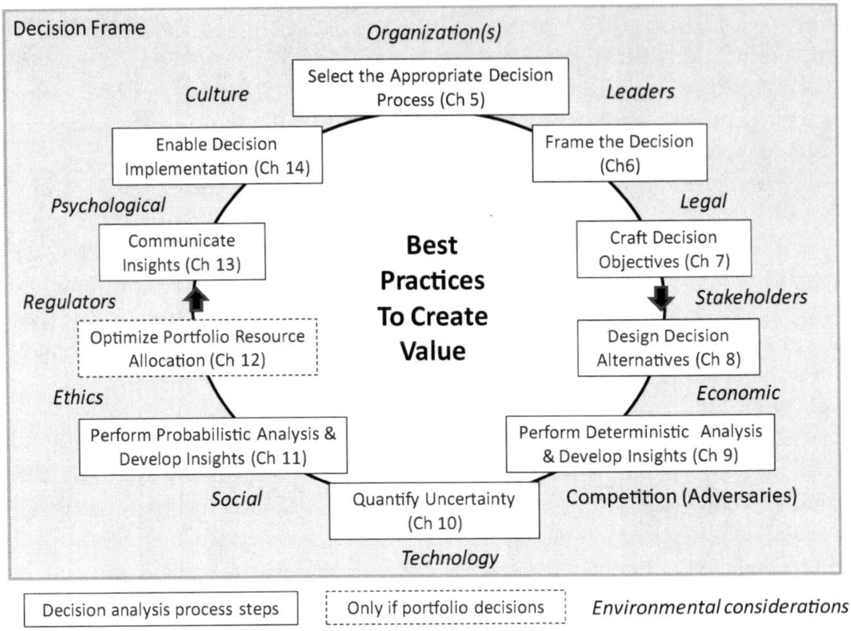

FIGURE 1.1 Decision analysis process.

steps may not be required in a particular application. Twelve environmental factors are placed in the decision frame of Figure 1.1, but outside the decision analysis process cycle to highlight the important considerations that apply in many of the steps of the decision analysis process. The location of a factor is not necessarily an indication of alignment with a particular step in the process. The 12 factors are meant to be illustrative and not all inclusive.

Next, we use this decision analysis process to identify the decision analysis technical products and soft skills that are essential for the decision professional. We identify these skills in Table 1.1 with steps in the process. Soft skills include personal and interpersonal skills.

Based on our experience and the above analysis, we aggregate the soft skills into the following nine categories.

- **Thinking strategically** about the client organization, the problem domain, and the role of the decision analysis in achieving the current strategy or, when appropriate, developing a new strategy, and new decision opportunities
- **Leading teams**, including developing team goals, motivating individuals to achieve team goals, and guiding the client organization to achieve the most value from the study
- **Managing decision analysis projects**, including developing analysis plans; identifying and scheduling activities; and managing the completion of tasks

TABLE 1.1 List of Technical Products and Soft Skills

Steps	Engagements	Technical Products	Soft Skills
Select the appropriate decision process (Chapter 5)	DM, SH	Decision process diagram	Strategic thinking Lead teams Manage teams
Fame the decision (Chapter 6)	DM, SH, SME	Vision statement Decision hierarchy Issue identification matrix Qualitative influence diagram Study schedule	Strategic thinking Lead teams Manage teams Research problem domain Interview DM & SH Facilitate group sessions Communicate insights
Craft decision objectives (Chapter 7)	DM, SH	Objectives hierarchy Functional value hierarchy	Strategic thinking Interview DM, SH Survey SH, SMEs Facilitate group sessions Communicate insights Aggregating expertise
Design decision alternatives (Chapter 8)	SME, SH, DM	Strategy generation table Strategic alternatives Real options Means ends network	Strategic thinking Use individual and group creativity techniques Research problem domain
Perform deterministic analysis and develop insights (Chapter 9)	SME, SH, DM	Deterministic influence diagram Quantitative deterministic value models (NPV and multiple objective models) Swing weight matrix Value components Value over time Value by business unit Waterfall chart Direct and delta tornado diagrams Sensitivity analysis Cost vs. value chart	Research data Interview SME to develop model structure Aggregating expertise Facilitate group sessions to develop model structure Elicit value curves Elicit swing weights Assess alternative scores Communicate insights
Quantify uncertainty (Chapter 10)	SME, SH, DM	Influence diagram Probability assessments of uncertain variables	Elicit probabilistic relationships Aggregating expertise Elicit probability distributions

TABLE 1.1 ***(Continued)***

Steps	Engagements	Technical Products	Soft Skills
Perform probabilistic analysis and develop insights (Chapter 11)	SME, SH, DM	Decision tree Monte Carlo simulation Net present value distribution Dominance analysis Value component chart Direct and delta tornado chart Perform sensitivity analysis Value over time Value by business unit Risk preference Utility	Strategic thinking Elicit risk preference Use individual and group creativity techniques to reduce risk and increase value
Portfolio resource allocation (Chapter 12)	DM, SH, SME	Portfolio constraints Portfolio model Efficient frontier Football chart	Strategic thinking Manage teams Interview DM, SH, SME to identify constraints Facilitate group session(s) to develop value model and evaluate alternatives Communicate insights
Communicate insights (Chapter 13)	SH, DM	Communication objectives Analysis story Key insights Executive summary Presentation(s) Technical report(s)	Develop the story and key insights Present analytical results Communicate to DM, SH
Enable decision implementation (Chapter 14)	SH, DM, SME	Implementation schedule Implementation success measures Implementation risks and risk mitigation plan	Manage teams Interview DM, SH Facilitate group sessions to identify success measures and risks

- **Researching the problem domain, modeling approaches, and data sources**
- **Interviewing individuals** (DMs, SH, and SMEs) to frame the decision problem and obtain modeling information
    - Interact with senior leaders and SMEs (listening, learning, and discovery)
    - Elicit knowledge (preferences [value, time, and risk], probabilities, and alternatives)
- **Surveying stakeholders and experts** can be a efficient way to collect knowledge for geographically dispersed individuals
- **Facilitating groups** of DMs, SH, and SMEs to frame the decision problem and obtain modeling information (also includes focus groups)
    - Frame decision opportunity (initial and updated)
    - Elicit knowledge (preferences [value, time, and risk], probabilities, and alternative)
    - Use individual and group creativity techniques (values, sources of risk, strategy design, strategy improvement) to generate better alternatives
- **Aggregating expertise** is needed to combine different views of SHs and SMEs
- **Communicating with DMs, SH, and SMEs** (see Chapter 13).
    - Communicate the story, analytic results, and the key insights in ways that are understandable to the audience.

In the subsequent chapters, we present in more detail both the technical skills and the soft skills that are essential to decision analysis.

1.3 Decision Analysis Applications

Decision analysis has been used in many important corporate and public applications. These decision analysis applications typically have four features in common: difficult decisions, multiple (possibly conflicting) objectives of SH, significant uncertainties, and important consequences. One of the first compendiums of decision analysis applications was published in 1983 (Howard & Matheson, 1983). In addition to applications, this two-volume set also includes some important early foundational technical articles on decision analysis. Two more recent applications summaries are Corner and Kirkwood (1991) and Keefer et al. (2004). These two papers list several published applications in a wide variety of problem domains. These applications summaries greatly underestimate the number of applications since practitioners generally do not publish their work due to the confidentiality of the results, the lack of time for writing publications, and lack of incentives for publication.

Three important enduring areas of decision analysis applications have been oil and gas, pharmaceuticals, and military.[2] Table 1.2 (modified from Burk & Parnell [2011]) compares these three significant decision analysis application areas using several factors: organizational objectives, key SH, major environmental uncertainties, technological development uncertainties, schedule uncertainties, cost uncertainties, operating environment, strategic partnerships, intraorganizational resource competition, and decision reviews. The primary organizational objective of private firms (e.g., oil and gas and pharmaceuticals) is to increase shareholder value, while public organizations (e.g., military) provide products and services that are not easily measured in terms of dollars. The three examples illustrate the difficulty of decisions, the conflicting preferences of SH, and the major uncertainties.

All three domains have a significant number of private and public SH with complex and, many times, conflicting objectives. Clearly, each application area has a significant number of environmental, technical, schedule, and cost uncertainties. The operating environments and adversaries are different. Finally, the resource competition and decision review processes are significantly different for public and private problem domains.

There are many decision analysis success stories. Next, we describe decision analysis success stories in each of the three major application areas.

1.3.1 OIL AND GAS DECISION ANALYSIS SUCCESS STORY: CHEVRON

Over the past 20 years, Chevron has used decision analysis for its major decisions (Menke et al., 2011). The Chevron Vice Chairman, George Kirkland, summarizes the use of decision analysis to create value and manage risk on over 40 projects with investments of over $1B.[3] According to Mr. Kirkland, Chevron "uses decision analysis because it works." Chevron's Larry Neal estimated the benefit of decision analysis as $100B over 10 years, and highlighted the additional benefits of decision framing (see Chapter 6) and improvements in thinking.[4] Chevron's Frank Koch noted the added confidence decision analysis gives DMs to pursue projects and accept risk (see Chapters 11 and 12). In addition, Koch stated that the marginal cost of doing decision analysis is small and the cost of training and learning software is significantly outweighed by the benefits.[5]

> Chevron uses decision analysis because it works.

---

[2]Perhaps not surprisingly, the authors have worked in these application areas.

[3]http://www.youtube.com/chevron#p/u/12/JRCxZA6ay3M, recorded December 1, 2010.

[4]http://www.blip.tv/file/4567268

[5]Op. cit.

TABLE 1.2 Comparison of Three Decision Analysis Application Areas

Factor	Oil and Gas	Pharmaceuticals	Military
Illustrative organizational objectives	Increase shareholder value Provide energy for the nation and energy consumers Protect the environment	Increase shareholder value Improve health and quality of life of patients Minimize potential side effects	Provide defense capabilities for the national command authority to achieve national objectives Minimize causalities if a conflict occurs Reduce collateral damage during a conflict Be cost-effective
Key stakeholders	Board of directors Shareholders Government regulators Environmental groups Nation where resources are located Energy distributors and retailers Consumers Employees	Board of directors Shareholders Current and future patients Health care providers Government regulators Employees	Citizens Department of Defense Congress Federal budget organizations Defense contractors Military, civilian, and contractor employees International security groups Allies
Environmental uncertainties	Existence and quantity of resources at particular locations Actions of competitors Actions of governments Actions of environment groups	Causes of diseases Efficacy of competitor and company's drugs Prevalence of future diseases	Future national, regional, and terrorist threats to national interests Economic resources devoted to defense Political constraints on military actions
Technology development uncertainties	Effectiveness and efficiency of location, extraction, and processing technologies Impact of operations and products on the environment	Efficacy of drugs Unwanted side effects of new drugs	Technology readiness to develop and produce future systems R&D test failures in potential operational environments Battlefield conditions impact on weapons systems
Schedule uncertainties	Local, state, national, and international approvals to extract and operational restrictions	Success of trials National and international regulatory approvals	Testing success Acquisition approvals Congressional funding authorizations

TABLE 1.2 (*Continued*)

Factor	Oil and Gas	Pharmaceuticals	Military
Cost uncertainties	Drilling rig availability, technology development problems, environmental protection requirements, and schedule slips	Size of clinical trials required	Changes by adversaries, immature technologies, and schedule changes
Operating Environment	Natural environment Competition	Pharmaceutical laboratories Human body Competition	Hostile natural and adversarial environment
Strategic partnerships	Mergers and acquisitions	Mergers and acquisitions	Foreign military sales to offset costs and support international security objectives Joint deterrence activities
Intra-organizational resource competition	Divisions Other corporate programs	Divisions Other corporate programs	Services and defense agencies Other government programs
Decision reviews	Corporate Board of directors	Corporate Board of directors	Military hierarchy, defense agency, Department of Defense, Office of Management of Budget, Congress

1.3.2 PHARMACEUTICAL DECISION ANALYSIS SUCCESS STORY: SMITHKLINE BEECHAM

Research and development decisions are the lifeblood of any pharmaceutical company. SmithKline Beecham (now GlaxoSmithKline) used decision analysis to make better resource allocation decisions (Sharpe & Keelin, 1998; Menke et al., 2011). SmithKline Beecham selected decision analysis because it was technically sound and organizationally credible. In their article, Sharpe and Keelin describe the benefits of decision analysis as follows:

> The new process not only reduced the controversy in the resource allocation process, it also led the company to change its investment strategy. Although top management had set out to cut back on the company's development budget, they now saw their investment decision in a new light; they believed the new portfolio to be 30% more valuable ($2.6B) than the old one without any additional investment. Furthermore, the marginal return on additional investment had tripled from 5:1 to

> 15:1. To exploit this opportunity, the company ultimately decided to increase development spending by more than 50%.

The results of this analysis were a dramatic increase in shareholder value.

1.3.3 MILITARY DECISION ANALYSIS SUCCESS STORIES

Public organizations use multiple objective decision analysis to evaluate the stakeholder value of alternatives and make defensible decisions.

1.3.3.1 U.S. Army Installations. In 2001, Congress enacted legislation that required a 2005 Base Realignment and Closure (BRAC) round to realign military units, remove excess facility capacity, and support defense transformation. This BRAC round was the fifth round of base closures. The U. S. Army used multiple objective decision analysis with 40 value measures to determine the military value of installations and an installation portfolio model to develop the starting point for identification of potential unit realignments and base closures and provide the basis for evaluating all recommendations (Ewing et al., 2006). The BRAC 2005 Commission accepted 95% of the Army's recommendations.[6] According to Army estimates, the approved recommendations will create a 20-year gross savings of \$20.4B for a one-time cost of \$12.8B and generate 20-year net savings of \$7.6B, which are 1.2 times the net Army savings of the first four BRAC rounds combined. After completion of the 5-year BRAC implementation, the Army estimated that the recommendations would create a recurring savings of \$1.5B annually. In addition, the Army leadership believes that the transformation realignments have made the Army more effective.

1.3.3.2 Data Center Location. Organizations with large computing needs have used data centers to help meet the demand for processing capabilities. The data centers can cost around \$0.5B per center (without the computers and software costs!). There are typically many groups of SH involved in the decision to select the best locations for these data centers, with highly diverse objectives. Multiple objective decision analysis has been successfully used four times in the intelligence community to select the best location that provides the highest value data center at an affordable life cycle cost.[7] The success of these projects led us to develop the IT illustrative example used throughout this handbook.

1.4 Decision Analysis Practitioners and Professionals

This handbook is intended for decision analysis practitioners. Some decision analysis practitioners may only occasionally use one or more of the decision

[6]The overall acceptance rate for all defense agencies was 86%. The Army had the highest acceptance rate.

[7]The first author facilitated the development of the first multiple objective value and life cycle cost model and mentored the analysts performing the studies.

analysis techniques to help DMs. Other decision practitioners, whom we call decision professionals, are individuals who, for a significant portion of their professional careers, seek to learn and apply proven decision analysis technical and soft skill best practices to help senior leaders create value for their organizations. To be effective and credible to DMs and SH, the decision professional must have knowledge about decision making and decision analysis techniques. Some decision professionals use their decision analysis techniques and soft skills to help groups solve problems in domains where they do not have significant knowledge or expertise (See Appendix C, Decision Conferencing). Other decision professionals acquire deep domain knowledge by working for extended periods in the field (e.g., oil and gas, pharmaceuticals, or military).

> A decision professional is an individual who seeks to learn and apply proven decision analysis technical and soft skill best practices to help senior leaders create potential value for their organizations.

To support their continual learning, many decision professionals belong to two types of professional societies. The first are societies that focus on decision analysis methods, education, and professional development. The second are professional societies that focus on particular problem domains.

1.4.1 EDUCATION AND TRAINING

Some decision professionals learn decision analysis in undergraduate or graduate degree programs. A listing of the graduate decision programs can be found on the Decision Analysis Society website (see the next section). Many decision professionals begin their education with a degree in engineering, science, or business. Some even begin with a liberal arts degree. Many individuals become decision analysts after working in a particular application domain by taking professional decision analysis training courses. All four of the authors took graduate courses in decision analysis and later taught undergraduate, graduate, and/or professional training courses. All of us have supplemented our formal education with reading to better understand our application domains and human and organizational decision making.

1.4.2 DECISION ANALYSIS PROFESSIONAL ORGANIZATIONS

The oldest decision analysis professional organization (founded in 1980) is the Decision Analysis Society (DAS) of the Institute for Operations Research and Management Science (INFORMS). DAS "promotes the development and use of logical methods for improving decision-making in public and private enterprise . . . members include practitioners, educators, and researchers with backgrounds in engineering, business, economics, statistics, psychology, and other social and applied sciences."[8] The DAS is a subdivision of INFORMS, which is world's largest organization of operations researchers and management scientists, with over 10,000 members. The DAS is among the largest of INFORMS' subdivisions, with more than 1000 members. Historically, a large percentage of the members have been consultants and students. DAS conducts its annual meeting and sponsors one or more tracks at the annual INFORMS meeting in the fall of each year. DAS has also organized decision analysis tracks in other INFORM sponsored meetings, including international meetings.

INFORMS and international operations research societies publish decision analysis articles in their technical journals. In addition, INFORMS and DAS publish *Decision Analysis*, which focuses on decision analysis theory and applications.

The Decision Analysis Affinity Group (DAAG) is a group of corporate and consulting decision analysis leaders who meet once a year for 2 or 3 days to share decision analysis insights, challenges and successes. It is more "practitioner" oriented than INFORMS DAS, which has a heavier "academic" and theoretic focus. The attendance at these meeting usually ranges from 30 to 80 individuals.

The Society of Decision Professionals (SDP) is a newer organization devoted to helping "decision professionals become the trusted advisors of choice for DMs facing important and complex decisions. The Society fosters collaboration, continual learning, and networking amongst its members and other professional societies and organizations so that as a growing community, we can bring clarity and insight to DMs."[9] The SDP wants to reach both DMs and decision professionals. Established in 2010, the society held its first meeting in the spring of 2011 at the annual Decision Analysis Affinity Group meeting.

1.4.3 PROBLEM DOMAIN PROFESSIONAL SOCIETIES

Many problem domains have professional societies that include decision analysis applications in their meetings and publications. As an example, the Military Operations Research Society (MORS) is a professional society devoted to furthering the development and use of operations research techniques for national security problems. Since the late 1980s, MORS has had a decision analysis working group at their annual meeting. In addition, INFORMS also has a Military Applications Society that has many military decision analysts, including the authors of this chapter.

The Society for Petroleum Engineering publishes many journals about oil and gas exploration and production, including some that address the decision analysis involved in the effort.

---

[8]Homepage of the Decision Analysis Society of INFORMS, http://www.informs.org/Community/DAS accessed July 29, 2011.

[9]Society of Decision Professionals, http://www.decisionprofessionals.com/, accessed July 29, 2011.

The Society for Medical Decision Making holds annual meetings and publishes a journal that has decision analysis approaches to guide the choice of medical treatment, at both the individual and societal level.

1.4.4 PROFESSIONAL SERVICE

Decision professionals perform professional service by taking leadership positions in professional societies and serving on national, regional, and local public service activities. Decision analysts have been president of many professional societies, including INFORMS, MORS, Society for Risk Analysis, and, of course, DAS and SDP. Many decision analysts have served on committees of the National Research Council where they use decision analysis expertise to help solve some of our nation's most significant challenges. As another example, decision professionals volunteer their time and talents to teach decision analysis concepts to youth through programs such as the Decision Education Foundation (DEF).[10]

1.5 Handbook Overview and Illustrative Examples

The handbook is organized as follows. Chapters 2–4 provide essential information that all decision analysis practitioners should know. Chapters 5–14 describe the decision analysis best practices in a sequential order. Chapter 15 provides a summary of these decision analysis best practices.

Chapter 2 describes the decision-making challenges in organizations and the cognitive and motivational biases from the behavioral decision analysis literature. Chapter 3 provides the theoretical foundations of decision analysis. Chapter 4 describes the soft skills that are the key to success of the decision analysis practitioner.

Chapters 5–14 are aligned with the steps in our decision analysis process (Fig. 1.1). Chapter 5 addresses the important issue of tailoring the decision process for the organization. Chapter 6 describes the use of soft skills to develop the decision frame. Chapter 7 describes techniques to craft the decision objectives. Chapter 8 introduces the creative process of designing the decision strategies. Chapter 9 focuses on the technical skills of model building and the soft skills of getting credible data for the models. We introduce single- (e.g., net present value) and multiple-objective value models. Chapter 10 focuses on the techniques for assessing uncertainty. Chapter 11 describes probabilistic modeling and analysis techniques to improve value and better manage risk. Chapter 12 introduces and describes the important techniques of portfolio decision analysis. Chapter 13 focuses on communicating the analysis results and insights to DMs to help them select the best alternatives. Chapter 14 addresses the implementation of the decision to achieve the potential value identified at the time of the

[10]Decision Education Foundation homepage, http://www.decisioneducation.org/, accessed July 31, 2011.

decision. Chapter 15 provides a summary of the decision analysis best practices that have been described in the book.

Each chapter has several standard features. First, we begin the chapter with a quotation to capture an important theme of the chapter. Second, we present the chapter material and illustrate the material with the three illustrative examples. Third, we list and define the key words introduced in the chapter. Fourth, we provide a list of the references we have used in the chapter.

One of the key features of this handbook is the integration of illustrative examples in almost all chapters of the book to illustrate the key concepts and techniques, to show the diversity of applications, and to demonstrate how the techniques are tailored to different problems. The first example is an oil and gas problem that we use to illustrate a single objective decision analysis using net present value. The second problem is the development and commercialization decision of a personalized medicine for breast cancer that also illustrates the use of net present value. The third example involves a government agency's decision about data center location and, in Chapter 12, an IT portfolio decision problem. We use this example to illustrate multiple objective decision analysis techniques.

Since the three illustrative examples are used throughout the book, we provide Table 1.3 as a reference to where to find the material for each of the examples. The table is also referenced in subsequent chapters.

1.5.1 ROUGHNECK NORTH AMERICAN STRATEGY (RNAS) (by Eric R. Johnson)

The title of the first illustrative example is the Roughneck North American Strategy (RNAS). The example is based on a specific decision analysis consulting engagement, but content is changed to preserve client confidentiality. Roughneck is the fictitious name of an international oil and gas operator, with headquarters and sizeable holdings in North America. Typical revenues were $1.5B a year. Market cap was roughly $5B. In the years preceding the strategic decision-making process described here, Roughneck had viewed North America as a mature market that was largely played out, and had focused its plans for growth on international assets. This was found to be less promising than initially hoped, due to ever-rising prices for development assets being paid by other international bidders, particularly developing countries with large populations and high aspirations for economic growth. Accordingly, Roughneck wanted to take another look at the growth potential of its North American properties.

1.5.2 GENEPTIN PERSONALIZED MEDICINE FOR BREAST CANCER (by Sean Xinghua Hu)

Our second illustrative example is a decision in the field of personalized medicine. Most medicines today are intended for a broad patient population, and many are effective in only 30–50% of patients. Personalized medicine, sometimes referred to as stratified medicine (Hu et al., 2005) (Trusheim et al., 2007),

TABLE 1.3 Section Locations of Illustrative Examples in Each Chapter

Chapter	Roughneck North American Strategy	Geneptin	Data Center
Introduction (Chapter 1)	1.5.1	1.5.2	1.5.3
Decision-making challenges (Chapter 2)	2.8.1	2.8.2	2.8.3
Select the appropriate decision process (Chapter 5)	5.4.1	5.4.2	5.4.3
Fame the decision (Chapter 6)	6.5.1	6.5.2	6.5.3
Craft decision objectives (Chapter 7)	7.8.1	7.8.2	7.8.3
Design decision alternatives (Chapter 8)	8.7.1	8.7.2	8.7.3
Perform deterministic analysis and develop insights (Chapter 9)	9.6 Spreadsheet Model 9.8 Analysis	9.11.1	9.11.2
Quantify uncertainty (Chapter 10)	10.2 Influence Diagram 10.3 Elicit and Document Assessments	10.4.1	NA
Perform probabilistic analysis and develop insights (Chapter 11)	11.3 Value Dialogue	11.5.1	11.5.2
Optimize portfolio resource allocation (Chapter 12)	12.3.3 RNAS Portfolio and 12.3.5 Tradeoffs	NA	12.4.3 Application to the Data Center Portfolio
Communicate insights (Chapter 13)	13.5.1	13.5.2	13.5.3
Enable decision implementation (Chapter 14)	14.5.1	NA	14.5.2

uses a diagnostic test (often referred to as "companion diagnostic tests") based on a molecular biomarker to "preselect" (or "stratify") the patients for whom the drug is most suitable. There have been only a few dozen personalized medicine drugs developed to date (Frueh et al., 2008; Laing et al., 2011) (Hu et al., 2012; FDA, n.d.).

One of the first successful personalized medicine products is Herceptin[11] (trastuzumab), which was marketed for cancer patients whose bodies make too much of the growth factor HER2, that is, they "overexpress" it. It is approved for treating HER2-overexpressing breast cancer patients, both for metastatic stage, and as an adjuvant therapy for early-stage patients. It is also approved for

[11]Herceptin, http://www.herceptin.com/breast/herceptin/, accessed May 25, 2012.

HER2-overexpressing metastatic gastric cancer. Herceptin was the first targeted medicine whose regulatory approval relied upon the use of a "companion diagnostic" to identify patients with a biomarker (in this case, HER2 overexpression). Herceptin was developed and marketed by Genentech (now owned by Roche).

Our "Geneptin" case is based on the development of Herceptin, but modified, simplified, and fictionalized to demonstrate some general considerations of personalized medicine development decision making.

Our Geneptin case is set in 1994, when the hypothetical Geneptin manufacturer, DNA Biologics, was designing the large, expensive "Phase III" clinical trial aimed at demonstrating safety and efficacy in metastatic breast cancer to secure FDA approval. DNA Biologics needed to decide whether to use a traditional all-comers approach, or to restrict the trial to patients who overexpress HER2. Previous Phase II studies had given some indication that HER2-overexpressing patients would likely respond better to Geneptin, though the evidence from these small trials was far from definitive.

The VP of Clinical Development at DNA Biologics believed that stratification could result in an enhanced benefit/risk ratio to patients and, therefore, a higher probability of technical and regulatory success (PTRS) of the drug development effort. However, it was not clear how to implement the stratification, because HER2 expression was measured on a continuous scale. The key question was where to draw the line defining HER2 overexpression?

Meanwhile, the VP of Commercialization believed that patient stratification would mean a smaller addressable patient population; based on the internal opinion on where to draw the line defining HER2 overexpression, HER2 overexpression-positive patients likely comprise only 25–35% of metastatic breast cancer patients. Would the reduction of addressable patient population still allow Geneptin to be a commercially viable product? Or, should the company proceed with an all-comers approach to maximize the number of patients Geneptin could serve, and avoid the risk of unnecessarily denying access to a somewhat arbitrarily defined population of HER2-negative patients?

1.5.3 DATA CENTER LOCATION AND IT PORTFOLIO (by Gregory S. Parnell and Terry A. Bresnick)

Our third decision example focuses on information technology (IT) for a large government organization. IT is critical to the ability of many organizations' ability to perform their missions. This government agency collects and processes large amounts of data. Due to the expanding variety and rapidly increasing volume of data, the agency required significant increases in data analytics. In addition, technology advances had resulted in supercomputers and servers becoming smaller, consuming more power, and requiring more cooling. The agency uses large data centers to process the collected and stored data. "A **data center** is the department in an enterprise that houses and maintains back-end information technology (IT) systems and data stores—its mainframes, servers and databases. In the days of large, centralized IT operations, this department and all the systems resided in one physical place, hence the name data center."[12] At the time of the case, all of the agency's data centers were located in one metropolitan area. Senior leaders viewed the need for a new data center as an opportunity to make data center operation more secure by selecting a new location outside of the metropolitan area. There would be multiple approval levels to obtain the funds and approve the location decision within and outside of the agency. The agency needed to select the best data center location and justify the decision to budget approvers in the executive branch and Congress.

We also illustrate an IT portfolio decision (Chapter 12) for the same agency as an example of portfolio decision analysis. Again, the agency needed to justify the IT portfolio decisions to budget approvers in the executive branch and the Congress.

1.6 Summary

This chapter provides an introduction to the discipline of decision analysis. Decision analysis is a socio-technical process that must have sound quantitative theoretical underpinnings, but also must be done in the context of organizational and environmental considerations. We introduce the iterative decision analysis process that we use in the handbook and use this process to list the key technical products and the nine categories of soft skills (strategic thinking, leading teams, managing teams, researching, interviewing individuals, surveying individuals, facilitating groups, aggregating expertise, and communicating) necessary to help organizations create value for their SH. We compare three important application areas (oil and gas, pharmaceuticals, and military) and summarize several decision analysis success stories (Chevron, SmithKline Beecham, Army base realignments and closures, and government data centers). Next, we define a decision professional, discuss the education and training of decision professionals, identify some of their major professional societies, and described some of their professional service activities. We conclude with an overview of the handbook and an introduction to the three substantive illustrative examples (RNAS, Geneptin, and Data Center/IT).

KEY TERMS

Decision An irrevocable allocation of resources.

Decision analysis Decision analysis is a philosophy and a social-technical process to create value for DMs and SH facing difficult decisions involving multiple SH, multiple (possibly conflicting) objectives, complex alternatives, important uncertainties, and significant consequences.

Decision analyst An individual who uses the technical and soft skills of decision analysis.

[12]http://www.gartner.com/it-glossary/data-center/, accessed November 24, 2012.

Decision maker The leader vested with the responsibility and the authority to make organizational decisions.
Decision professional An individual who wants to learn and apply the decision analysis technical and soft skills that have been proven to help senior leaders create value for their organizations.
Probability A mathematical theory of uncertainty based on three axioms. See Appendix A.
Stakeholder An individual or an organization with a significant interest in a decision under consideration.
Risk Risk is the probability and consequence of a bad outcome.
Uncertainty The potential outcome of an event or events, which is not known with certainty.
Value A fair return or equivalent in goods, services, or money for something exchanged.[13]

REFERENCES

Burk, R.C. & Parnell, G.S. (2011). Portfolio decision analysis—Lessons from military applications. In A. Salo, J. Keisler, & A. Morton (eds.), *Advances in Portfolio Decision Analysis Methods for Improved Resource Allocation*, pp. 333–358. New York: Springer.

Clemen, R.T. & Reilly, T. (2001). *Making Hard Decisions with Decision Tools*. Belmont, CA: Duxbury Press.

Corner, J.L. & Kirkwood, C.W. (1991). Decision analysis applications in the operations research literature, 1970–1989. *Operations Research, 39*(2), 206–219.

Ewing, P., Tarantino, W., & Parnell, G. (2006). Use of decision analysis is the army base realignment and closure (BRAC) 2005 Military Value Analysis. *Decision Analysis, 3*(1), 33–49.

FDA (n.d.). Table of valid genomic biomarkers in the context of approved drug labels. http://www.fda.gov/Drugs/ScienceResearch/ResearchAreas/Pharmacogenetics/ucm083378.htm, accessed June 19, 2012.

Frueh, F.W., Amur, S., Mummaneni, P., Epstein, R.S., Aubert, R.E., DeLuca, T.M., et al. (2008). Pharmacogenomic biomarker information in drug labels approved by the United States Food and Drug Administration: Prevalence of related drug use. *Pharmacotherapy, 28*, 992–998.

Howard, R.A. (1966). Decision analysis: Applied decision theory. *Proceedings of the Fourth International Conference on Operations Research*, pp. 55–71. Wiley-Interscience.

Howard, R.A. & Matheson, J.E. (eds.). (1983). *Readings on the Principles and Applications of Decision Analysis*. Vols. I and II. Palo Alto, CA: Strategic Decisions Group.

Hu, S., Berndt, M.R., Aitken, E.R., & Epstein, A.M. (2012). Identifying personalized medicine therapeutics and quantifying their utilization. Submitted for reveiw.

[13]Definition of Value, http://www.merriam-webster.com/dictionary/value, accessed 29 July 2011.

Hu, S.X., Foster, T., & Kieffaber, A. (2005). Pharmacogenomics and personalized medicine: Mapping of future value creation. *BioTechniques*, *39*(10 Suppl), s1–s6.

Keefer, D.L., Kirkwood, C.W., & Corner, J.L. (2004). Perspectives on decision analysis applications: 1990–2001. *Decision Analysis*, *1*(1), 5–24.

Laing, R.E., Hess, P., Shen, Y., Wang, J., & Xu, S. (2011). The role and impact of SNPs in pharmacogenomics and personalized medicine. *Current Drug Metabolism*, *12*, 480–486.

Menke, M., Spetzler, C., & Keelin, T. (2011). The value of decision quality/decision analysis: Building a compelling case for decision makers. http://www.decisionprofessionals.com/news_events.html, accessed September 4, 2011.

Phillips, L.D. (2005). Decision analysis in 2005. In A. Robinson & J. Parkin (eds.), *OR47 Keynotes/Tutorials*, pp. 115–132. Birmingham: Operational Research Society.

Phillips, L.D., Humphreys, P., Embrey, D., & Selby, D. (1990). A social-technical approach to human reliability. In R.M. Oliver & J.Q. Smith (eds.), *Influence Diagrams, Belief Nets, and Decision Analysis*. Chichester: John Wiley & Sons.

Raiffa, H. (1968). *Decision Analysis: Introductory Lectures on Choices Under Uncertainty*. Reading, MA: Addison-Welsey Publishing Co.

Sharpe, P. & Keelin, T. (1998). How SmithKline Beecham makes better resouce allocation decisions. *Harvard Business Review*, *76*, 45–57.

Trusheim, M.R., Berndt, E.R., & Douglas, F.L. (2007). Stratified medicine: Strategic and economic implications of combining drugs and clinical biomarkers. *Nature Reviews Drug Discovery*, *6*, 287–293.

CHAPTER TWO

Decision-Making Challenges

TERRY A. BRESNICK and GREGORY S. PARNELL

Two Perspectives on Decision Making

Nothing is more difficult, and therefore more precious, than to be able to decide.
—Napoleon, "Maxims," 1804

Nothing good ever came from a management decision. Avoid making decisions whenever possible. They can only get you in trouble.
—Dogbert, *Dogbert's Top Secret Management Handbook*, 1996

Handbook of Decision Analysis, First Edition. Gregory S. Parnell, Terry A. Bresnick, Steven N. Tani, and Eric R. Johnson.

2.1 Introduction

In this chapter, we describe decision-making challenges and introduce some reasons why decision analysis is valuable to decision makers, but also why decision analysis may be difficult to apply. The axioms of decision analysis (See Chapter 3) assume rational decision makers operating in efficient organizations, but that may be the exception rather than the rule. Although decision analysis is mathematically sound, it is applied in the context of human decision making and organizational decision processes. As decision analysts, we interact with decision makers, stakeholders, and subject matter experts to build models in environments where objective data may be scarce and we have to rely on eliciting knowledge from decision makers, stakeholders, and subject matter experts (SMEs) who are prone to many cognitive biases. We must develop our soft skills as well as our technical skills. The challenges introduced in this chapter include understanding organizational decision processes, understanding decision traps, and understanding cognitive and motivational biases. Soft skills, such as strategic thinking, leading teams, managing projects, researching, interviewing, and facilitating group meetings, are covered in Chapter 4. Even if we do a superb technical analysis, the most important part of our job remains—communicating results of the analysis to decision makers and stakeholders as discussed in Chapter 13.

The rest of the chapter is organized as follows. Section 2.2 discusses the decision-making processes that humans typically employ. Section 2.3 introduces the factors that make decision making a challenge. Section 2.4 introduces the social and organizational decision factors that impact how a decision analysis can be conducted, including the organizational culture, the impact of stakeholders, and the level at which decisions are made. Section 2.5 discusses issues involved in obtaining credible domain knowledge for making the decision and the role of experts in providing this data. Section 2.6 focuses on the behavioral aspects of decision making to include decision traps and barriers and cognitive biases that affect the decision-making process. Section 2.7 provides two anecdotes of success and failure of supporting the human decision-making process. Section 2.8 sets the stage for our illustrative decision problems used throughout the handbook by setting the decision-making context.

2.2 Human Decision Making

Decision analysis practitioners sometimes take it for granted that people have little difficulty in making decisions. We often assume that thinking about alternatives, preferences, and uncertainty comes naturally to those we are trying to help, and that rational thought is the norm. The reality is that regardless of how many well-documented methodologies with thoroughly proven theorems are provided to them, human decision makers are inconsistent. Roy Gulick, a decision analyst with Decisions and Designs, Inc., coined the term "anatomical decision making" to describe the answers he had gathered from decision makers over the years to the question, "How do you make your decisions?" Why "anatomical decision making"? The responses he received typically included things such as "*seat* of the pants," "*gut* reaction," "rule of *thumb*," "top of the *head*," "*knee*-jerk reaction," "pulled it out of my . . ."—almost every part of the body was mentioned except "the brain"! These responses serve to remind us that no matter how well-honed our analytical methods are, and no matter how much objective data we have, human decision making cannot be overlooked. Humans are sometimes inconsistent, irrational, and subject to cognitive biases. Nonetheless, their subjective judgments, tenuous though they may be, must be included in the decision analysis.

All that said, decision analysts believe that the human decision-making process can be studied systematically, and that coherent, structured, and formal processes are better than purely "anatomical" decision-making processes.

2.3 Decision-Making Challenges

To achieve effective decision making, it is desirable to bring rational decision makers together with high-quality information about alternatives, preferences, and uncertainty. Unfortunately, the information is not always the high quality we would like. While we want to have factual information, very often, erroneous and biased data slip in as well. While we would like to have objective information based upon observed data, sometimes, the best we can obtain is opinion, advice, and conjecture. Similarly, the rational decision makers are not always as rational as we would like. Often, the goals of an organization are ambiguous and conflicting. In many cases, the decision-making environment is characterized by time pressures that impose additional constraints. As a result, the effective decision making that we seek is often less attainable than we desire. One key role of decision analysis thus becomes providing an effective link between the decision makers and the best available information.

Decision problems are complex, and this complexity can be characterized in three dimensions, as shown in Figure 2.1 (Stanford Strategic Decision and Risk Management Decision Leadership Course, 2008). Content complexity ranges from few scenarios with little data and a relatively stable decision-making setting, to many scenarios with data overload, many SMEs involved, and a dynamic decision context. Analytic complexity ranges from deterministic problems with little uncertainty and few fundamental and means objectives to problems with a high degree of uncertainty, many alternatives, and a complicated value hierarchy with many dependencies (see Chapter 7). Organizational complexity ranges from a single decision maker with a homogeneous set of stakeholders to multiple decision makers requiring consensus and a diverse set of stakeholders with conflicting perspectives. The best time to address the organizational complexity is when we are setting up the project structure by engaging the right people in the right way (see Chapter 4).

As we discuss in Chapters 9, 10, and 11, when we are eliciting expertise about a decision situation, modeling its consequences, and analyzing the results,

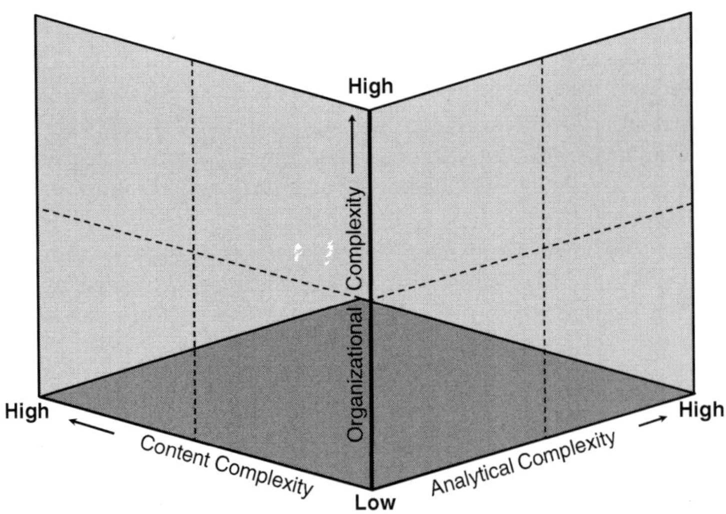

FIGURE 2.1 Dimensions of decision complexity. (Adapted from Stanford Strategic Decision and Risk Management Decision Leadership Course, 2008, used with permission.)

it can be helpful to further decompose the analytical and content complexity into five more specific dimensions: value components, uncertainty, strategy, business units, and time.

2.4 Organizational Decision Processes

As decision analysis practitioners, we are asked to go into organizations, whether in the public or private sector, and to work within the existing decision processes. The key thing to remember is that one size does not fit all. Each decision opportunity has its own unique characteristics, and we must be willing and able to adapt to the individuals and the organizational decision-making environment. There are many examples of analytically sound studies that sit on bookshelves or in trashcans because the processes used and the conclusions reached did not "fit" with the existing organizational decision processes. All too often, analysts tend to think of processes used to "solve" the client problems that they face as "technical" processes. Properly applying decision trees, or influence diagrams, or Monte Carlo simulations, may provide a superb technical solution to the problem, but, by themselves, they can miss what may be the most important part of the solution—the social aspects of the solution. Larry Phillips of the London School for Economics describes what decision analysts do as a "socio-technical process." (Phillips, 2007) The way that the technical solution fits into the organizational culture, structure, decision-making style, and other factors may determine the acceptability of the technical solution. These factors are discussed in the next section.

2.4.1 CULTURE

Decision analysis cannot be performed in isolation in any organization. The approach must be tailored to the context of the problem and the culture and environment of the organization. Culture can include many aspects that must be considered. Some of the major factors to consider include:

Public versus private sector. Decision making in a public sector environment can be very different than in a private sector environment. Many public sector decisions are made in a setting of openness and transparency, while others are made in parts of the public sector, such as the Department of Defense (DoD) and the Intelligence Community (IC), which are very security conscious and where information protection and "need-to-know" are the guiding principles. Private sector decisions in some domains are often proprietary and protected as well.

Geographical. Decision-making practices and approaches can vary greatly from country to country. Understanding the value systems, legal systems, moral and ethical underpinnings, and cultural mores are critical for success. What is an accepted practice in one country can be a terrible *faux pas* in another. For example, according to O'Boyle, people brought up in the U.S.

have difficulty understanding those who prefer identity as a group rather than as an individual; Japanese find it unsettling to deal with U.S. companies whose policies change as management changes; and Americans consider the Dutch unassertive in their business approach, while the Dutch consider the standard American business résumé to be so boastful as to be unreliable (O'Boyle, 1996).

Leadership style. The decision-making process and the role that decision analysts can play is highly dependent on leadership style. Some of the most significant aspects of leadership style that impact the nature of the analysis that can be performed include:

- **Degrees of authoritativeness.** Some organizations have highly authoritative leaders, while others practice a more democratic style of leadership.
- **Degree of delegation.** Some leaders are more willing than others to delegate both decision-making responsibility and authority.
- **Decision-maker engagement.** Some decisions makers will provide initial guidance and will not want to be involved again until results are ready, while others will want to be involved in every step of the process.
- **Number of decision makers.** In some rare cases (especially in public decisions), there is a single decision maker, while in others, the decisions are made by committee, which involves aggregating across decision makers.
- **Degree of formality.** Some organizations have very formal decision-making and leadership styles that can make it very difficult to get access to the decision maker without having to go through "gatekeepers" who fiercely protect schedules. Other organizations provide easier access to the decision maker.
- **Openness to new ideas and innovation.** Some organizations are highly innovative and are willing to accept new and better approaches to problem solving, while others prefer their current approaches to doing business. "Not invented here" can be a significant barrier to being able to perform a sound decision analysis.
- **Comfort with outside consultants versus use of insiders.** Some organizations are more comfortable getting most of their analytical support from inside the organization using analysts who are experts in both the process and the subject matter. In such an environment, it may be difficult for an "outsider" to have an impact.

The most important thing to remember is that as decision analysts, we must be prepared to adapt our techniques and processes to the culture of the organization, especially to the style of the decision maker. Keep in mind the "golden rule" of consulting—"the one with the gold makes the rules." As we develop our analytical solutions, we must offer a process that matches the organizational culture.

Decision analysis is a social-technical process: we must design a process that uses the right people (broad and deep knowledge of the problem), the right forum (conducive to discussion and interaction), the right balance of modeling and challenging the model with intuition, and the right duration (meet needed deadlines but enable information gathering and socializing the results).

2.4.2 IMPACT OF STAKEHOLDERS

> A stakeholder is a person, group, or organization that has direct or indirect stake in an organization because it can affect or be affected by the organization's actions, objectives, and policies. Key stakeholders in a business organization include creditors, customers, directors, employees, government (and its agencies), owners (shareholders), suppliers, unions, and the community from which the business draws its resources. (BusinessDictionary.com, 2011)

Stakeholders comprise the set of individuals and organizations that have a vested interest in the problem and its solution (Sage & Armstrong, 2000). Understanding who is affected by a solution to a decision problem provides the foundation for developing a complete definition of the problem. Stakeholders collectively perform many functions. They help frame the problem and specify constraints; they participate in the alternative generation and solution process to include evaluation and scoring; they provide data and subject matter expertise; and they identify and often execute tasks for implementing recommended solutions (Parnell et al., 2011).

A straightforward taxonomy of stakeholders is offered in Decision Making in Systems Engineering and Management (Parnell et al., 2011). Stakeholders can choose to be active or passive when it comes to participating in the decision process. Stakeholders are listed in typical order of relative importance:

- **Decision authority.** Person or persons with ultimate authority and responsibility to accept and implement a solution to a decision opportunity;
- **Client.** Person or organization that initiated the request for decision support; often, the client defines the requirements and holds the purse-strings for the effort;
- **Owner.** Person or organization responsible for proper and purposeful operations surrounding the decision;
- **User.** Person or organization accountable for conducting proper operations of systems related to the decision;
- **Consumer.** Persons and organizations with intentional dependencies on the implications of the decision.

Stakeholder analysis is a key technique to ensure that the problem has been fully described before we attempt to obtain a solution to the problem. The three most common techniques for stakeholder analysis are interview, focus groups, and surveys. Several techniques are available for soliciting input from diverse stakeholders as shown in Table 2.1 (Trainor & Parnell, 2007). The techniques are characterized and compared on five attributes—time commitment, ideal stakeholder group, preparation, execution, and analysis.

Stakeholder analysis is critical, since the fundamental and means objectives (see Chapter 7) are built upon the needs of the stakeholders. Without a clear understanding of the different perspectives and different success criteria upon which alternatives will be judged, the analysis can easily be built upon a shaky foundation that will not withstand the pressures of intense scrutiny and organizational implementation. The best practices for the use of these techniques are presented in Chapter 4.

2.4.3 DECISION LEVEL (STRATEGIC, OPERATIONAL, AND TACTICAL)

Decision analysis can be applied at a variety of decision levels in an organization. A common characterization of decision levels includes strategic, operational, and tactical. A good analysis must balance concerns across all three and must consider the dependencies across levels.

2.4.3.1 Strategic Decision Making. This level is focused on the long-term goals and directions of the organization, which are often expressed in the organization's strategic plan. Strategic decision making is oriented around the organization's mission and vision for where it wants to be in the future. It addresses very fundamental issues, such as what business the organization *is* in versus what business it *should* be in? What are the core values? What products and services should it deliver? Who are the customer sets? What is management's intent about how the organization will evolve and grow? From the decision analyst's perspective, this level of decision making typically involves the fewest viable alternatives, the greatest degree of uncertainty since it is future oriented, and the greatest need for fleshing out the fundamental objectives since statements of strategic goals are often broad and vague. In order to help an organization, the decision analyst must be a strategic thinker. We identify this as one of the soft skills required for decision analysts (Chapter 4).

2.4.3.2 Tactical Decision Making. This level focuses on turning the broad strategic goals into achievable, measurable objectives (Eyes Wide Open: Tips on Strategic, Tactical and Operational Decision Making, 2011). It requires developing actions and allocating resources that will accomplish the objectives. It includes the set of procedures that connects the strategic goals with the day-to-day operational activities of the organization, and its primary purpose is to enable the organization to be successful as a whole rather than as independent parts (Tactical Decision Making in Organizations, 2011). From the decision analyst's perspective, it is important to identify redundancies and synergies in

TABLE 2.1 Techniques for Stakeholder Analysis

	Time Commitment of Participants	Ideal Stakeholder Group	Preparation	Execution	Analysis
Interviews	30–60 minutes	Senior leaders and key stakeholder representatives	Develop interview questionnaire(s) and schedule/reschedule interviews	Interviewer has conversation with senior leader using questionnaire as a guide. Use a separate note taker.	Note taker types interview notes. Interviewer reviews typed notes. Team analyzes notes to determine findings, conclusions, and recommendations.
Focus groups	Shortest—60 minutes Typical—4–8 hours	Mid-level to senior stakeholder representatives	Develop meeting plan, obtain facility, and plan for recording inputs. May use Group Systems Software to record (Group Systems 2006).	At least one facilitator and one recorder. Larger groups may require breakout groups and multiple facilitators.	Observations must be documented. Analysis determines findings, conclusions, and recommendations.
Surveys	5–20 minutes	Junior to mid-level stakeholder representatives	Develop survey questions, identify survey software, and develop analysis plan. Online surveys are useful.	Complete survey questionnaire, solicit surveys, and monitor completion status.	Depends on number of questions and capability of statistical analysis package. Conclusions must be developed from the data.

the alternatives, to conduct value of information analysis (see Chapter 11) to avoid modeling uncertainties that do not affect decisions, to fully understand how to decompose fundamental objectives of the organization into manageable means objectives, and to avoid suboptimization (see Chapter 7).

2.4.3.3 Operational Decision Making. This level focuses on day-to-day operational decisions, particularly on how the organization allocates scarce resources. Decisions are short term, and the decision context can change rapidly (Eyes Wide Open:Tips on Strategic, Tactical and Operational Decision Making, 2011). In a business context, it can be highly reactive since much depends upon the competitive environment. From the decision analyst's perspective, it frequently involves rapid response, "quick turn" analyses with little data other than that of SMEs. Benefit/cost analysis, to include net present value (NPV) analysis, and high-level multiple-objective decisions analyses (MODA), are frequently used tools that are appropriate for longer time horizons.

Identifying the decisions to be made in a decision analysis is a nontrivial task. Knowing the decision level is one useful technique. In Chapter 6, we introduce the decision hierarchy, which is a decision framing tool to help define the decisions in sufficient detail to perform a decision analysis.

2.5 Credible Problem Domain Knowledge

2.5.1 DISPERSION OF KNOWLEDGE

In theory, it is easy to think of the decision analyst as working directly with the decision maker to build models and solve problems. In practice, it is rarely that straightforward. Most decision analyses rely on knowledge that is dispersed among many experts and stakeholders. The views of individual decision makers are subject to both cognitive and motivational biases, and often, these views must be interpreted and implemented by groups of stakeholders in multiple organizations. The decision analyst is often asked to take on a complicated role other than model builder—the analyst must be the facilitator who translates management perspective to others, who balances conflicting perspectives of stakeholders, who elicits knowledge from dispersed SMEs, and who combines these varied ingredients into a composite model that organizes and integrates the knowledge (see Chapter 9). While unanimity among participants is a noble goal, it is exceptionally rare. Consensus is a more achievable goal—if it is defined as developing a solution that everyone can "live with" rather than as any form of unanimity. But even to achieve consensus, the decision analyst must use modeling and facilitation skills to bring together technical knowledge (often the purview of scientists and engineers) with business knowledge (often the purview of managers and financial personnel) in the particular domain at hand. This is no easy task as sources of such knowledge can be varied and uneven in quality.

2.5.2 TECHNICAL KNOWLEDGE: ESSENTIAL FOR CREDIBILITY

For a decision analysis to be credible, it must be based upon sound technical knowledge in the problem domain. In some cases, the decision maker may have such knowledge by coming up through the ranks. In others, the decision maker is more focused on the business side of the organization and relies upon the technical staff of scientists, engineers, and others to provide such knowledge. In some ways, it is easier to reconcile conflicting opinions on technical matters than on business matters since technical matters tend to be more factually based and objective. That said, sometimes, it is difficult to establish which "facts" to believe, particularly on controversial issues, such as global warning!

There can be a huge difference in the level of domain technical knowledge required of the decision analyst. Specific technical knowledge may be demanded of a decision analyst who is internal to an organization. This is typical, for example, in the oil and gas industry and in the pharmaceutical industry. For a decision analyst external to an organization, there may be less expectation of technical knowledge, but rather the expectation is that the decision analyst can work with a group of technical experts to identify and model the key concerns. In fact, in many consulting firms, it is considered to be an advantage for the decision analyst to not be burdened by having to be an expert in the technical aspects of the organization; this allows the decision analyst to focus on the decision process.

For either the internal or the external decision analyst, it is essential to help the client develop a clear set of objectives, a range of possible outcomes, and probability distributions, to flesh out the key assumptions and constraints, to understand the factors that could create extreme outcomes, and to document how the technical knowledge obtained from others is used.

2.5.3 BUSINESS KNOWLEDGE: ESSENTIAL FOR SUCCESS

While a firm grasp on domain technical knowledge is essential for credibility, a firm grasp on business knowledge is essential for success. Such knowledge includes analysis of the competition, analysis of the economic environment, analysis of the legislative environment, and analysis of required rates of return, among other business environment areas. As with technical knowledge, it would not be unusual for the decision analyst to not be a SME in these areas, but rather, obtain such knowledge required for decisions from business experts internal or external to the organization. Familiarity by the decision analyst with corporate financial reports, benefit/cost analysis, costing approaches, net present value calculations, and portfolio theory may be essential for success.

2.5.4 ROLE OF EXPERTS

The role of experts in decision analyses is not always as straightforward as one might think. Clearly, they provide factual, objective information in the areas of their expertise. Whether it is actual hard technical data, actual performance data, or projected performance data on proposed systems, most technical experts are comfortable providing both point estimates and uncertainty ranges on such data. Where it becomes murkier is when SMEs or technical experts are asked to provide "preference" or "value judgment" estimates that may require access to stakeholders and decision makers.

2.5.5 LIMITATIONS OF EXPERTS

As one would expect, all experts are not created equal. Some have more technical or business knowledge than others, and it is not always easy for the decision analyst to determine their limitations. Even experts are subject to motivational and cognitive biases—in fact, as will be pointed out in Section 2.6.2, experts may be even more subject to biases, such as failing to spread probability distributions widely enough. As indicated above, some experts are very uncomfortable doing anything other than "reporting" on what they know and may be unwilling to provide value judgments. Some experts will attempt to dominate the group by citing that their knowledge is more recent or more authoritative than others, and this may have the effect of "shutting down" other experts that are present. Some will "stretch" their areas of expertise to go far beyond their actual areas of expertise, often providing a mix of very good quality information with less credible information. One of the greatest challenges for the decision analyst is to determine the *bona fides* of both the experts *and* of the expertise they provide and to determine what can and cannot be used.

2.6 Behavioral Decision Analysis Insights

This section provides insights into the decision traps and barriers that get in the way of good decision analysis practice, as well as into the cognitive and motivational biases that can impact the quality of the knowledge we elicit from decision makers and SMEs.

2.6.1 DECISION TRAPS AND BARRIERS

An excellent summary of behavioral decision insights and barriers to good decision making can be found in *Decision Traps* (Schoemaker & Russo, 1989). The authors describe the ten most dangerous decision traps as follows:

1. **Plunging in.** Starting data gathering and problem solving before fully understanding the complete nature of the problem and the organizational decision processes.
2. **Frame blindness.** Setting out and solving the wrong problem because the mental framework we are using is incomplete or incorrect. If the framework is wrong, it is difficult to develop a complete set of alternatives or to properly specify the values used in the decision.

3. **Lack of frame control.** Failing to consciously define the problem in more ways than one or being unduly influenced by the frames of others.
4. **Overconfidence in our judgment.** When we rely too heavily on our assumptions and opinions, it is easy to miss collecting the key factual information that is needed.
5. **Short-sided shortcuts.** Relying inappropriately on "rules of thumb," and failing to avoid well-known cognitive biases.
6. **Shooting from the hip.** Believing we can informally keep track of all information gathered in our head, and "winging it" rather than relying on systematic procedures.
7. **Group failure.** Assuming that a group of many smart people will automatically lead to good choices even without managing the group decision-making process.
8. **Fooling ourselves about feedback.** Failing to interpret the evidence from past outcomes, either through hindsight biases or ego issues.
9. **Not keeping track.** Failing to keep systematic records of results of key decisions and failing to analyze them for lessons learned. This is sometimes called not being a learning organization.
10. **Failure to audit our decision process.** Failing to develop an organized approach to understanding our own decision making, thus exposing ourselves to the other nine decision traps.

The first author of this chapter has compiled a similar list of barriers to good decision analysis based upon his experiences over the last 35 years as follows:

1. **Inadequate problem formulation.** This is related to Schoemaker and Russo's frame blindness. Analysts frequently overconstrain or underconstrain the problem statement, thus leading to alternatives that do not really make sense, or eliminating alternatives prematurely. James Adams, in *Conceptual Blockbusting* (Adams, 1979), cites an example of how millions of dollars were spent to reduce damage to crops being caused by mechanical tomato pickers. The original problem statement was to "develop a better mechanical tomato picker to keep from damaging the tomato crops." Once the problem statement was reframed as "reduce damage to the tomato crop that is currently caused by mechanical tomato pickers," a totally new class of solutions emerged, and the problem was solved by developing a thicker-skinned tomato that was more resistant to damage.
2. **Decision paralysis by waiting for "all of the data."** Analysts and decision makers frequently fail to finish their work because they continually search for more data. In reality, it is rare that "all" the data are ever available, and it is more effective to use a "requisite" approach to data gathering—gather what is needed to make the decision and no more (Phillips, 2007).
3. **Looking for a 100% solution.** The analyst has to know when to stop modeling. Similar to the notion of requisite data as described above, a

requisite decision model is one that is sufficient in form and content to resolve the issues at hand (Phillips, 2007). As the French author Voltaire said, "the perfect is the enemy of the good".

4. **Ineffective group decision-making processes.** This relates to Schoemaker and Russo's Group Failure. Too often, analysts assume that group processes are automatically better than individual decision processes. This view has been particularly magnified with the advent of group decision-making software. Such software manufacturers often tout the advantage of the anonymity that is provided to increase willingness to participate. However, the opinion of many experienced decision analysts runs counter to this, and anonymity can degrade the group process by providing a shield to hide behind in allowing participants to remain parochial in their views, thus decreasing open exchange of information.
5. **Lack of access to the decision maker.** All too often analysts do not get the access to the decision maker that is essential to properly frame the problem and to understand the preferences essential for value-focused thinking. Frequently, this is the result of "gatekeepers" to the decision makers who are afraid to have questions asked of the decision maker, either by themselves or by outside consultants, lest the decision maker think that his or her analysts do not know what they are doing. It is the collective experience of all authors of this handbook that it is the very rare decision maker who does not appreciate the opportunity to participate in the process. Rather than fearing an opportunity to communicate with the decision makers face-to-face, decision analysts should seek to meet with the decision makers as early and as often as possible without wasting their valuable time (see Chapter 5).
6. **Insensitivity to deadlines.** Analysts often get so enraptured with the cleverness and sophistication of their techniques that they lose sight of the decision timelines. It is essential that analyses be "designed to time" so as to be responsive to decision-maker needs. A timely 80% solution is usually better than a 100% solution that is too late. Managing projects is one of our essential soft skills (Chapter 4).
7. **No plan to implement the decision.** It is essential to have a workable plan to implement the solution that is produced by the analysis. A mathematically correct alternative that cannot be executed within time, budget, and other organizational constraints is of little use to the client organization. See Chapter 14 for more discussion on implementation.

2.6.2 COGNITIVE BIASES

"Cognitive biases are mental errors caused by our simplified information processing strategies. A cognitive bias is a mental error that is consistent and predictable" (Heuer, 1999). As analysts attempt to elicit value judgments and probabilities from decision makers and SMEs, they must confront many of the cognitive biases that are well documented in the behavioral decision analysis literature.

Some of the biases are related to decision making, some to probability or value assessment, and some to personal motivation (motivation to have positive attitudes toward oneself). This section provides a quick overview of the most common biases. The letters DM after the name of the bias refer to decision-making and behavioral biases, P to probability or belief biases, and M to motivational biases.

- **Framing effect (DM).** Drawing different conclusions from the same information, depending on how that information is presented. For example, assume a military commander has 500 troops under his command and is about to undertake a military operation. If he takes alternative A, he is certain to lose 250 people (the rest will survive unscathed). If he takes Alternative B, there is a 50% chance 150 will die, and a 50% chance 350 people will die. Which should he choose? When framed in this manner and presented to subjects, an overwhelming majority select alternative B. Note that the expected losses for A and B are the same—250 deaths. Now consider the question reframed as "if he takes alternative A, he is certain to save 250 people. If he takes Alternative B, there is a 50% chance he will only save 150 people, and a 50% chance he will save 350 people." When presented this way, the vast majority of subjects select option A. When couched as certain lives lost, they choose the "lottery" alternative; when couched as certain lives saved, they choose the certain alternative even though the choices are the same (Mellers & Locke, 2007).
- **Bandwagon effect (DM).** The tendency to act or perceive things in a certain way merely because many other people act or perceive them in the same manner. The bandwagon effect is related to groupthink and herd behavior, and is often seen in election polling where people want to vote for the candidate they perceive will be the winner. We also see the bandwagon effect in personal finance where, for example, many people "jump on the bandwagon" and buy the mutual funds that performed best the previous year, yet it is rare that the same funds are top performers year after year.
- **Information bias (DM).** The tendency to seek information believing that it will help the decision process even when it cannot affect action (Baron, 2000). For example, people expend resources to gather information without performing a value of information analysis.
- **Confirmation bias (also known as selective search for evidence) (DM).** The tendency to search for or interpret information in a way that confirms one's preconceptions (Oswald & Stefan, 2004). It is often manifested by early information leading to misinterpretation of later information. For example, when the U.S. Navy ship U.S.S. Vincennes erroneously shot down a commercial Iranian Airbus, the Tactical Control Officer (TCO), responsible for identifying the aircraft as friend or foe, initially believed it to be a foe based upon its flight path. As additional information came in, such as altitude, speed, and location within safe passage corridors, the TCO misinterpreted each new clue in a way that supported his initial hypothesis

of foe (Silverman, 1992). It is a natural tendency to seek confirming information since we all like to be proven correct—yet disconfirming information often has far more value than confirming information.

- **Anchoring and adjustment bias (DM, P).** The common human tendency to rely too heavily, or "anchor," on one trait or piece of information when making decisions and fail to sufficiently adjust from that anchor (Tversky & Kahneman, 1974). When focusing on the initial estimate in attempting to describe a complete probability distribution, there is a tendency to stay too close to the anchor and not adjust the extremes of the distribution enough. For example, in trying to put a probability distribution on the number of McDonalds in the United States, most people will first select their "best guess" (the number where it is equally likely to be over as under) and tend to underestimate the spread between the 1% and 99% points of the distribution. Probability assessors do better if they first estimate the point where they are 99% sure that the right answer is less than the number they have specified at the high end and the point where there is a 99% chance that the right answer is greater than the number they have specified at the low end. (There were 13,381 McDonalds in the United States as of May 2009!)
- **Availability bias (DM, P).** People predict the frequency of an event, or a proportion within a population, based on how easily an example can be retrieved from memory (Heuer, 1999). In our decision-making processes, our estimates are influenced by the information that is most available to us either through personal experiences, dramatic or easily imagined events, or through the normal workings of the mind. For example, when asked if the letter "k" is more likely to be the first letter of a word or the third letter, most people will say the first letter even though it is three times as likely to be the third letter. Words with "k" as the first letter are far more "available" in our minds than words with the letter "k" in the third position (Tversky & Kahneman, 1974).
- **Base rate bias (P).** The tendency to base judgments on specifics, ignoring general statistical information (Baron, 2000). Base rate bias occurs when the conditional probability of some hypothesis given some evidence fails to take into account the "base rate" or "prior probability" of the hypothesis and the total probability of the evidence (Tversky & Kahneman, 1982). The base rate bias is one example of the broader category of biases known as representativeness biases. For example, many people would say that a person who is described as quiet, introverted, and orderly is more likely to be a librarian than a salesperson even though a person chosen at random is much more likely to be a salesperson than a librarian (Tversky & Kahneman, 1974).
- **Certainty illusion (P).** Prior expectations of relationships lead to correlations that do not exist. Many people use correlation interchangeably with causality, but they are different concepts (Heuer, 1999). For example, in experiments posing the multiple choice question "Potatoes are native to (A) Peru or (B) Ireland," the vast majority of respondents will say Ireland. When

asked to put a probability on the likelihood that they have the right answer, most will say more than 90%, with many saying 100%. Yet the correct answer is Peru. Potatoes are not native to Ireland, but many people associate potatoes with Ireland from their prominence during the well-known famine. The association of potatoes to Ireland during the famine leads to a correlation that is false in determining the native source of potatoes. This is also referred to as the illusion of validity, where the confidence that people express in their predictions has little or no regard to factors that limit predictive accuracy (Tversky & Kahneman, 1974).

- **Hindsight bias (P).** Sometimes called the "I-knew-it-all-along" effect, the tendency to see past events as being predictable at the time those events happened (Pohl, 2004). For example, researchers asked college students to predict how the U.S. Senate would vote on the confirmation of Supreme Court nominee Clarence Thomas. Prior to the vote, 58% of the participants predicted that he would be confirmed. When the students were again polled after the confirmation hearings, 78% of the participants said they thought he would be approved. In another study, potential voters were asked for whom they would vote in an upcoming election. In after-the-fact polling of the same voters 1 month later, a far greater percentage said they voted for the winner than originally said they would vote for that candidate (Myers, 1994).
- **Representativeness bias (P).** Occurs when people judge the probability or frequency of a hypothesis by considering how much the hypothesis resembles (or is representative of) available data as opposed to using a Bayesian calculation (see Appendix A). In causal reasoning, the representativeness heuristic leads to a bias toward the belief that causes and effects will resemble one another (Tversky & Kahneman, 1974). Humans tend to ignore sample sizes when making probability judgments, and they expect small samples to mirror population statistics. Local representativeness is when people perceive that small samples represent their population to the same extent as large samples (Tversky & Kahneman, 1982). The Gambler's fallacy is an example of this thinking pattern—when a sequence of randomly generated trials repeatedly strays in one direction (e.g., a roulette wheel comes up red three times in a row) people mistakenly expect the opposite to be more likely over the next few trials (e.g., the wheel landing on black) (Goodie, 2011).
- **Recency effect (P).** The tendency to weigh recent events more than earlier events. For example, people will generally overestimate the likelihood of a shark bite or of a person being hit by lightning if they have recently read or heard about such an event. This is closely related to the availability bias, where time frame proximity makes the information more readily accessible (Tversky & Kahneman, 1974).
- **Self-serving bias (M).** Occurs when people attribute their successes to internal or personal factors but attribute their failures to situational factors beyond their control. The self-serving bias can be seen in the common human tendency to take credit for success but to deny responsibility for

failure (Miller & Ross, 1975). Another version occurs when individuals evaluate ambiguous information in a way that is the most beneficial to their self-interests. This bias is related to the better-than-average effect (also known as the superiority illusion) in which individuals believe that they perform better than the average person in areas significant to their self-esteem (Kruger, 1999).

These and other biases can lead to common assessment mistakes to include the following:

- Innate overconfidence makes us assume that what we know is correct—even when facts are limited, or when dealing with SMEs.
- We rely on data that should not count to make important decisions.
- We confuse memorable events with important and meaningful ones.
- We tend to ignore the odds, even when they are heavily against us.
- We underestimate the role of "lady luck" in everyday events.
- We do not always treat a dollar as a dollar; for example, we feel differently if we lose a theater ticket that we paid for than one we receive as a prize, yet the loss is the same—a ticket that could be used for the theater!
- We mentally put money into categories that do not make much sense;, for example, we think about spending money differently if we receive it as a gift verse having earned it, even though both go in the same bank account and can buy the same things.
- Our tolerance for risk is inconsistent.
- We often throw good money after bad because we do not ignore sunk costs
- We often overlook the opportunity costs of not making a decision.

The first step in overcoming these biases is to be aware of and to recognize them when they occur. Additional ways to counter biases and faulty assessment heuristics include the following:

- Carefully define what is being estimated.
- Use multiple assessment methods as consistency checks.
- Postulate multiple hypotheses and set up a list of pros and cons for each.
- Use the "crystal ball" test: "assume a crystal ball said your most unlikely hypothesis was true; how could this possibly have happened?"
- Seek disconfirming information as well as confirming information.
- When anchors may be present, seek several different anchors.
- Seek other opinions from individuals known or likely to have different opinions.

2.7 Two Anecdotes: Long-Term Success and a Temporary Success of Supporting the Human Decision-Making Process

We highlight the key points of this chapter with two anecdotes that show the importance that human decision making plays in determining the success or failure of a decision analysis.

In 1978, decision analysts from a decision analysis consulting firm began working with the U.S. Marine Corps (USMC) to develop a new approach for prioritizing items to be funded in the annual budget cycle. They had a "champion" in the form of a young Colonel who was fighting a "that's not how we do it here" attitude in trying to implement an innovative, decision analysis approach. By developing a sound prioritization method based on benefit/cost analysis, tailoring it to the organizational culture and demands, and evolving it as the organizational considerations changed, the decision analysts developed a facilitated process that is still being used today. It is still being facilitated by some of the same decision analysts more than 30 years later. Over the years, the Marines have tried other approaches to prioritizing budgetary items, but they continue to return to the decision analysis framework. In a presentation at INFORMS, the following reasons were given for why the decision process has been so successful (Leitch et al., 1999):

- The Marines (not the decision analysts) own and control the process.
- It forces professional discussions about what is best for the USMC.
- It allows all relevant voices to be heard at the right time.
- The process permits rapid analysis and modification.
- It supports decisions based upon the best-available information.
- It creates an effective synergy between:
  - a quantitative framework and qualitative judgments
  - a people-oriented process and automation
  - rational and irrational input
  - complex thinking and simple modeling.
- It has adapted over time to changing organizational needs.
- It works!

As a point of interest, the young Colonel who took the risk of innovating and who made it all happen was P.X. Kelly, who later became a four-star general and served as Commandant of the Marine Corps.

In 2000, decision analysts from a consulting firm were asked by the Chief Systems Engineer of a major intelligence agency to develop a methodology and process for putting together the annual budget. The existing process was stovepiped, highly parochial, and there was little collaboration among those fighting for shares of the budget, thus leading to suboptimization. The guidance from the decision maker was to put in place a process that would allocate resources efficiently, but more importantly, that would break down internal barriers, encourage cross-discipline discussion, and foster shared purpose. A facilitated process was established that evolved over 7 years that included detailed stakeholder analysis (both internal and external), prioritization of fundamental goals and objectives, a detailed multiple objective value model with scenario-specific value curves, and clear communication tools. The process worked well over the first few years, and collaborative analysis was greatly enhanced. Few argued with the technical correctness of the analytical model. However, the more the process did, the more it was asked to do. Instead of addressing the strategic and tactical decisions for which it was designed, it was being used for day-to-day operational decisions for which it did not have the right degree of sensitivity and granularity to discriminate value among the alternatives. The data demands grew more than the stakeholders could tolerate and accommodate, stakeholder support for the process waned, and the process began to collapse under its own weight. Later, a subsequent decision maker determined that a simpler process was needed, and mandated that no value curves or weights be used and that facilitated processes were not necessary since SMEs internal to the organization could prioritize initiatives on their own. The organization is currently engaged in developing a new process "from scratch" that is "based on only objective data" and better fits the evolved culture and constraints of the organization and the decision-making style of the decision maker. After all, that is the golden rule of management!

2.8 Setting the Human Decision-Making Context for the Illustrative Example Problems

This chapter discusses decision-making challenges in general. We now describe those challenges further in the context of the three illustrative example problems introduced in Chapter 1. See Table 1.3 for further information on the illustrative examples.

2.8.1 ROUGHNECK NORTH AMERICAN STRATEGY (by Eric R. Johnson)

Roughneck Oil and Gas is a global oil and gas company with operations in North America. They had an advocacy-based approach to decision making that led to narrowly framed decisions and only incremental change from one year to the next. Decision makers were not familiar with creating distinctly different alternatives, and analysts had no experience accounting for the actual range of outcomes that might be encountered. The organization had many silos, with little communication between business areas. Within this situation, there was a desire to take a broad strategic look at the assets in North America.

2.8.2 GENEPTIN PERSONALIZED MEDICINE (by Sean Xinghua Hu)

Compared with traditional pharmaceuticals, personalized medicine decision-making creates additional complexity and challenges for a drug development team, particularly for a young biotech company like DNA Biologics.

To make the Geneptin decision, DNA Biologics had to wrestle with a number of cultural and organizational issues. How formal or informal should the decision process be? Should the approach be driven more by data and analysis or more by intuition and experience? To what extent should the senior management be involved in the decision process—fully engaged or a "just show me your final recommendation" approach? Which of the different organizational "habits" or cultures for building consensus should be employed —a top-down, hierarchical approach or a collaborative approach based on preagreed upon criteria? To what extent should the company's decision be driven by science and innovation versus by commercial value?

These organizational and cultural aspects are not specific to personalized medicine, but because of inertia, lack of sufficient knowledge, and higher level of uncertainty associated with personalized medicine, these challenges were amplified at DNA Biologics.

Additional complexity and uncertainty associated with personalized medicine can make the decision analysis more challenging. There are additional decisions to make regarding design, pricing and positioning of diagnostic biomarker tests. There are additional variables to consider, including biomarker prevalence, addressable patient population, market share within the addressable patient population, and probability of success of both the drug and companion diagnostic test. Personalized medicine brings different tradeoffs; for example, a biomarker reduces addressable patient population but can increase the drug's market share among the stratified patient segment. Some variables have a different impact on value—for example, personalized medicine R&D costs may be higher or lower than traditional costs; while other variables serve as new drivers of value—for example, a personalized medicine can offer a better benefit/risk ratio, allowing patients to potentially take it for longer duration and drug manufacturers to provide a more compelling value proposition to payers for reimbursement and to physicians and patients for clinical adoption, increasing commercial value to the drug company.

2.8.3 DATA CENTER DECISION PROBLEM (by Gregory S. Parnell)

A major government agency had an expanding mission that required significant increases in data analysis. The agency's headquarters data centers were already operating at capacity, and the existing data centers lacked additional floor space, power, and cooling. The agency had identified the need for a large new data center and had already ordered a significant number of new servers, but had no process to verify that sufficient data center capacity would be provided to support mission demands. They also had no process for determining the best location of the data center.

There were several complicating factors in the decision. Technology advances had resulted in large servers becoming smaller, consuming more power, and requiring more cooling. Multiple stakeholder organizations within the agency were involved in the decision, including the mission-oriented operational users, the information technology office responsible for procuring and operating, and the logistics office responsible for facilities, power, and cooling. Some of their objectives conflicted with each other. Each believed that it should be the final decision authority, and each believed it had the expertise to make the decisions without using SMEs from the others. All of the organizations appeared to be biased toward solutions with which they were already familiar and comfortable. Life cycle costs were a major factor in the selection of the best data center alternative, and budgets for IT had been shrinking. Multiple approval levels were necessary to obtain the funds from both within and outside the agency. This included the requirement to communicate the need for funds to Congress. The agency had a history of getting what it asked for, but it had been challenged more and more to justify requests for additional funds.

2.9 Summary

For a decision analysis to be useful, it must be theoretically sound, performed with accepted analytical techniques, and be methodologically defensible. But that is not enough. It must work in the environment of the organizational culture and be compatible with the decision making style of the decision makers. It must be based on sound objective data when they are available and on sound subjective information from credible experts when objective data are not available. It must be as free as possible from cognitive and motivational biases of the participants. It frequently must reconcile stakeholders positions based on conflicting perspectives and, often, conflicting information. It often must reach a single conclusion that can be agreed upon by participants who at times have little or no motivation to reach consensus. And finally, it must be communicated clearly and effectively.

These challenges demand that decision analysts be far more than technical experts. We must be knowledge elicitors to gain the information required. We must be facilitators to help overcome group decision-making barriers. We must be effective communicators to prepare presentations and reports that will be read, understood, and accepted. Most of all, we must recognize that decision analysis is not just a science, but an art, and we must be fluent from both perspectives.

KEY TERMS

Anatomical decision making a description given to how people have been observed to make decisions, citing every body part but the brain! (rule of thumb, seat of pants, gut reaction, etc.)

Business knowledge knowledge of the business environment, practices, and procedures of an organization
Cognitive bias a pattern of deviation in judgment that occurs in particular situations.
Decision level the level in an organization at which decisions are made; may be strategic, tactical, or operational
Decision trap a behavior or barrier that impedes effective decision making
Human decision making the process by which we make decisions; it is essentially a technical process that includes the quantitative frameworks that we use coupled with a social process that includes organizational culture, human biases, and intuition used in decision making.
Motivational bias the effect that motivation to have positive attitudes to oneself has on the way a person perceives or acts upon information in the decision process.
Organizational decision making the process of choice through which organizational leaders and managers select among alternatives, allocate resources, or implement strategic goals and objectives.
Technical knowledge substantive knowledge of the specific domain of interest
Soft skills the non-quantitative, "social" skills that complement the quantitative decision-making methodologies; they include strategic thinking, managing, leading, facilitating, interviewing, researching, networking, and communicating.

REFERENCES

Adams, J. (1979). *Conceptual Blockbusting.* Stanford, CA: Stanford Alumni Association.

Baron, J. (2000). *Thinking and Deciding,* 3rd ed. New York: Cambridge University Press.

BusinessDictionary.com. Stakeholder Definition. 2011.

Eyes Wide Open. Tips on Strategic, Tactical and Operational Decision Making (2011). http://www.smallbusinesshq.com.au/factsheet/20305-tips-on-strategic-tactical-and-operational-decision-making.htm.

Goodie, F. (2011). Cognitive distortion as a component and treatment focus of pathological gambling: A review. *Psychology of Addictive Behaviors, 26*(2), 298–310.

Heuer, R. Jr. (1999). *Psychology of Intelligence Analysis.* Mclean, VA: Central Intelligence Agency.

Kruger, J. (1999). Lake Wobegon be gone! The "below-average effect" and the egocentric nature of comparative ability judgments. *Journal of Personality and Social Psychology, 77*(2), 221–232.

Leitch, S., Kuskey, K., Buede, D., & Bresnick, T. Of princes, frogs, and marine corps' budgets: Institutionalizing decision analysis over 23 years. *INFORMS.* Philadelphia, 1999.

Mellers, B. & Locke, C. (2007). What have we learned from our mistakes? In W. Edwards, R. Miles, & D. von Winterfeldt (eds.), *Advances in Decision Analysis*, pp. 351–371. New York: Cambridge University Press.

Miller, D.T. & Ross, M. (1975). Self-serving biases in the attribution of causality: Fact or fiction? *Psychological Bulletin*, *82*(2), 213–225.

Myers, D. (1994). *Did You Know It All Along? Exploring Social Psychology*. New York: McGraw-Hill, 15–19.

O'Boyle, J.G. The culture of decision-making. *R&D Innovator*, 1996.

Oswald, M. & Stefan, G. (2004). Confirmation bias. In R.F. Pohl (ed.), *Cognitive Illusions: A Handbook on Fallacies and Biases in Thinking, Judgement and Memory*. Hove, UK: Psychology Press.

Parnell, G.S., Driscoll, P., & Henderson, D. (eds.) (2011). *Decision Making in Systems Engineering and Management*. Hoboken, NJ: John Wiley & Sons.

Phillips, L. (2007). Decision conferencing. In W. Edwars, R. Miles, & D. Von Winterfeldt (eds.), *Advances in Decision Analysis*, pp. 375–398. New York: Cambridge University Press.

Pohl, R.F. (2004). Hindsight bias. In R.F. Pohl (ed.), *Cognitive Illusions: A Handbook on Fallacies and Biases in Thinking, Judgement and Memory*. Hove, UK: Psychology Press.

Sage, A. & Armstrong, J. (2000). *Introduction to Systems Engineering*. New York: John Wiley & Sons.

Schoemaker, J.E. & Russo, P. (1989). *Decision Traps*. New York: Bantam/Doubleday/Dell Publishing Group.

Silverman, B.G. (1992). Modeling and critiquing the confirmation bias in human reasoning. *IEEE Transactions on Systems, Man, and Cybernetics*, Sept/Oct: 972–982.

Society for Decision Professionals. (2012). *SDP Home Page*. http://www.decisionprofessionals.com, accessed April 2012.

Stanford Strategic Decision and Risk Management Decision Leadership Course (2008).

Tactical decision making in organizations. (2011). http://main.vanthinking.com/index.php/20080910125/Tactical-Decision-Making-in-Organizations.html, accessed 2011.

Trainor, T. & Parnell, G. Using stakeholder analysis to define the problem in systems engineering. *INCOSE*, 2007.

Tversky, A. & Kahneman, D. (1974). Judgment under uncertainty: Heuristics and biases. *Science*, *185*, 1124–1131.

Tversky, A. & Kahneman, D. (1982). Evidential impact of base rates. In P. Slovic, A. Tversky, & D. Kahneman (eds.), *Judgment under Uncertainty: Heuristics and Biases*. Cambridge: Cambridge University Press.

CHAPTER THREE

Foundations of Decision Analysis

GREGORY S. PARNELL and STEVEN N. TANI

Information is a source of learning. But unless it is organized, processed, and available to the right people in a format for decision making, it is a burden, not a benefit.

—William Pollard

Handbook of Decision Analysis, First Edition. Gregory S. Parnell, Terry A. Bresnick, Steven N. Tani, and Eric R. Johnson.

3.1 Introduction

In Chapter 1, we state that decision analysis is a social-technical process to create value. In Chapter 2, we identify the challenges that decision makers face in making difficult decisions for organizations and stakeholders. We also discuss the decision traps, cognitive biases and motivational biases faced by individuals and groups. Decision analysis is an operations research/management science (OR/MS) discipline that is designed to help decisions makers faced with difficult decisions, multiple stakeholders with many (possibly conflicting) objectives, complex alternatives, significant uncertainties, and important consequences. In this chapter, we examine the technical foundations of decision analysis. In Chapter 4, we describe the soft skills that are essential to successfully perform complex decision analyses in large organizations.

This chapter if organized as follows. We start with the historical foundations of decision analysis. We then define the tenets of decision theory, embodied in five simple rules of behavior that together provide a complete framework for determining how decisions should be made in the face of uncertainty that are consistent with our preferences. Next, we describe the scope of decision analysis using four dimensions: interacting with decision makers, alternatives, preferences, and uncertainty. We then describe the variations in the practice of decision analysis that have developed during the last half-century, pointing out both differences and similarities in those variations. Finally, we review an important philosophical perspective on decision analysis called value-focused thinking (VFT), created by Ralph Keeney.

3.2 Brief History of the Foundations of Decision Analysis

How did the discipline of decision analysis begin? The foundations of decision analysis come from the disciplines of probability (see Appendix A), mathematical decision theory, behavioral decision theory, and soft skills.[1] We use probability to assess subjective beliefs about potential outcomes and analyze uncertainty. We use mathematical decision theory to analyze decisions under uncertainty using our preferences. We use the research findings of behavioral decision theory to understand individual and group decision making and the heuristics and biases that are used when assessing uncertainty and value (Chapter 2). Using these research findings, we are able to better assess probability distributions for uncertain variables. Finally, we develop and use soft skills to develop models and elicit information from and communicate insights to decision makers, subject matter experts, and other stakeholders. This section focuses on subjective probability, mathematical decision theory, and behavioral decision theory.

The origins of probability theory date back to the seventeenth and eighteenth centuries. Based on the work of Blaise Pascal and Pierre de Fermat, Christiaan Huygens published *On Reasoning in Games of Chance* in 1657. In 1713, Jacob Bernoulli published *The Art of Conjecturing*, which included combinations, permutations, and the law of large numbers. In 1738, Daniel Bernoulli introduced diminishing marginal utility and risk averse behavior as a stimulant to insurance (Bernoulli, 1954). Bernoulli also proposed maximizing the expected value of something other than wealth, which he referred to as "moral expectation." The Reverend Thomas Bayes derived Bayes' law, the equations to update the probability of an outcome given new data. This work was published posthumously (1763) in *Essay towards solving a Problem in the Doctrine of Chance*. In 1933, Kolmogorov published *Foundations of the Theory of Probability*, which defined the axioms of probability theory.

While probability is one of the foundations of decision analysis, the development of the *subjective* view of probabilities was essential. The *objective* view of probability is that probabilities are a state of the world. The subjective view (sometimes called the Bayesian view) is that the probabilities we assign to potential outcomes depends on our state of information. In 1812, Pierre-Simon Laplace published *Theorie Analytique des Probabilities*, which adopted a subjective view of probability and used the Bayes approach. This work was expanded in 1931, when Frank Ramsey published *Truth and Probability*. In 1937, Bruno de Finetti continued the theoretical foundations for subjective probability.

The next foundational discipline is mathematical decision theory. In 1944, John von Neumann and Oskar Morgenstern published *Theory of Games and*

[1]Ronald Howard views the four pillars of decision analysis as systems analysis, decision theory, epistemic probability, and cognitive psychology (Howard, 2007).

Economic Behavior (von Neumann & Morgenstern, 1944). They formalized game theory and developed decision theory, including the maximization of expected utility, based on four axioms and the concepts of risk aversion and risk-seeking behavior. In 1954, Leonard J. Savage published *Foundations of Statistics* (Savage, 1954), which integrated decision theory with subjective and personal probability.

The third foundational discipline is behavioral decision analysis. Ward Edwards created behavioral decision research as a new field in psychology with two seminal articles: "The Theory of Decision Making" (Edwards, 1954) and *Behavioral Decision Theory* (Edwards, 1961) (Howard, 2007). Amos Tversky and Daniel Kahneman published "Judgment Under Uncertainty: Heuristics and Biases" (Tversky & Kahneman, 1974), which described three heuristics that people use to make judgments about uncertainty: representativeness, availability, and anchoring. While generally effective, these heuristics can lead to systemic bias. We must consider these biases when we elicit probabilities (see Chapter 10). In 1981, Tversky and Kahneman published *The Framing of Decisions and the Psychology of Choice* (Tversky & Kahneman, 1981), which described the importance of decision frames on our perception of the decision problem. This work has significantly influenced how we frame a decision opportunity. Although not an economist, Kahneman received the Nobel Prize in Economics in 2002 for his work on prospect theory (Kahneman & Tversky, 1979). Prospect theory uses behavioral decision analysis insights to develop a descriptive decision-making theory that is different than expected utility theory.

Decision analysis uses the axioms of probability and utility theory (Howard, 1966). The first decision analysis book, written by Howard Raiffa (Raiffa, 1968), used subjective probability and a single objective, net present value. The first multiple objective decision analysis book was published in 1976 by Ralph Keeney and Howard Raiffa (Keeney & Raiffa, 1976). Decision analysts are Bayesians since we believe that probability is a personal assessment of our belief in the outcome of an event based on our state of information, and we use Bayes' law to update our beliefs as we learn new information. In addition to the mathematical foundations of decision theory, decision analysts have adopted lessons from behavioral decision theory research about the cognitive processes, heuristics and biases people use to reason with uncertain information and make decisions. Decision analysts have used behavioral decision theory research and effective soft skills to develop effective protocols for the elicitation of probability, value, and utility.

3.3 Five Rules: Theoretical Foundation of Decision Analysis

The entire discipline of decision analysis stands on a strong theoretical foundation that was first stated in the 1940s by John von Neumann and Oskar Morgenstern (von Neumann & Morgenstern, 1944), who showed that making decisions based on expected utility is consistent with four axioms of rational behavior. Since then, several others, notably Leonard J. Savage (Savage, 1954) and Ronald A. Howard (Howard, 2007), have formulated alternative sets of axioms that also lead to decision making based on expected utility.

As formulated by Howard, decision theory starts with five rules of behavior (i.e., axioms). The five rules are so transparent and embody so much common sense that it would be considered an irrational act for anyone to express a desire to violate any of them.

The statement of the five rules uses two terms defined here (Howard, 2007):

Prospect.[2] A possible future. An example of a prospect is winning a prize of \$100 (or, more accurately, living your future life after having won \$100).

Deal.[3] A complete set of prospects, each with a probability of occurrence. An example of a deal is the opportunity to win \$100 with 50% probability and to win nothing with 50% probability.

The Five Rules

1. The *probability rule* requires that you can fully describe any deal in terms of possibilities and probabilities. A possibility is a clear description of an event that may or may not occur. A set of possibilities (also called an outcome space) is complete if they are *mutually exclusive* (only one may occur) and *collectively exhaustive* (one of the set must occur). A probability is a number between 0 and 1 that expresses your degree of belief that a possibility will occur. Note that in the realm of decision analysis, a probability does not describe a characteristic of the physical world that we can discern only through repeated experimentation. Rather, a probability is the quantification of your (or someone else's) belief about the uncertainty of a future event.
2. The *order rule* requires that you can rank any set of prospects in order of preference from best to worst. Indifference (i.e., equal preference) between two prospects is allowed. This rule implies transitivity of preference. If you prefer A to B and you prefer B to C, then you must prefer A to C (i.e., C cannot be both below A and above A in the preference ranking).
3. The *equivalence rule* requires that you can always create an uncertain deal involving two prospects such that you would be indifferent between receiving that deal and receiving a third prospect that is intermediate in your preference ranking between the two prospects in the deal. So, if you prefer A to B and prefer B to C, then the rule requires that there is a probability p such that you are indifferent between (1) a deal that gives you A with

[2]A similar term for a prospect is an outcome. The difference is that "outcome" has connotations that all pertinent uncertainty has been resolved, while prospect does not.

[3]Another term for a deal is a lottery.

probability p and C with probability (1 – p) versus (2) receiving B for sure. The probability p is called a preference probability because it is defined by your preferences rather than by your beliefs about the likelihood of any real events. Prospect B is said to be the *certain equivalent* of the deal involving A and C with the preference probability that you specify.

4. The *substitution rule* requires that your preference for a prospect will not change if an uncertain deal contained in the prospect is replaced by its certain equivalent, or vice versa.
5. The *choice rule* requires that if given the choice between two deals involving the same two prospects but with different probabilities, you must prefer the deal having the higher probability of receiving the more preferred prospect. Suppose that you prefer A to B and that you are offered two different deals. In Deal 1, you would receive either A or B with probabilities 40% and 60%, respectively. In Deal 2, you would receive either A or B with probabilities 25% and 75%, respectively. The rule requires that you must prefer Deal 1 because it offers the higher probability of the more preferred prospect A.

In any decision situation, the choice that is consistent with the five rules can be found through direct application of those rules, as follows:

1. Identify all possible outcomes of the decision. Identify the best and worst outcomes (i.e., the two that you prefer the most and the least).
2. For each possible outcome, assess the probability that expresses your degree of belief that it will occur.
3. For each possible outcome, identify the preference probability of best versus worst that you deem to be equivalent to it.
4. Substitute these best–worst deals for all the outcomes. This results in defining for each alternative an equivalent deal involving only the best and worst outcomes.
5. In the equivalent deal for each alternative, use probability calculations to find the probability of receiving the best outcome.
6. Choose the alternative whose equivalent deal has the highest probability of the best outcome.

We can view the preference probabilities from Step 3 as a metric that indicates the relative attractiveness of the outcomes. The calculations in Step 5 amount to finding the probability-weighted average (also called expected value, or EV) of that metric for each alternative, and in Step 6 we choose the alternative with the highest EV of that metric. We give to that metric the name "utility." It is straightforward to show that a linear transformation of the utility metric (i.e., adding and/or multiplying by a constant) does not change its essential property—that the alternative with the highest EV of the transformed utility metric is the alternative with the highest EV of the preference probabilities.

Thus, the acceptance of the five rules implies the existence of a utility function (Howard uses the term "u-curve") that maps any prospect to a utility metric. Decision theory tells us that the only way to act in accordance with the five rules is to make decisions that maximize the probability-weighted average of this utility metric.

An Example of the Direct Use of the Five Rules

The following highly simplified example illustrates the direct use of the Five Rules to find the best alternative in a decision situation. Please note that this method is not generally used in practice. Instead, more efficient methods are used that ensure consistency with the Five Rules.

Suppose that a company is planning the production of a commemorative T-shirt linked to a particular sporting event. It has only two choices of the number of T-shirts produced—Many and Few. The demand for the T-shirt will either be High or Low, depending on several factors, such as which teams actually participate in the event. The following figure shows the decision tree.

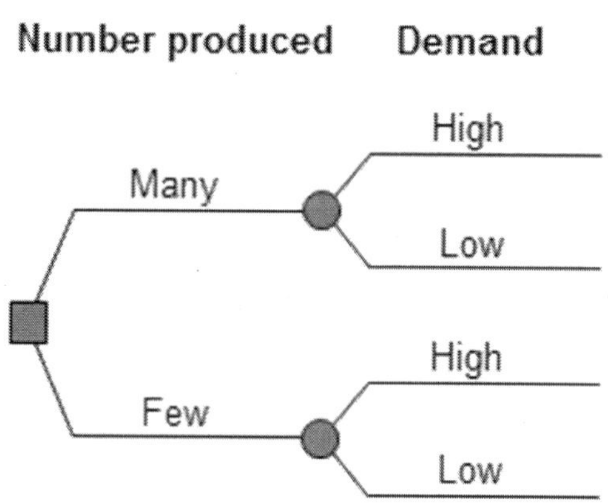

In Step 1, the possible outcomes are listed and ranked from best to worst, as follows:

1. **Many, High.** High sales, many happy customers, no waste (Best)
2. **Few, Low.** Low sales, few customers, but all are happy, no waste
3. **Few, High.** Low sales, many disappointed would-be customers, no waste
4. **Many, Low.** Low sales, much wasted cost (Worst)

In Step 2, the probability of High demand is assessed as 40%. In Step 3, an assessment is made of the probability of getting the best versus worst that makes the deal equivalent to the Few, Low outcome (= 50%). The same assessment is made for the Few, High outcome (= 30%), as shown in the figure below.

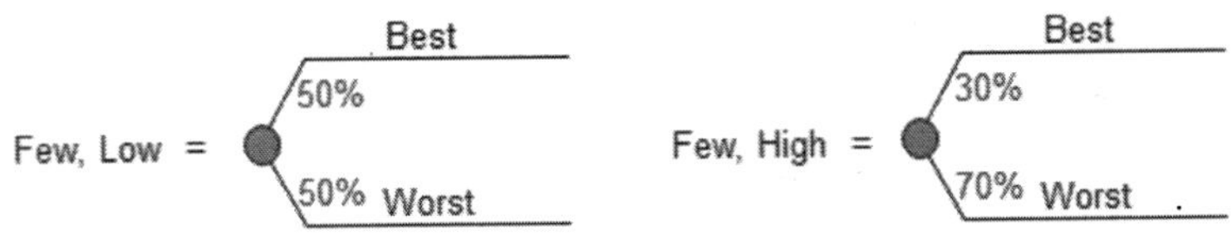

In Step 4, each outcome in the decision tree is replaced by its equivalent best–worst deal, as shown in the left-hand part of the figure below.

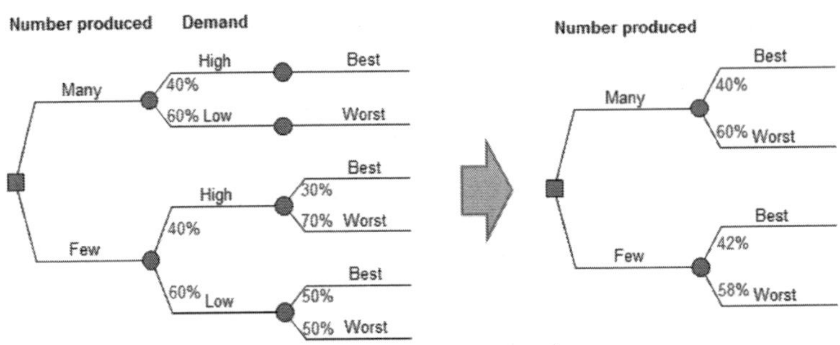

In Step 5, the probability of the best outcome is calculated for each alternative, as shown in the right-hand part of the figure above. Finally, in Step 6, producing Few T-shirts is identified as the better alternative because it has the higher probability of the best outcome (42% vs. 40%).

In a complex decision situation, finding the best alternative by direct application of the five rules can be cumbersome and time-consuming. In practice, we apply more streamlined methods (described throughout this handbook) that are designed to be consistent with the five rules. We take as given that a good decision professional will always use a decision process that results in choices that are consistent with the five rules. Be aware that there are popular decision methods (such as the Analytic Hierarchy Process) that can lead to violations of the five rules.

3.4 Scope of Decision Analysis

The purpose of decision analysis is to create value for decision makers and stakeholders. Decision analysis is an operations research/management science (OR/MS) technique that is appropriate for modeling decisions with preferences (value, time, and risk), uncertainties about future consequences, and complex alternatives.

Many OR/MS techniques have been used to model complex decisions. There are several OR/MS techniques that use probability to model uncertainty, including simulation, applied statistics, queuing theory, stochastic game theory, and stochastic optimization, to name a few. The distinction of decision analysis is that it provides an axiomatic theory for normative decision making using the alternatives we perceive, our probabilistic belief about uncertain consequences, and our preferences for the potential consequences.

Figure 3.1 displays a depiction of the scope of decision analysis (modified from (Parnell, 2009)). The diagram has four dimensions: interaction with decision makers and stakeholders; value and time preference; uncertainty and risk preference; and decisions. The simplest techniques are in the middle of the diagram. The complexity increases as we go outward on the spokes. The decision analysis concepts are shown as levels on the spokes. The decision analysis techniques are shown in black. The "none" level under decision maker and stakeholder interaction and the "importance weights" on the value spoke are not recommended (see swing weight discussion in Chapter 9). The interactions with decision makers are shaded gray to indicate their importance (see Chapter 5).

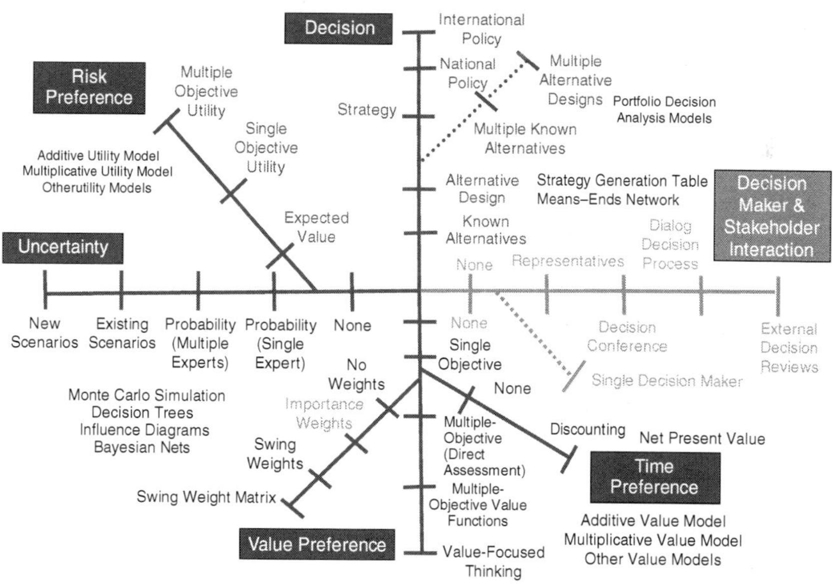

FIGURE 3.1 The scope of decision analysis.

The concepts and techniques in this figure will be developed and illustrated in Chapters 5–12.

3.5 Taxonomy of Decision Analysis Practice

The practice of decision analysis takes on a number of forms that differ from each other in the treatment of objectives and values. The choice of form is driven in part by the nature of the decision problem to be addressed and in part by the training and inclination of the decision practitioner. We present a taxonomy (see Fig. 3.2) of the various approaches, pointing out both how the forms differ from each other and how they are similar.

3.5.1 TERMINOLOGY

For clarity, we use in this section the following terminology:

Objective. A specific goal whose achievement is desired.

Performance score. A quantitative scale that measures the degree to which an objective is achieved.

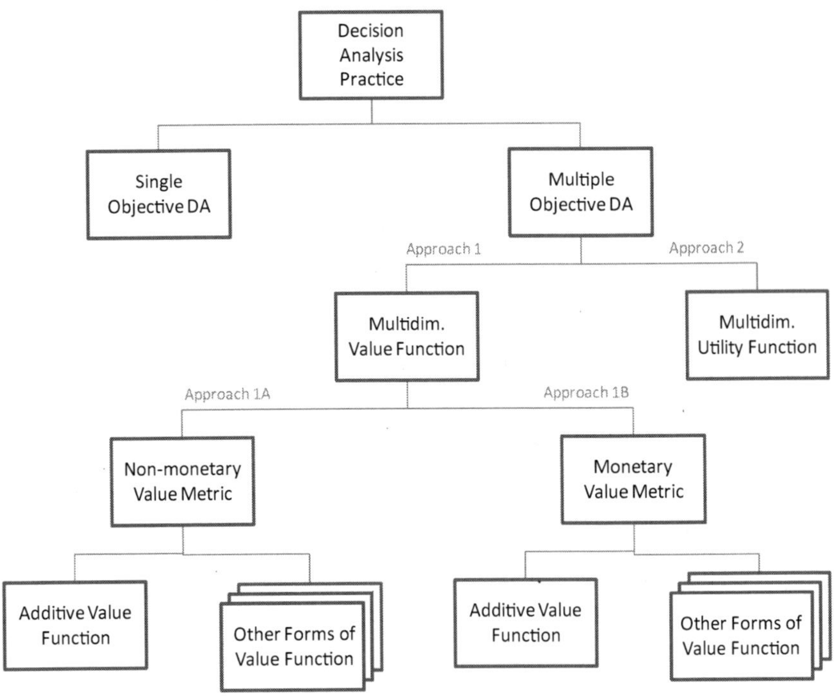

FIGURE 3.2 A taxonomy of decision analysis practice.

Risk preference. A description of a decision maker's attitude toward risk; an individual can be risk averse (the typical risk preference), risk neutral (often seen in Government decision making); or risk preferring.

Value metric. A quantitative scale that measures the value to the decision makers and stakeholders of the degree to which objectives are achieved.

Value function. A mapping of performance scores to the value metric. It can be a mapping from one performance score or from many performance scores to the value metric.

Utility metric. A quantitative scale that expresses the decision maker's attitudes toward risk-taking for the value metric. Optimizing the probability-weighted average of the utility metric (expected utility) is ideally the sole criterion for decision making. There is only one utility metric in a decision analysis.

Utility function. A mapping to the utility metric from the value metric in the case of a single-dimensional utility function or from all of the performance scores in the case of a multidimensional utility function.

3.5.2 TAXONOMY DIVISION: SINGLE OR MULTIPLE OBJECTIVES

The first division in the taxonomy is defined by the number of objectives that are addressed in the decision analysis. There are just two possibilities—one objective and more than one objective. This choice is driven by the nature of the decision situation. If the decision makers and key stakeholders believe that there is only one objective to be optimized in the decision, a *single-objective decision analysis* is appropriate. This is very often the case for decisions made within private sector companies, where usually the only objective of interest is to maximize shareholder value. However, if the decision makers and key stakeholders believe that there are several objectives to be optimized in the decision, a *multiple-objective decision analysis* is the right choice.

3.5.3 SINGLE-OBJECTIVE DECISION ANALYSIS

The single objective form of decision analysis is illustrated in Figure 3.3. For a business decision, the objective is usually to maximize shareholder value. The performance score used to measure shareholder value is the net present value of future cash flows, discounted at a rate representing the company's time value of money. In this case, the one performance score, which is on a monetary scale, can serve directly as the value metric, so the single dimensional value function is simply the identity function. In other cases, the performance score may be expressed in a way that makes it inappropriate to be used directly as a value metric. In these cases, a single dimensional value function is defined to map the performance score to a suitable value metric.

When appropriate, using a monetary scale rather than a unitless scale for the value metric is desirable in a decision analysis because it allows comparisons that are more meaningful to the decision makers. Each of us has acquired an entire lifetime of experience in using a monetary scale to measure relative values.

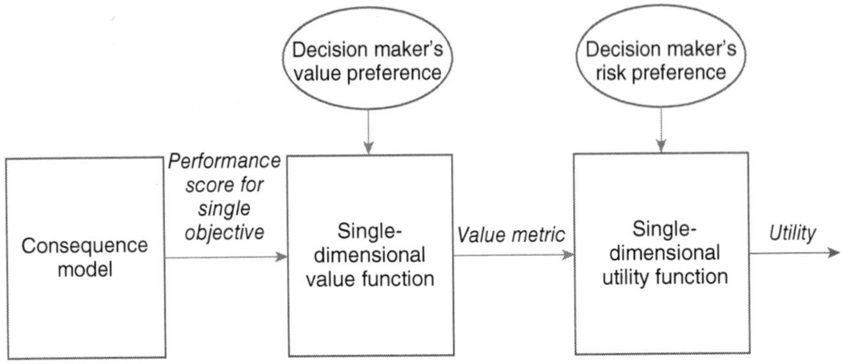

FIGURE 3.3 Single objective decision analysis.

So, the statement "Alternative A is $5 million better than Alternative B" is much more meaningful and insightful to us than "Alternative A is 50 value units better than Alternative B". In addition, using a monetary scale allows us to determine if the value of gathering additional information to improve the decision is worth the cost of doing so.

The desirable properties stated for a monetary value metric are shared by some other value metrics. One example is a metric defined on the amount of time expended or saved to accomplish an objective. Another, useful in some military decisions, is defined on the number of lives lost or saved. The important characteristic is that the value metric can be used by the decision makers to measure both benefits and costs.

One may ask: Is it always possible to define a value function that translates a nonmonetary performance score into a monetary value metric? The answer is yes. One of the Five Rules requires that any set of prospects can be put in order of preference, where equal preference (i.e., indifference) between two prospects is allowed. We define the set of prospects of achieving various levels of the nonmonetary performance score in question. We can add to that set of prospects another set defined by the receipt of various sums of money. All of these prospects are then ranked in order of preference, and additional monetary prospects can be added so that each level of the nonmonetary performance score has a monetary prospect that is equally preferred. Through this process, we define the desired value function that maps the nonmonetary performance score to a monetary value metric. However, although it is always possible in theory to define a monetary value metric, it might not be possible in practice because the decision makers and stakeholders are unable or unwilling to make the necessary assessments. For example, some governmental agencies are prohibited by law from putting a monetary value on a human life.

If the level of risk in the decision is significant, single-objective decision analysis includes the assessment of a single-dimensional utility function that embodies the decision maker's preferences regarding risk-taking relative to the value metric. Assessment of the utility function requires that the decision maker state preferences between pairs of hypothetical choices involving different levels of the value metric, where at least one choice in each pair is an uncertain deal.

For example, the decision maker may be asked the question, "Would you prefer to receive a sure $10 million gain or to receive an uncertain deal in which you have a 50% chance of a $30 million gain and a 50% chance of no gain?" A similar assessment can be made for a nonmonetary value metric.

For decision situations where decision makers are risk neutral or the level of risk is not significant (e.g., the range of possible outcomes is quite small relative to the possible outcomes of the enterprise), we do not bother to assess the utility function because we know that it will be approximated quite well by a straight line. That is, the decision maker is essentially risk neutral in such a situation and the decision can be based on maximizing the probability-weighted average (expected value) of the value metric.

The Roughneck North American Strategy illustrative example in this handbook utilizes the single objective decision analysis approach.

3.5.4 MULTIPLE-OBJECTIVE DECISION ANALYSIS

Some decision situations are characterized by multiple objectives. For example, a company may want to choose a production strategy that maximizes shareholder value while simultaneously minimizing the damage to the environment caused by its operations. Or a governmental agency may want to design a space program that accomplishes several different objectives, such as enhancing national pride, strengthening national defense, fostering international cooperation, and advancing scientific knowledge.

An important distinction to make is that between *fundamental* (or *ends*) objectives and *means* objectives (Keeney, 1992). Howard uses the terms *direct* values and *indirect* values for the same distinction (Howard, 2007). Fundamental objectives (direct values) are those that are valued by the decision makers as ultimate goals, whereas means objectives (indirect values) are valued only because they contribute to fundamental objectives. So, for example, if increasing shareholder value is a fundamental objective for a company, decreasing production costs would be a means objective that is valued only because it contributes toward the ultimate goal. In a systems decision, two common objectives are to increase availability and to increase reliability of a system. Since reliability is included in the calculation of availability, availability is a fundamental objective and reliability is a means objective. In a multiple objective decision situation, it is very important for the decision practitioner to make sure that the set of objectives defined for the decision comprise only fundamental objectives, since this will simplify the mathematical form of the value function and help avoid double-counting value.

3.5.5 TAXONOMY DIVISION: ADDRESSING VALUE TRADE-OFFS AND RISK PREFERENCE SEPARATELY OR TOGETHER?

Every decision analysis with multiple objectives must ultimately produce a single utility metric that provides the criterion for decision making—the alternative with the highest expected utility is the preferred alternative. This single utility metric must incorporate two different types of preferences held by the decision makers and key stakeholders:

1. Preferences on the trade-offs among the multiple objectives ("How much do we value achieving this objective relative to achieving the others?")
2. Preference regarding risk-taking ("How much potential value are we willing to forgo to reduce risk?").

The next division in the taxonomy is defined by whether the two types of preferences are addressed separately or together. If the decision analysis addresses the preferences separately, we call it Approach 1. Otherwise, we call it Approach 2 (see Fig. 3.4).

Approach 1 comprises two steps. In the first step, the performance scores representing the multiple objectives are mapped to a single value metric by a

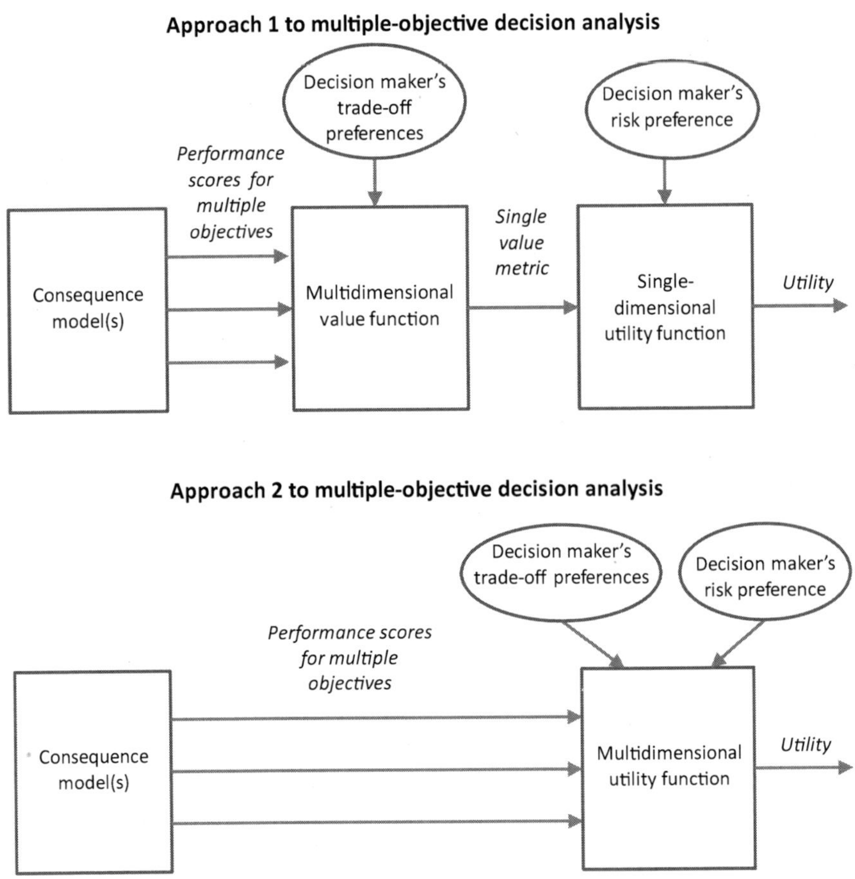

FIGURE 3.4 Two approaches to multiple objective decision analysis.

multidimensional value function, which embodies the preferences regarding the value trade-offs among the performance scores. In the second step, the single value metric is mapped to the utility metric in a separate single-dimensional utility function that expresses the preferences regarding risk-taking.

By contrast, in Approach 2, there is just one step in which the multiple performance scores are mapped to the utility metric by a multidimensional utility function that embodies both the preferences on the trade-offs among the performance scores and the preferences regarding risk-taking.

Because the one function in Approach 2 must represent both types of preferences, directly assessing this function can be a daunting task. In the general case, it would involve stating preferences between uncertain deals involving combinations of levels of the performance scores. "Would you prefer to receive for sure 10 units of score X, 20 units of score Y, and 40 units of score Z, or a deal in which you have a 50% chance of receiving 5 units of score X, 15 units of score Y, and 30 units of score Z vs. a 50% chance of receiving 15 units of score X, 25 units of score Y, and 0 units of score Z?" Daunting indeed.

To avoid this difficulty, a simplified form of multidimensional utility function can be used. First, a single-dimensional utility function is assessed separately for each of the performance scores. Then, the overall utility function is defined as a simple combination (such as additive or multilinear) of these single-performance score utility functions. This simplification of the multidimensional utility function is valid only if certain independence conditions hold among the performance scores (Keeney, 1992).

The great majority of multiple objective decision analysis applications use Approach 1 rather than Approach 2 because it is generally easier to implement. In Approach 1, the first step is to define a single value metric that can be used to rank in order of preference different combinations of multiple performance scores. Creating this value metric can be done in a variety of ways. For example, one frequently used method is to map each performance score to its own value scale and then to combine these single-performance score values into one overall value metric.

Suppose that there are *N* performance scores representing *N* objectives. We can imagine an *N*-dimensional space defined by those performance scores, in which indifference curves (or surfaces) exist, each comprising equally preferred combinations of levels of the *N* performance scores. One of the Five Rules requires the existence of these indifference curves. Figure 3.5 shows example indifference curves for the case in which there are only two objectives (and, hence, only two performance scores). These curves can be continuous or discrete.

It is important to note that the indifference curves are defined without regard to uncertainty since they are determined by comparing value preferences for sure prospects.

In essence, a multidimensional value function associates a quantity of value with each indifference curve. By convention, the value function is generally defined so that greater amounts of value are preferred to lesser amounts.

In Approach 1, if the decision maker is risk neutral or the range of possible outcomes in the decision poses very little risk, there is no need to assess a utility

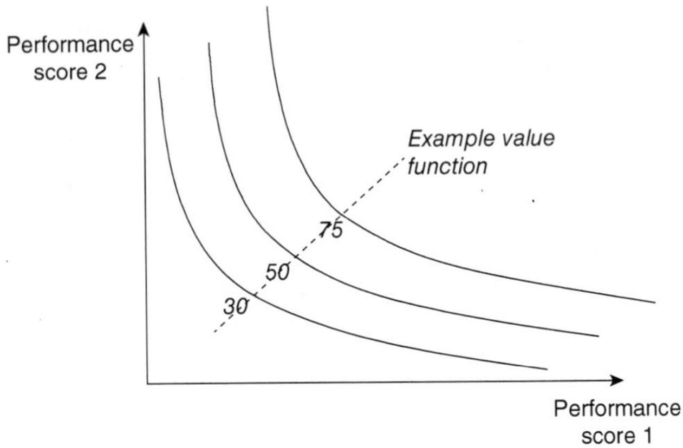

FIGURE 3.5 Example of indifference curves.

function to represent the preferences regarding risk-taking. In this case, the single value metric serves as the utility metric, so that the alternative with the greatest expected value is preferred. However, if the level of risk in the decision is important, a utility function is assessed that maps the single value metric to the utility metric and the decision criterion is to maximize the expected utility.

3.5.6 TAXONOMY DIVISION: NONMONETARY OR MONETARY VALUE METRIC?

The next division in the taxonomy applies only to Approach 1 and is defined by whether the multi-dimensional value function produces a value metric on a scale that is nonmonetary or monetary. This choice is often driven by the nature of the decision situation. Some decision makers, particularly in the public sector, are unable or unwilling to assess the value of achieving various objectives in monetary terms. In this case, using a nonmonetary value metric would be appropriate. We label as Approach 1A a multiple objective decision analysis using a nonmonetary value metric and as Approach 1B one using a monetary value metric.

In either case, the multidimensional value function represents the decision maker's preferences for trade-offs between the performance scores, and hence may take on any functional form. However, it is quite common to construct a value function that meets the mathematical assumptions required for a simple form.

3.5.7 TAXONOMY DIVISION: DEGREE OF SIMPLICITY IN MULTIDIMENSIONAL VALUE FUNCTION

The final division in the taxonomy is defined by the level of simplicity in the multidimensional value function. This choice applies to both Approaches

1A and 1B. One form of simplification is of particular interest because it is used quite frequently—the additive value function. In an additive value function, total value is calculated as the weighted sum of contributions of value from the performance scores, one contribution from each performance score. For an additive value function to be valid, every objective must be *preferentially independent* of all other objectives. That means that in the opinion of the decision makers, the value of achieving one objective does not depend on the level of achievement of any other objective. Keeney and von Winterfeldt state (Keeney & von Winterfeldt, 2007) that if the objectives used in the analysis are all *fundamental* rather than *means* objectives, and if those objectives collectively satisfy certain criteria, notably that they are nonredundant (i.e., they do not include overlapping concerns), then an additive value function is likely to be valid.

When an additive value function is used in Approach 1A, each performance score is mapped to a unitless value scale by a single-dimensional value function. Also, a swing weight (see Chapter 9) is assessed for each performance score. Total value is then calculated by multiplying the single-dimensional value for each performance score by that performance score's weight and summing across all performance scores. The Data Center illustrative example used throughout this handbook uses Approach 1A. An additive value function is defined in Approach 1B by mapping each performance score to a monetary value scale. Total value is then calculated by summing the monetary values of all performance scores.

Chapter 9 presents a more detailed discussion of the types of value functions.

3.6 Value-Focused Thinking

An important philosophical approach to the analysis of decisions was introduced by Ralph Keeney in his book entitled *Value-Focused Thinking* (Keeney, 1992). To be successful for professional decisions, VFT must interactively involve decision maker(s), stakeholders, subject matter experts, and decision analysts in seeking alternatives that can create higher shareholder and/or stakeholder value.

3.6.1 FOUR MAJOR VFT IDEAS

VFT has four major ideas: start first with values, use values to generate better alternatives, create decision opportunities, and use values to evaluate the alternatives.

3.6.1.1 Start with Values. VFT starts with values and objectives before identifying the alternatives. Keeney describes the contrasting approach as alternative-focused thinking (AFT), which starts with the alternatives and seeks to differentiate them. He lists three disadvantages of AFT. First, if we start with known alternatives this will limit our decision frame (our understanding of the opportunity or problem, see Chapter 6). Second, if we try to evaluate only known alternatives, we may not identify important new values and objectives that are relevant for future solutions (see Chapter 7). Finally, since we may not have a full understanding of the opportunity and the decision frame, we may not have a sound basis for the generation of alternatives.

3.6.1.2 Generate Better Alternatives. Our decision can only be as good as the best alternative that we identify. If we have several crummy alternatives, the best analysis will only identify a crummy alterative! Once we have identified the values and objectives for the decision problem, we can use them to generate better alternatives. We do this by qualitatively and quantitatively defining the value gaps. We can then focus our creative energy on developing alternatives that better meet our objectives (see Chapter 8).

> If we have several crummy alternatives and fail to find improvements to any of them, the best analysis will only identify a crummy alterative!

3.6.1.3 Create Decision Opportunities. Many of us have been taught to look for potential problems, carefully define the problems, and then look for solutions. Whereas problem definition is reactive, opportunity definition is proactive. Keeney encourages us to not wait for problems to appear but to focus our energy and creativity on identifying decision opportunities to identify potential value for an organization's shareholders and stakeholders.

3.6.1.4 Use Values to Evaluate Alternatives. Finally, we should use the values to qualitatively and quantitatively evaluate our alternatives. The mathematics of single- and multiple-objective decision analysis can be used to evaluate of the alternatives, and once completed, we can use this information to improve existing alternatives or develop new alternatives (see Chapters 9 and 11).

3.6.2 THE BENEFITS OF VFT

Keeney has identified several benefits of VFT (Fig. 3.6).

1. Since strategy is a choice among alternatives based on strategic objectives, VFT can support the organization's strategic planning by helping to identify strategic objectives and creative strategies for evaluation in strategic planning.
2. VFT is broadly applicable to both decision opportunities and to problems. The difference is that decision opportunities are proactive and problem solution is reactive.
3. Since VFT takes a broad perspective of the decision opportunity, it can help identify hidden objectives. The objectives may be hidden because they result from new opportunities and challenges faced by the organization.

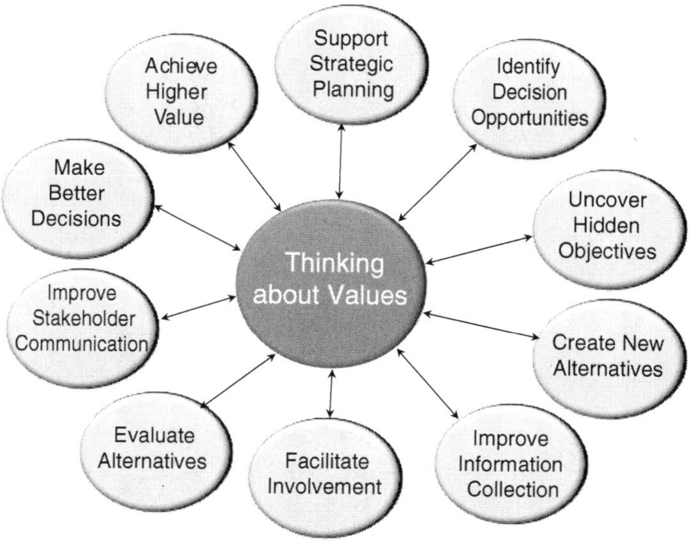

FIGURE 3.6 Benefits of value-focused thinking (modified from Keeney, 1992).

4. VFT can identify creative new alternatives with potential to achieve value for shareholders and stakeholders. This is a critical part of any decision analysis, since the quality of the alternatives limits the potential value that can be identified.
5. Since VFT identifies important objectives and new alternatives, it can focus information collection on what is important and NOT just what is available. The prioritization of the objectives and measures is critical to establishing what is important.
6. The focus on stakeholder participation to identify their objectives and measures is important to achieve their involvement in the analysis, decision making, and strategy implementation. When stakeholders see that their objectives are included, they are more willing to participate.
7. The definition of objectives and measures can improve the evaluation of alternatives. By using the mathematics of decision analysis, we can develop a value model directly from the objectives and measures that can be used to evaluate the alternatives.
8. A clear understanding of the organization's values, objectives, and value measures can improve communications between stakeholders. This understanding is critical for the development of alternatives, the analysis of alternatives, and the implementation of the selected alternative.
9. Better alternatives and better evaluation can lead to better decisions. The identification of better alternatives offers the potential for higher value

outcomes, and better analysis provides insights and decision clarity for decision makers.

10. Early stakeholder involvement and better communication can increase the probability of successful solution implementation, which increases the likelihood of converting potential value to actual value.

3.7 Summary

This chapter provides an introduction to the discipline of decision analysis that has its roots back in the seventeenth and eigtheenth centuries. Modern decision analysis was "born" in the 1960s, and its four foundational disciplines are: probability theory, decision theory, behavioral decision theory, and soft skills. Decision analysis is an operations research technique, with a technical foundation of five simple rules of behavior that together provide a complete framework for determining how decisions should be made in the face of uncertainty that are consistent with our preferences—to maximize our expected utility. We describe the scope of decision analysis using four dimensions: interacting with decision makers, alternatives, preferences, and uncertainty. We then describe the variations in the practice of decision analysis that have developed during the last half century, pointing out both differences (number of objectives and monetary/nonmonetary value metric) and similarities (consequence models, performance scores, value functions, and utility functions). Finally, we review the four key ideas and the 10 benefits of Keeney's value-focused thinking.

KEY TERMS

Decision Theory Five basic axioms of choice that call for assessment of probabilities and utilities of possible outcomes and selection of the alternative with the highest expected utility.

Objective A specific goal whose achievement is desired.

Objective probability Our knowledge of probability is only based on a large series of identical trials.

Performance score A quantitative scale that measures the degree to which an objective is achieved.

Possibility A clear description of an event that may or may not occur

Preference probability The probability specified for a deal between two prospects that makes the deal have the same value as a third intermediate prospect.

Probability Theory A mathematical theory of uncertainty based on three axioms.

Subjective probability The quantification of your (or someone else's) belief about uncertainty about a future event.

Utility function A mapping to the utility metric from the value metric in the case of a single-dimensional utility function or from all of the performance scores in the case of a multidimensional.

Utility metric A quantitative scale that expresses the decision maker's attitudes toward risk-taking for the value metric. Optimizing the probability-weighted average of the utility metric (expected utility) is the sole criterion for decision making. There is only one utility metric in a decision analysis.

Value-focused thinking A philosophy of decision making that advocates the central role of values in our decision making.

Value function A mapping of performance scores to the value metric. It can be a mapping from one performance score or from many performance scores to the value metric.

Value metric A quantitative scale that measures the value to the decision makers and stakeholders of the degree to which objectives are achieved.

ACKNOWLEDGMENTS

We would like to acknowledge course notes on the history of decision analysis developed by Terry Bresnick's colleague Roy Gulick. We used these notes as a reference for our section on the history of the foundations of decision analysis.

REFERENCES

Bernoulli, D. (1954). Exposition of a new theory on the measurement of risk. *Econometrica, 22*(1), 23–36.

Edwards, W. (1954). Theory of decision making. *Psychological Bulletin, 41*, 380–417.

Edwards, W. (1961). Behavorial decision theory. *Annual Review of Psychology, 12*, 473–498.

Howard, R.A. (1966). Decision analysis: Applied decision theory. *Proceedings of the Fourth International Conference on Operations Research*, pp. 55–71. Wiley-Interscience.

Howard, R.A. (2007). The foundations of decision analysis revisited. In W. Edwards, R.F. Miles, & D. Winterfeldt (eds.), *Advances in Decision Analysis: From Foundations to Applications*, pp. 32–56. Cambridge: Cambridge University Press.

Kahneman, D. & Tversky, A. (1979). Prospect theory: An analysis of decisions under risk. *Econometrica, 47*(2), 263–291.

Keeney, R.L. (1992). *Value-Focused Thinking: A Path to Creative Decision Making*. Cambridge, MA: Harvard University Press.

Keeney, R.L. & Raiffa, H. (1976). *Decisions with Multiple Objectives: Preference and Value Tradeoffs*. New York: John Wiley & Sons.

Keeney, R.L. & von Winterfeldt, D. (2007). Practical value models. In W. Edwards, J.R. Miles, & D. von Winderfeldt (eds.), *Advances in Decision Analysis: From Foundations to Applications*, pp. 232–252. Cambridge: Cambridge University Press.

Parnell, G.S. (2009, May). Decision analysis in one chart. *Decision Line—Newsletter of the Decison Sciences Institute.*

Raiffa, H. (1968). *Decision Analysis: Introductory Lectures on Choices Under Uncertainty.* Reading, MA: Addison-Welsey Publishing Co.

Savage, L.J. (1954). *Foundations of Statistics.* New York: Dover Publications.

Tversky, A. & Kahneman, D. (1974). *Judgment under Uncertainty: Heuristics and Biases.* Cambridge: Cambridge University Press.

Tversky, A. & Kahneman, D. (1981). The framing of decisions and the psychology of choice. *Science, 211*, 453–458.

von Neumann, J. & Morgenstern, O. (1944). *Theory of Games and Economic Behavior.* Princeton, NJ: Princeton University Press.

CHAPTER FOUR

Decision Analysis Soft Skills

TERRY A. BRESNICK and GREGORY S. PARNELL

Leadership is communicating to people their worth and potential so clearly that they come to see it in themselves.

—Stephen Covey

The mind is like a parachute. It only works when it is open.

—Tony Robbins

Handbook of Decision Analysis, First Edition. Gregory S. Parnell, Terry A. Bresnick, Steven N. Tani, and Eric R. Johnson.

4.1 Introduction

In Chapter 1, we identify the decision analysis soft skills, and in Chapter 2, we discuss decision-making challenges and note that the decision analyst must have the soft skills necessary to implement decision analysis as a socio-technical process. Soft skills are the nonquantitative, "personal," "social," or "interpersonal" skills that complement the quantitative hard skills of the mathematics of decision making methodologies. Soft skills, such as leadership and facilitation, include the personal skills to prepare for the analysis, such as strategic thinking and researching, and the ability to interact with and manage people, while hard skills involve technical tasks, mathematical modeling, and analysis techniques that we learn and perform individually. Our business and personal success relies on many traits, including honesty, integrity, dependability, friendliness, communications, optimism, etc. However, we focus on the soft skills that are directly related to a decision analysis project and help us interact with decision makers, stakeholders, and experts. For example, good communications is a general soft skill, interviewing an expert to assess a probability distribution is a decision analysis soft skill, and solving a decision tree for the best alternative with the probability information is a decision analysis hard skill. Every experienced decision analyst knows that soft skills are critical, but we have not found a comprehensive list in the decision analysis literature. Based on discussions with colleagues and our experience, we present and describe the following nine decision analysis soft skills: thinking strategically, leading teams, managing projects, researching, interviewing individuals, conducting surveys, facilitating groups, aggregating across experts, and communicating analysis results. We are not attempting here to provide a "cookbook" of steps that has been codified with rules that should be followed for every project, but rather, a collection of diverse ideas, general approaches, and specific techniques that the authors have found extremely useful over the years. The specific soft skills that decision analysts will use in a particular study will depend upon the nature of the decision analysis study, their background and training, their experience level, and their judgment about what is required for success in that study.

The remainder of the chapter is organized as follows. Sections 4.2–4.10 describe the nine decision analysis soft skills listed above. Section 4.11 provides a summary of the chapter.

4.2 Thinking Strategically

Strategic thinking is a personal skill that involves taking a long-term, systematic view of one's environment, goals, and opportunities. Strategic thinking is recognized as an important attribute for our professional and personal lives (Dixit & Nalebuff, 1993). Liedtka (Liedtka, 1998) identifies five attributes of strategic thinking in practice: systems perspective, intent focused, thinking in time, hypothesis driven, and intelligent opportunism. We add a sixth attribute: speaking truth to power. We will describe each of these attributes using a decision analysis perspective.

- **Systems perspective.** "A strategic thinker has a mental model of the complete end-to-end system of value creation, his or her role within it, and an understanding of the competencies it contains" (Liedtka, 1998). Strategic thinkers understand the "big picture," the decision environment, the strategic objectives of the organization, and the challenges that senior leaders face. A systems perspective includes a sound understanding of the strategic planning space of the organization, as well as the decision frame (Chapter 6) of the current study.
- **Intent focused.** Liedtka describes strategic intent as "the focus that allows individuals within an organization to marshal and leverage their energy, to focus attention, to resist distraction, and to concentrate for as long as it takes to achieve a goal." A decision analyst should maintain a determined focus on the purpose of the decision analysis study and not be distracted by smaller issues and challenges along the way that might distract from the goal of their study.
- **Thinking in time.** "Strategy is not driven by future intent alone. It is the gap between today's reality and intent for the future that is critical." Strategic thinkers must be able to understand the past, envision the future, and act in the present. Decision analysis must understand the past and know which elements are important to the future. Decision analysts can use scenario planning to creatively think about future opportunities and challenges (Chapter 6). However, they must take actions during the study that position

the organization for successful implementation once the decision is taken (Chapter 14).

- **Hypothesis driven.** Strategic thinking requires a balance of ensuring both creative and critical thinking. The creative thinking is especially important in the design of new alternatives (Chapter 8) and the development of new models (Chapters 9 and 11). Critical thinking requires listening to individuals with different views, collecting data to confirm or reject hypotheses, and using the best available data and sound mathematical techniques to evaluate alternatives.
- **Intelligent opportunism.** "The dilemma involved in using a well-articulated strategy to channel organizational efforts effectively and efficiently must always be balanced against the risks of losing sight of alternative strategies better suited to a changing environment." The decision analyst must always be on the "lookout" for significant changes in the environment that could signal major new risks or game-changing opportunities with the potential to make dramatic improvements in value through improved strategies.
- **Speaking truth to power.** Being a strategic thinker is easy when the decision analyst is presenting exciting new ideas with great potential for value creation. Unfortunately, decision analysts sometimes are the bearers of bad news. Decision analysts must have the integrity and courage to present the unvarnished truth to senior leaders even if there are undesirable potential professional and personal consequences.

4.3 Leading Decision Analysis Teams

Decision analysts must be effective decision analysis team leaders. Leading teams includes identifying the correct decision issues; developing common team goals; motivating individuals to achieve team goals; guiding the team during the study; and helping the client organization to achieve the most value from the study. J. Richard Hackman (Hackman, 2002) describes five conditions necessary for successful teamwork: The team must be a real team, rather than a team in name only; it has compelling direction for its work; it has an enabling structure that facilitates teamwork; it operates within a supportive organizational context; and it has expert teamwork coaching. We describe each of these in the context of a decision analysis.

- **Form a team.** Assigning the name "team" to a new decision analysis group does not make them act like a real team or perform effectively as a team. The decision analysis team leader must motivate the team to perform as a team. An important first step is establishing team objectives for the study. The team that the decision analysts may have to develop may be the facilitators and experts supporting a project, or it may even be the internal decision board itself (see Chapter 13).

- **Provide compelling direction.** Groups that have a shared understanding of the importance of their decision analysis project to the organization are more likely to perform effectively. The team leader must develop and communicate this compelling direction.
- **Delegate responsibility and authority.** The leader must be careful not to try to do everything himself/herself. Key tasks should be delegated to team members along with the authority to make things happen and the responsibility and accountability for results.
- **Enable teamwork structure.** The leader must design and implement a team structure that facilitates team work. This structure should include an understanding of the responsibilities and authorities of each team member.
- **Obtain organizational support.** The team leader must insure that the key leaders and stakeholders support the decision analysis project and provide the necessary resources for success. A critical issue is the access to decision makers, stakeholders, and experts. The decision process must be designed to provide access to senior decision makers (see Chapter 5).
- **Obtain expert coaching.** Taking a college course in decision analysis or performing technical work on a decision study does not prepare one to be a decision analysis team leader. Leadership skills are developing by doing. Studying good leadership traits and examples can be very helpful. Having a coach or mentor can also be very helpful.

4.4 Managing Decision Analysis Projects

There is a difference between leadership and management—"management is doing things right; leadership is doing the right things" (Covey, 1989). In addition to leading teams as described in Section 4.3, leaders must also manage the decision analysis project. Project managers are concerned with effectiveness and efficiency. Managing decision analysis projects includes developing analysis study plans; identifying and scheduling activities; monitoring progress; managing the completion of tasks; and transitioning project activities to individuals and teams with authority and responsibility for decision implementation.

- **Develop a project plan.** For large projects with many participants, a project plan is essential to identify the responsibilities of all contributors. Depending on the scope and duration of the effort, the plan can range in size from a few pages to a large document.
- **Identify and schedule activities.** Decision analysis is not a standard set of "cookie-cutter" techniques that are used identically every time. The decision analysis techniques and process must be developed uniquely for each decision analysis project. A common schedule is required to assign and monitor the activities.

- **Monitor progress.** Every project and every team is different. The project manager needs to monitor progress and, usually, needs to revise activities or reallocate resources to keep on schedule and within cost.
- **Manage task completion.** The decision analysis study participants are key individuals in the organization. Once their tasks are complete, the project manager needs to reassign team members to other tasks or return them to their "day job" or another project.
- **Transition to decision implementers.** Many times the decision analysis team does the study up to the decision and then transitions project activities to individuals and teams with authority and responsibility for decision implementation. The greater the participation of the decision implementers in the study, the more likely they are to understand the decision, and effectively execute the decision implementation (see Chapter 14).

4.5 Researching

Researching is an important personal skill. As we discuss the important role of decision analysts in interacting with decision makers, stakeholders, and subject matter experts (SMEs), it is important to note that many times, the analyst must perform research prior to these meetings to better understand the problem domain, identify the key issues, identify objectives, and develop the key questions to ask. The best researchers look for a variety of credible sources from both the problem domain and the decision analysis literature that offer conflicting views on the problem. The amount of research depends on the decision analyst's prior understanding of the problem domain, knowledge of key terminology, and amount of domain knowledge expected of the decision analyst in the decision process.

Another important role of research is to find decision analysis techniques that are appropriate for the problem being addressed. For example, we may decide to use scenario planning or Monte Carlo simulation and decide that we need to review the best practices in the literature.

4.6 Interviewing Individuals

A frequently used approach for knowledge elicitation is the interview process. Interviews are especially appropriate for senior leaders who do not have the time to attend a longer focus group or the interest to complete a survey. Interviews can be highly effective, but there are many potential pitfalls. The time of high-level decision makers is valuable, so it is essential to keep the interview moving. The interviewer must probe for thorough responses to open-ended questions rather than simple yes/no or few-word answers and should use research and subject matter expertise to ask challenging questions. The questions asked must be clear, and the interviewer must quickly provide clarification as needed. The interviewer must not impose his or her emotions and feelings on the participant. While an occasional war story is okay to stimulate thought, the interviewer must avoid dominating the discussion. The interviewer should be aware of his or her own personal gestures and body language—participants pick up on them easily. Finally, and most importantly, the interview team must keep good notes.

There following are a variety of formats that can be used to frame good questions:

- **Fact-finding questions.** These are designed to gather verifiable data or facts.
- **Feeling-finding questions.** Theses are designed to gather subjective information that gets at feelings, opinions, beliefs, and so on.
- **"Tell me more" questions.** These are designed to follow up or to expand upon what participants are saying.
- **Best/least questions.** These are designed to test limits of wants and needs; questions often take the form of "what would be best in this circumstance?" or "what is the most critical factor?"
- **Third-party questions.** Theses are designed to uncover sensitive information in an indirect manner. Often, questions are asked about what other persons would say or do or think about a specific topic.
- **"Ruler for a day" questions.** These are designed to remove constraints that inhibit or limit responses; they may take the form of "what things would you change if you were in charge?"

Regardless of the questioning technique, it is essential to:

- **Obtain the facts.** When asking factual questions, ascertain the level of confidence that the participant has in his or her ability to answer.
- **Distinguish among facts, feelings, and emotions.** With open-ended questions in particular, find out what is known, what is assumed, and what is opinion.
- **Frame questions carefully.** How an open-ended question is framed and probed can greatly affect the answer. Develop a script and stick to the script unless a potentially interesting issue is raised that has not been thought of before.
- **Identify biases** from the managers and subject matter experts that could skew judgmental assessments.

Kayser, in *Mining Group Gold* (Kayser, 2011), indicates that interviewees can be classified by the way they react to interview questions. He describes different types of interviewees:

- **Expressers (open words, open emotions).** They easily represent their words and feelings and are quick to get things off their chests. The interviewer must pay careful attention to the specific words behind the emotions that the participant chooses.

- **Directors (open words, guarded emotions).** They tend to use strong positive words and speak quickly and confidently, and have little patience for time wasting. The interviewer must keep things moving; avoid extended side discussions.
- **Reactors (guarded words, open emotions).** They get easily upset and their emotions are clear. They make heavy use of body language and gestures, their responses are indirect and obscure, and their reasoning is sometimes hard to clarify. The interviewer must be sensitive to emotions and must not let the participant think that his or her emotions are being ignored.
- **Stoics (guarded words, guarded emotions).** They tend to keep facts and emotions to themselves. The interviewer must be sensitive to body language which may give clues to when to probe further; do not let participants' discomfort fester and explode.

Since interviews take time, it is important to get the best information possible. The following are best practices from the authors' personal experiences for each phase of the interview process: planning, scheduling, conducting, documenting, and analyzing interviews.

4.6.1 BEFORE THE INTERVIEW

For interviews with senior leaders and key stakeholder representatives, it is important to prepare a questionnaire to guide the interview discussion. The following are best practices for interview preparation:

- Develop as broad and diverse a list of interviewees as possible. Identify one or more interviewee for each stakeholder group. Review the interview list with the decision maker(s) to ensure that all key stakeholder groups are represented on the list of potential interviewees.
- Begin the questionnaire with a short explanatory statement that describes the reason for the interview, the purpose of the decision analysis, and the stakeholders being interviewed.
- Begin the interview with an "unfreezing question" that encourages the interviewee to think "outside the box" about the future and how that will impact the key elements of the decision.
- Tailor the questionnaire to each interviewee. Make the questions as simple as possible.
- Do not use leading questions that imply the answer is known in advance and that guide the interviewee to agree with the preconceived answer.
- Do not ask a senior leader a detailed question that can be obtained by research.
- End the questionnaire with a closing question, for example, "Is there any other key issue that we have not discussed?"

- Arrange to have an experienced interviewer and a recorder for each interview.
- Decide if the interviews will be for attribution or not for attribution.

4.6.2 SCHEDULE/RESCHEDULE THE INTERVIEW

Interviews with senior leaders require scheduling, and frequently, rescheduling. The more senior the leader, the more likely schedule will be a challenge. The following are best practices for interview scheduling:

- It is usually best to have the interviews scheduled by team members who are representatives of the decision makers and stakeholders since they have better access.
- Depending on the importance of the decision and the difficulty of scheduling, we usually request 30–60 minutes for the interview.
- It is usually best to conduct interviews individually to obtain each interviewee's ideas on the topic at hand. Additional attendees change the interview dynamics. The senior leader may be reluctant to express ideas in front of a large audience or may defer to staffers to let them participate.
- While some facilitators prefer to provide the interview questionnaire to the interviewees in advance, there is a danger that the recipient will have his or her staff fill it out and we will not get the recipient's perspective. This can result in the interviewee reading "staff answers." If we want staff answers, we should consider a focus group or a staff interview.
- The interviews can be done in person or over the phone. In-person interviews are the most effective since interaction is easier; however, sometimes they are not possible and the only practical choice is a phone or video interview.

4.6.3 DURING THE INTERVIEW

The interview team's execution of the interview creates an important first impression with the senior leader about the decision team. Some thoughts for conducting interviews:

- The best number of people to conduct the interview is one interviewer and one note taker. An alternative to the note taker is a recorder. Some interviewees may be reluctant to be recorded. If considering using a tape recorder, request permission first.
- Conduct the interview as a conversation with the interviewee. Use the interview questionnaire as a guideline. Take the questions in the order the interviewee wants to discuss them.
- Make the interview interesting to the interviewee. Be flexible, follow-up on an interesting observation even if it was not on the questionnaire. Many times, one interviewee will identify an important issue that no one else has mentioned.

- Ask clear, open-ended, questions that require the interviewee to think and respond. Avoid complex questions that might confuse the interviewee.
- Respect the interviewee's time. Stay within the interview time limit unless the interviewee wants to extend the interview period. When the interviewee's body language signals that they have finished the interview (e.g., fold up paper and look at their watch), go quickly to a closing question and end the interview.

4.6.4 AFTER THE INTERVIEW

Documentation of the interview is important to capture the key decision issues since they provide important data for all of the steps that follow. Best practices for documenting the interviews include the following:

- As soon as possible after the interview, type the interview notes. The questions and the answers should be aligned to provide proper context for the answers. The documentation should be consistent with the decision to use the notes with or without attribution.
- The interviewer should review the typed notes and make revisions as required.
- Once the interview notes are complete, the interviews should be provided to the decision team.
- It is good practice to follow-up the session with an individual thank you note (or email)

Freeman Marvin et al., in their "Soft Skills Workshop," suggest additional best practices for interviewing as shown in the boxed text (Klimack et al., 2011):

Seven Steps to a Successful Interview

1. Establish rapport.
2. Motivate by establishing legitimacy and purpose.
3. Structure the interview using an "hourglass" approach—go easy early in the interview, challenge in the middle, and gently wrap up and exit.
4. Condition the interviewee in order to impart the right perspective.
5. Be flexible in the order of questions; follow up on important new information.
6. Verify and confirm with questions, such as "Did I hear correctly that. . . .".
7. Conclude on time.

4.7 Conducting Surveys

Surveys are a useful technique for collecting information from a large group of individuals in different locations and are most appropriate for junior to mid-level stakeholders. Surveys can be used to gather qualitative and quantitative data on the decision objectives, uncertainties, and alternatives. A great deal of research exists on techniques and best practices for designing effective surveys. Decision analysts can distribute and collect survey data via mail, email, or the Internet. This section provides an overview of survey design and methods for conducting surveys. Surveys require detailed planning to accomplish the team's goals (The Survey System—Survey design, n.d.). These steps can be followed to plan, execute, and analyze surveys:

- Establish the goals of the survey.
- Determine the sample of stakeholders that will be targeted with the survey.
- Determine how the survey data will be distributed and collected.
- Develop the survey questions.
- Test the survey.
- Distribute the survey and collect data.
- Analyze the survey data.

4.7.1 PREPARING AN EFFECTIVE SURVEY: DETERMINE THE GOALS, SURVEY RESPONDENTS, AND MEANS OF DISTRIBUTING AND COLLECTING SURVEY DATA

The decision team needs to clearly articulate the goals of the survey and the target sample of stakeholders whom they want to answer the survey. Often, surveys for decision problems will be used to collect potential decision objectives and value measures. If the team plans to collect and analyze data from questions with standard answer scales (e.g., "Yes/No" or multiple choice answer scales), it is important to determine the appropriate sample size needed to draw valid statistical conclusions from the survey data. Sample size calculations are described in basic statistics books, and online tools are available to do these calculations (The Survey System—Sample size calculator, n.d.). The team needs to work with the decision maker(s) to determine the appropriate stakeholders to survey.

The method for implementing a survey should be selected before the survey is designed. Popular methods are mail, electronic mail, and web surveys. Table 4.1 provides a listing of some of the advantages and disadvantages of these survey methods.

The ability to collect survey responses in a database when using a web survey instrument can be beneficial if subsequent analysis is required. Several online programs now exist to help teams design web surveys, collect responses, and analyze the results. Some popular programs include SurveyMonkey.com

TABLE 4.1 **Advantages and Disadvantages of Surveys**

Survey Method	Advantages	Disadvantages
Mail	Respondents have flexibility in completing the survey	Takes the most time Hard to check compliance and conduct follow-up with respondents Response data will have to be transformed by the analysis team into a format for analysis
Electronic mail	Fast to distribute Low cost Easy to check compliance and do follow-up	Need to obtain email addresses for the survey sample Response data will have to be transformed by the analysis team into a format for analysis
Internet web survey	Extremely fast Can include special graphics and formatting Can collect responses in a database to facilitate analysis	May be hard to control who responds to the survey due to worldwide Internet access Respondents can easily provide only a partial response to the survey

(SurveyMonkey), InsitefulSurveys.com (Insiteful Surveys), and the SurveySystem.com (The Survey System).

4.7.2 EXECUTING A SURVEY INSTRUMENT: DEVELOPING THE SURVEY QUESTIONS, TESTING, AND DISTRIBUTING THE SURVEY

Surveys should be designed to obtain the information to meet the goals of the survey. To maximize response, the survey should be short with clearly worded questions that are not ambiguous from the respondent's perspective. When constructing questions, care must be taken not to overly constrain the answers. Start the survey with an overview of the purpose of the survey and the goals the decision analyst hopes to achieve from the information provided by the respondents. Here are some general principles that can be followed in developing effective survey questions (Fowler, 1995):

- Ask survey respondents about their first-hand experiences so they can provide informed answers.
- Ask only one question at a time.
- In wording questions, make sure that respondents understand the question. If the question includes terms that could be interpreted differently by respondents, provide a list of definitions to clarify any possible ambiguities. This list of definitions should precede the questions.

- Articulate to respondents the kind of acceptable answers to a question. For objective questions, answer scales can be setup as multiple choice answers from a rating scale or level-of-agreement scale. For open-ended text response questions, the question should be worded so respondents provide information germane to the question. Close the survey with a statement allowing respondents to provide any additional information pertinent to the goals of the survey.
- Format the survey so that it is easy for respondents to read the questions, follow instructions, and provide their answers. For example, answer scales should follow a similar pattern in terms of the order in which they are presented (e.g., the least desirable answer is the first choice ascending to the most desirable answer).
- Orient the respondents to the survey in a consistent way. This can be accomplished with a set of instructions that describe the goals of the survey, the method for completing their responses, and the means for submitting the completed survey.
- Once the survey questions are written, test the survey instrument with a few individuals outside the team. Ask them to complete the survey using the same medium that respondents will use (e.g., by e-mail, mail, or on the web). Ask for input from the test sample regarding the instructions and wording of the questions and answer scales. If a web survey is used, test the method for collecting responses. Use the input from the test sample to improve the survey.
- Once improvements are made, distribute the survey to respondents using the method chosen. Develop a plan for monitoring the response rate and establish when reminders will be sent to respondents who have not completed the survey. The team should also have a standard way to thank respondents for their time and efforts, for example, a thank you note or e-mail.

4.8 Facilitating Groups

Group facilitation is a process in which a person whose selection is acceptable to all members of the group, who is substantively neutral, and who has no substantive decision-making authority, diagnoses and intervenes to help a group improve how it identifies and solves problems and makes decisions, to increase the group's effectiveness.

A facilitator is a person who helps a group free itself from internal obstacles or difficulties so that it may more efficiently and effectively pursue the achievement of desired outcomes

—Roger Schwarz (Schwarz, 2002)

4.8.1 FACILITATION BASICS

Facilitating groups is a critical decision analysis soft skill. When planning to facilitate a meeting, one size does not fit all. Facilitation can run the gamut from totally noninterventional, "touchy-feely" group processes, to more interventional, structured, and highly quantitative group processes and modeling. Facilitation sessions may be for *Groups* that get together on a one-time or infrequent basis, have no sense of group identity, and may have little reason to seek or expect consensus. Or it may be for *Teams* that get together regularly, have a common sense of group identity and purpose, and often seek to achieve consensus.

One unique approach to facilitation is decision conferencing (Phillips, 2007). A decision conference is typically a 2- to 3-day session supported by a facilitation team consisting of a lead facilitator, a model builder, and a scribe (O'Connor, 1984). It is most useful when a decision must be made in a very short period of time, when the stakes are high, when little "hard data" is available, and when most of the subjective information needed to make the decision resides in a small group of experts. The lead facilitator moderates and controls the sessions, elicits information, asks questions, channels responses, and constructs analytical models in response to group input. The model builder operates a computer to build in real time the decision models developed by the group and displays them on a projection screen. The scribe acts as a conference recorder, documenting key decisions and providing a complete audit trail of rationale for the session. The decision conferencing approach to facilitation is described in detail in Appendix C.

The role of the facilitator is challenging and requires multitasking. The facilitator must ask questions carefully using clear vocabulary, clarify and elaborate, seek feedback, and be consistent. He or she must be able to listen to the responses carefully and use listening techniques, such as paraphrasing, repeating word for word, and so on. The facilitator must be cognizant of body language and expressions. The facilitator must be aware of dysfunctional group behaviors, such as late arrivers, silence, chronic objectors to the process, dominators, side conversations, rambling answers, off-the-wall remarks, and so on, and must be prepared to deal with them as they arise. Techniques for dealing with such behavior include (Schwarz, 2002; Kayser, 2011):

- Be friendly, but firm in confronting the behavior.
- Focus on the behavior, not the individual.
- Encourage group members to share responsibility in handling difficult members.
- Recognize acceptable behavior and highlight the impact of disruptive behaviors.
- Privately suggest more functional behaviors with disrupters.
- Avoid sustained one-on-one arguments or exchanges.

Finally, the facilitator must provide feedback to the group in real time.

Facilitator responsibilities include (Kayser, 2011):

- Stimulating the conversation and balancing group participation.
- Understanding (not evaluating) participants' feelings.
- Refocusing the group based upon desired outcomes.
- Staying neutral on content.
- Staying focused on the situation, not the group behaviors.

10 Commandments for Facilitators

1. You are there to help group process, not to solve their issues.
2. Always stay focused on the goal of the session.
3. Rational argument will not win over intense emotionalism!
4. Listen to what the group is telling you.
5. Be flexible—if one approach fails, shift to another.
6. Display a positive attitude—the group will feed off of it.
7. Never display your anger to the group, even if justified.
8. Maintain your sense of humor and use it with the group.
9. Stay aware of time and schedules.
10. Relax and have fun; enjoy your work!

On occasion, conflict occurs and must be managed by the facilitator. Some friction can be good if it stimulates critical thinking, enhances creativity, or minimizes stagnation. However, if the conflict becomes disruptive, the facilitator can manage it by clarifying opposing points of view, defining areas of agreement and disagreement, using effective problem-solving techniques, and in some cases, trying to resolve differences. Specific techniques include (Kayser, 2011):

- **Collaboration.** Looking for "win–win" solutions; extensive effort is typically spent on generating innovative solutions, on probing deeply into the problem, and on brainstorming.
- **Avoidance.** Often results in a "no winner, no loser," no solution, situation; participants can choose to withdraw, give up, or just leave the process.
- **Accommodation.** Often results in someone "winning" and someone "losing"; typically requires someone to give in, yielding to others' positions or smoothing over concerns with additional concessions.
- **Compromise.** The goal is "no winner, no loser"; tradeoffs are the norm as is looking for middle ground or the lowest common denominator. Solutions reached are rarely the best solution.

- **Domination.** Someone "wins" someone "loses." This is often achieved by overwhelming those with opposing positions, by unyielding defense of one's own position until others become weary or convinced, or by pulling rank. Frequently, arguments get heated and emotional.

As experienced facilitators, we strive for collaboration.

4.8.2 GROUP PROCESSES

Many of the inputs used to develop analytical models are elicited using well-documented group processes. Various methods are used to take advantage of the positive aspects of group dynamics and to minimize the negative effects as described later in the chapter.

4.8.2.1 Stages of Group Development. All collaborative groups evolve over time as members learn to trust and work with each other, whether the group meets for a one-day decision conference or a year-long project. Bruce Tuckman identified four stages of group development—a model that many facilitators still use today to understand and guide group process (Tuckman, 1965).

- **Forming.** When a new group forms, its members are consciously or unconsciously searching for trust, meaning, and a sense of belonging with the group. Facilitators must be aware of these needs and insure that each session begins with an activity or exercise that allows everyone to participate and get engaged in the work at hand.
- **Storming.** Inevitably, the problem-solving process begins to heat up under the pressures of conflicting perspectives and information. People jockey for influence. Patient and impatient people clash. Trust is tested, and confusion around goals and roles begins to surface. If there are deadlines and constraints, this stage can be tense. Facilitation can help a group move through this stage efficiently by turning the focus of the group members' anger and frustration on the emerging model of the problem and away from each other.
- **Norming.** As members get to know each other better and get a grip on the problem, they begin to reconcile differences and come to agreement on decision-making processes, resources, timing, and standards. Norms are the formal and informal rules that make up the operating system of productive work. Facilitators can suggest at this stage a wide range of group processes, decision aiding tools, and analytical techniques that will help the group move forward.
- **Performing.** The final stage of team development involves the coming together of all the experience and skills of the group members to get results for each other and the organization. If the decision conference or working session is to be repeated for another aspect of the problem, the group members will continue to build upon their newly acquired collaboration and problem solving skills.

4.8.2.2 Planning. Meeting management is all about being organized, and collaboration needs careful planning as well. As a best practice, we highly recommend conducting a dry run with a small group of colleagues to insure that terms are understandable and questions are interpreted as intended.

Kayser suggests that every facilitator use a cornerstone document on which all collaborative sessions rest—The PDORA document. PDORA stands for **P**urpose, **D**esired **O**utcomes, **R**oles, and **A**genda (Kayser, 2011). For the facilitator, the PDORA document will be an effective tool in keeping sessions on course and preventing breakdowns. In a similar vein, Marvin uses the mnemonic GRASP to help facilitators plan and prepare for a collaborative working session:

- **G**oals and outcomes
- **R**oom and logistics
- **A**genda and time available
- **S**upport team, tools, and techniques
- **P**articipants and observers.

4.8.2.3 Pulsing. Decision making and problem solving require two types of thinking: divergent thinking and convergent thinking. Groups that do not spend enough time on divergent thinking risk solving the wrong problem or falling into groupthink. Groups that spend all of their time on divergent thinking never come to conclusion and become nothing more than a debating society (Marvin, 2006).

The ideal group process spends an adequate amount of time in divergent thinking—gathering information, uncovering goals and objectives, and generating creative alternatives—and then switches to convergent thinking for evaluating, prioritizing, and reaching consensus. This creates a repeatable natural cycle or "pulse" by the group as it works through pieces of the problem. The facilitator helps maintain group rhythms of divergent and convergent thinking.

There is ample evidence that groups often fail to explore adequately the "problem space" or the "solution space." When a group convenes to make a decision or solve a problem, it is useful to follow the "tunnel—funnel" approach diagramed in Figure 4.1. This process encourages divergent thinking as a prelude to convergent thinking.

4.8.2.4 Pacing. The facilitator helps manage the group task and time available, spurring on the group when progress is too slow, reining back when the group is jumping to conclusions, and guiding the group toward its goals. This role is critical to success because groups usually try to accomplish a lot in a limited amount of time. Pacing is all about channeling the energy of the group.

There are three situations when the facilitator can use good group process techniques to channel group energy. These techniques are called Slow, Flow, and Go (Marvin, 2006).

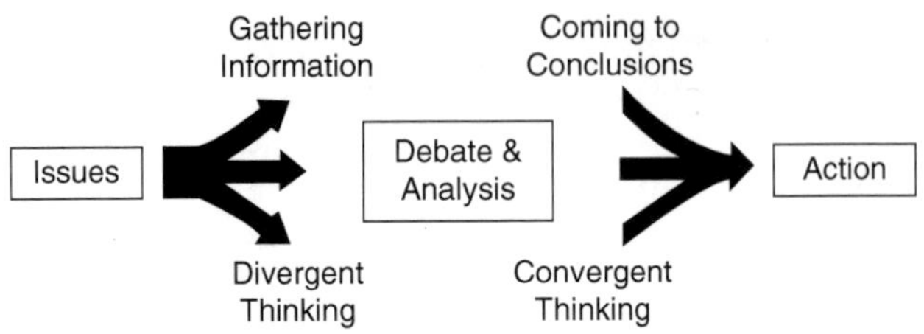

FIGURE 4.1 Divergent and convergent thinking.

Slow. When there is potential for too much conflict early in a group session, the facilitator may use the Nominal Group Technique (NGT) to slow down and control the rate of idea generation (Moore, 1987). NGT calls for all individuals to silently generate ideas in writing and then provide assessments to the group without discussion. Once these comments are made, the group then discusses all judgments for clarification, combination, elimination, and categorization. Discussions are carefully recorded in participant's own words or phrases, but there is no evaluation of ideas until the listing is complete. Once everyone understands the ideas, evaluation begins. Frequently, this involves rank-ordering ideas and aggregating the rankings.

Flow. Sometimes, it appears that there is not enough constructive conflict in the group. If the group cannot get a conversation flowing, the conflict probably has not been surfaced. The facilitator should start with a blank whiteboard and ask each group member to identify one aspect of the problem or decision. The resulting "cognitive map" will develop into a 360° View of the Problem and will lead the group to frame the problem and move on to solutions.

360° View of the Problem

- Use when there are a lot of confusing and interrelated aspects of the problem
- Identify all the issues, stakeholders, uncertainties, decisions, information sources, events, and conditions that impact this problem and its solution.
- Go around the room and get one item out loud from each participant.
- Write each item on a large sticky note and post on the wall.
- Draw links and make clusters to form a graphical representation of the problem space.

Go. When there is the right amount of constructive conflict and the energy level of the group is high, the facilitator should use techniques that encourage interaction among members. One such set of techniques are the classic rules for brainstorming developed by Alex Osborn in 1953 (Osborne, 1953). The facilitator simply asks the group for ideas and writes them on a whiteboard, flipchart, or computer screen. After about 30 minutes of brainstorming, ideas may be consolidated, sorted, and evaluated. In brainstorming, participants do not criticize or evaluate; wild ideas are encouraged; quantity of ideas leads to quality ideas; and participants often piggyback on each others' ideas.

4.8.3 FOCUS GROUPS

Facilitation is also required for focus groups. The term "focus group" is attributed to psychologist Ernest Dichter (Focus Groups, 2011). Focus groups are typically small groups of 6–12 people, and are primarily used to solicit qualitative opinions and attitudes about how satisfied users or customers of a service or product have been. They are also used to gather data on new or proposed products, services, or ideas. They can also be useful for identifying decision objectives, eliciting evaluation criteria, or identifying key uncertainties. They are typically shorter and more focused in scope than decision conferences or other facilitated working sessions. While interviews typically generate a one-way flow of information, focus groups create information through a discussion and interaction between the group members. As a general rule, focus groups should comprise between 6 and 12 individuals. Too few may lead to too narrow a perspective; too many will lead to some individuals not being able to provide meaningful input. As with interviews, the focus group facilitation team needs to devote time to prepare for, plan the execution of, and analyze data from focus groups.

4.8.3.1 Preparing for the Focus Group Session.

Developing the goals and objectives of the focus group session is critical to success. A few best practices for preparing for a focus group session:

- Develop a clear statement of the purpose of the focus group and what it hopes to achieve from the session.
- Develop a profile of the type of participant who should be part of the session and communicate that to the project client. Select a participant pool with the project client.
- Select and prepare a moderator who can facilitate a discussion without imposing his or her own biases on the group.
- Schedule a time and location during which this group can provide 60–90 minutes of uninterrupted discussion.

- Develop a set of questions that are open-ended and which will generate discussion. Do not use "Yes/No" questions that will yield little discussion. The most important information may come out of discussion resulting from a question posed to the group.
- Conduct a "dry run" with colleagues to insure questions convey what is intended.

4.8.3.2 Executing the Focus Group Session. The most important components of executing the session are the moderator and the recording plan. Here are some thoughts for the execution of a focus group session (McNamara, n.d.):

- The moderator should review the session goals and objectives, provide an agenda and the plan for recording the session.
- Ask questions and allow participants a few minutes to discuss their ideas. The moderator should ensure equal participation to prevent a few individuals from dominating the group.
- A good technology solution for facilitating focus groups is the Group Systems software (Group Systems Homepage, n.d.). This technology facilitates groups in brainstorming activities and generating ideas. It helps mitigate the impacts from individuals who tend to dominate discussions because participants type their ideas on a computer in response to questions generated by the moderator. It also helps record the data.
- Do a video and audio recording of the session if possible. If not, use multiple note takers.
- On closing, tell participants that they will receive a record of the decision objectives from the session.
- Follow up the session with an individual thank-you note for each participant.

4.9 Aggregating across Experts

When the decision analyst deals with multiple experts, it is often necessary to aggregate across experts to come up with a single position or assessment. Some of the more common techniques include:

- **Working toward consensus through discussion** (i.e., all can live with it)—the decision analyst/facilitator assists the group of experts in conducting a structured conversation designed to enable open exchange of information with the goal of informed consensus. Opportunities are made available to explore different perspectives, faulty assumptions, and hidden agendas. The ultimate goal is to reach consensus on a topic or assessment. The decision analyst must also recognize circumstances when consensus is just not achievable. There may be some differing perspectives that are parochial or political in nature, there may be just no middle ground positions, and achieving consensus may not be the goal of all participants. When this happens, it is useful to forego aggregating across *all* participants and instead, reporting a minority position that can be used as part of a sensitivity analysis to determine if it impacts the decision.

- **Voting across participants, majority rules.** This can be a quick way to determine how close or how far apart participants are. Whether it is voting on scores or weights or probabilities, each participant gets a vote, results are openly displayed (often using spreadsheets), opportunities are provided to understand and discuss differences, and often, revotes are allowed. Typically, a "majority rules" approach is used to reconcile the differences.
- **Mathematical aggregation (such as averaging, using copulas to develop joint distributions, etc.).** Mathematical approaches can be quick, relatively easy, and relatively common in aggregating across experts, but unfortunately, they are often applied inappropriately. The most common approach is averaging across voters. While this does represent the mathematical mean of the voters, much information can be lost, particularly when votes are highly bimodal. In such cases, the average does not represent what the group is saying at all. It is far better to look at the data to see if outliers are skewing results, and then try to explore reasons for outliers. If after discussion, outliers remain that skew results, it is better to initially eliminate the outliers, recalculate averages, and do sensitivity analysis on whether or not eliminating the outliers has an effect on the decision. The specific averaging approach should be a function of the types of scales being used.
- **Combined approaches.** Two or more of the above approaches can be used together to eliminate the shortfalls of a single approach. For example, the group can use the voting approach for a first pass to stimulate discussion, but then the facilitator, together with the group, can examine the data visually and determine which statistic (mean, mode, and median) best represents the holistic center of mass of the group.

Regardless of the technique used for aggregation, it is important to decide up front what the rule for achieving consensus will be; the group must agree on how the decision will be made.

4.10 Communicating Analysis Insights

The communication of analysis insights to decision makers and key stakeholders is such a critical decision analysis soft skill that we have devoted Chapter 13 to the topic.

4.11 Summary

In this chapter, we review the soft skills necessary for interacting with decision makers, stakeholders, and experts. The decision analyst should be a strategic thinker, a team leader, and a project manager. We describe the soft skill attributes for each of these three roles. Researching is the starting point for understanding the problem domain and understanding the key components of the decision. A frequently used approach for knowledge elicitation is interviewing. Interviews can be highly effective in capturing the preferences of senior decision makers and their staff, but can carry with it many pitfalls. In addition to research and interviews, surveys, when done properly, can provide a wealth of information by collecting information from large groups of individuals in different locations. In the chapter, we present best practices for all stages of the interview process and for preparing and conducting surveys.

In many cases, it is necessary to go beyond the techniques described above to bring together in one location a wide range of people, including decision makers, stakeholders, and subject matter experts. Facilitation of such groups is a critical soft skill. Whether in formal decision conferences or in less formal workshops, the role of the facilitator is challenging and requires an extensive set of skills. Facilitators must enable open exchange of information, encourage consensus building, deal with conflict, and deliver collaborative results. When we need to bring together small groups to gather information on well-focused topics, we can use focus groups to encourage discussion and interaction among group members. Familiarity with group processes, techniques for aggregating across experts, decision-making barriers, and cognitive biases found in Chapter 2 provides essential knowledge for the successful facilitator. The final soft skill is communicating analysis insights to decision makers and stakeholders, which we describe in Chapter 13.

KEY TERMS

Facilitation A process in which a person whose selection is acceptable to all members of the group, who is substantively neutral, and who has no substantive decision-making authority, diagnoses and intervenes to help a group improve how it identifies and solves problems, and makes decisions to increase the group's effectiveness.

Focus group Typically small groups of 6–12 people used primarily to solicit qualitative opinions and attitudes about how satisfied users or customers of a service or product have been. They are also used to gather data on new or proposed products, services, or ideas.

Interview The one-on-one process that a decision analyst uses to elicit information from decision makers and subject matter experts. The interview could be in person, by telephone, or by other electronic means.

Research The process that a decision analyst goes through at the start of an engagement to better understand the problem domain, identify the key issues, and develop the key questions to ask. The best researchers look for a variety of credible sources that offer conflicting views on the problem.

Soft skills The nonquantitative, "social" skills that complement the quantitative decision making methodologies; they include strategic thinking, leading teams, managing project, researching, interviewing, group facilitating, and communicating.

Strategic thinking Taking a long-term, systematic view of your environment, goals, and opportunities.

Subject matter expert (SME) Someone with credible substantive knowledge about the decision.

REFERENCES

Covey, S. (1989). *The 7 Habits of Highly Effective People*. New York: Simon & Schuster, Inc.

Dixit, A.K. & Nalebuff, B. (1993). *Thinking Strategically: The Competitive Edge in Business, Politics, and Everyday Life*. New York: Norton.

Focus groups (2011). Retrieved from Wikipedia.

Fowler, F. (1995). *Improving Survey Questions—Design and Evaluation*. Applied Social Reserach Methods Series. Thousand Oaks, CA: Sage Publications.

Group Systems Homepage (n.d.). Collaboration for leaders. http://www.groupsystems.com/, accessed October 17, 2011.

Hackman, J.R. (2002). *Leading Teams: Setting the Stage for Great Performances*. Boston: Harvard Business School Press.

Kayser, T. (2011). *Mining Group Gold*, 3rd ed. New York: McGraw-Hill.

Klimack, W., Marvin, F., Wicker, P., Buckshaw, D., Kloeber, J., Leonardi, D., et al. (2011). Soft Skills Workshop: Real World Skills for Decision Analysts and OR/MS Professionals. Vienna, VA.

Liedtka, J.M. (1998). Linking strategic thinking with strategic planning. *Strategy and Leadership*, *26*(4), 30–35.

Marvin, F. (2006). The Four Faces of Facilitation: A Guide for Facilitation of Problem-Solving and Decision-Making Groups.

McNamara, C. (n.d.). Basics of conducting focus groups. http://managementhelp.org/businessresearch/focus-groups.htm, accessed October 17, 2011.

Moore, C.M. (1987). *Group Techniques for Idea Building*. Thousand Oaks, CA: Sage Publications.

O'Connor, M. (1984). Methodology for corporate crisis decision making. In S. Andriole (ed.), *Corporate Crisis Management*. New York: Petrocelli Books.

Osborne, A. (1953). *Applied Imagination*. New York: Charles Scribner & Sons, Inc.

Phillips, L. (2007). Decision conferencing. In R. Miles, D. Von Winterfelt, & W.E. Edwards (eds.), *Advances in Decision Analysis*, pp. 375–398. New York: Cambridge University Press.

Schwarz, R. (2002). *The Skilled Facilitator.* San Francisco: Jossey-Bass.

The Survey System—Sample size calculator (n.d.). http://www.surveysystem.com, accessed October 17, 2011.

The Survey System—Survey design (n.d.). http://www.surveysystem.com/sdesign.htm, accessed October 17, 2011.

Tuckman, B. (1965). Developmental sequence in small groups. *Psychological Bulletin, 63*(6), 384–399.

CHAPTER FIVE

Use the Appropriate Decision Process

STEVEN N. TANI and GREGORY S. PARNELL

If you can't describe what you are doing as a process, then you don't know what you are doing.

—W. Edwards Deming

Handbook of Decision Analysis, First Edition. Gregory S. Parnell, Terry A. Bresnick, Steven N. Tani, and Eric R. Johnson.

5.1 Introduction

This chapter presents how to choose the appropriate process to provide decision support to senior decision makers who have the responsibility and authority to make an important decision for their organization or enterprise. We address several questions. What is a good decision? How much time and effort should be devoted to the decision? Who should be involved in making the decision? How should the decision process be structured?

We first discuss the goal of any decision process—making good decisions. We present a definition of a good decision using the six essential elements of decision quality. We describe four common decision processes: two that we believe are best practices and two that have major flaws. Then we discuss choosing a decision process that fits the decision situation. For the most difficult decisions, we recommend the dialogue decision process. We describe the decision processes used in the illustrative examples. We close the chapter with a discussion of making decision quality a way of life in organizations.

5.2 What Is a Good Decision?

The fundamental goal of any decision professional is to help others make good decisions. But how do we know if a decision is good or bad? It is tempting to define a good decision as one that turns out well—that is, one that has a desirable outcome. But this definition is not really helpful because of two deficiencies. First, it states that the quality of the decision may depend on factors beyond the control of the decision maker. Following this faulty definition, we would judge a decision to pay $1 for a 50–50 chance of winning $10 to be good if the prize is won but bad if the prize is not won. In other words, if two people make exactly the same choice in this situation, we might judge one choice to be good and the other to be bad. We want the definition of a good decision to be such that it depends only on factors that the decision maker can control. The second deficiency of the definition is that it requires us to wait for the outcome of the decision to be known in order to judge the quality of the decision. For major strategic decisions, the wait might be many years. We want a definition that allows us to judge the quality of the decision at the time that it is made.

5.2.1 DECISION QUALITY

In Chapter 1, we describe a good decision as one that is logically consistent with our preferences, our alternatives, and our assessment of the uncertainties. But

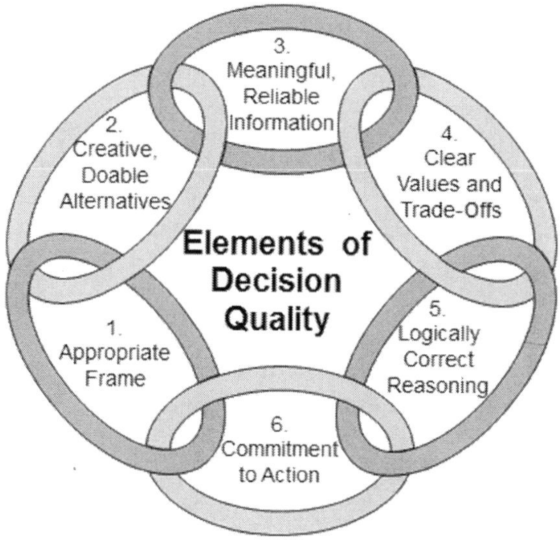

FIGURE 5.1 Six elements of decision quality.

how can a decision analyst know when such logical consistency is present? One approach for defining a good decision that has been used successfully over many years of decision consulting is embodied in the decision quality chain shown in Figure 5.1 (Matheson & Matheson, 1998). A good decision requires high quality in each of six essential elements:

1. Frame
2. Alternatives
3. Information
4. Values
5. Logical reasoning
6. Commitment to action.

5.2.2 THE SIX ELEMENTS OF DECISION QUALITY

1. **Frame.** A good decision requires an appropriate frame that clearly specifies its purpose, perspective, and scope. A good decision is timely, addresses the right problem, and involves the right people. See Chapter 6 for more information on framing.
2. **Alternatives.** A good decision requires a set of alternatives that offer the potential to create value for shareholders and/or stakeholders, each of which is feasible and compelling and which together span the full range of possible actions. See Chapter 8 for more information on designing alternatives.

3. **Information.** A good decision requires meaningful and reliable information, which usually includes properly-assessed probability distributions on key uncertain factors. The ideal here is not to collect *all* of the information that might be relevant, but rather, to collect information only up to the point where the cost of obtaining additional information exceeds the value of that information for making the decision. See Chapters 9–11 for more on information.
4. **Values.** A good decision requires that value metrics for comparing possible outcomes of the decision are clearly specified. The value metrics should faithfully represent the decision makers' preferences and the trade-offs between competing values should be explicitly stated. See Chapter 7 for more information on values.
5. **Logical reasoning.** A good decision has to make sense because it is based on logically correct reasoning. To have high quality in this element, the decision must be consistent with the Five Rules (see Chapter 3). See Chapters 9 and 11 for decision modeling using logically correct reasoning.
6. **Commitment to action.** A good decision is one which the decision makers are prepared to implement in a timely manner. Without commitment to action, we may have clarity of intention, but that is not really a decision. Commitment to action is enhanced by participation of the implementers in the decision process. See Chapter 14 for more information on implementation.

The six elements of decision quality are shown as links in a chain to emphasize that the decision is only as good as the weakest link. A good decision must be strong in all six elements.

5.2.3 INTUITIVE VERSUS DELIBERATIVE DECISION MAKING

Often, as discussed in Chapter 2, decisions are made not on the basis of careful deliberation but on the basis of intuitive "gut-feel." This can be a reasonable approach when the decision maker has made similar decisions previously, has received good timely feedback on their outcomes, and has no motivational involvement with the decision at hand (Campbell & Whitehead, 2010; Kahneman & Klein, 2010).

But for difficult decisions with complex preferences, major uncertainties, and important consequences, making decisions based entirely on intuition is not a useful process. For one thing, a person's intuition is fallible—a choice made intuitively might simply be a bad choice. Second, if the decision is to be made by a group of people within an organization, it is generally difficult for one person to convince the others of a choice based on "gut feel." Third, for complex problems, no one person has an intuitive understanding of the complete problem. It is useful to combine the intuition and experience of many experts into a composite understanding of the decision.

Although making decisions entirely on the basis of intuition is not a good idea, intuition can play a valuable role in decision making. If a choice that is about to be made does not "feel" right, it is worth exploring why that is so. Using the six elements of decision quality as a checklist can be very helpful. Perhaps the decision is framed inappropriately. Or perhaps the choice is being made with inadequate attention paid to exploring alternatives. Or the decision is being made on the basis of too little information or value measures that do not truly reflect the decision makers' preferences. Or there are logical flaws in the analysis. Or the decision makers are not fully committed to implementing the decision. Going through this checklist may reveal why the decision does not "feel" right and point the way to improving the decision. But if the checklist review does not reveal any weakness in the decision and yet it still feels wrong intuitively, it may be one of those times when one's intuition is incorrect and needs to be educated.

5.3 Selecting the Appropriate Decision Process

In this section, we begin by describing how to tailor the decision process to the decision. Next, we present two decision processes that are best practices—the dialogue decision process (along with the derivative system decision process) and decision conferencing. Then we describe two types of flawed decision processes that are nevertheless in common use—strictly analytical processes and advocacy processes.

5.3.1 TAILORING THE DECISION PROCESS TO THE DECISION

It is fair to ask whether it is appropriate to perform a full decision analysis effort on all decision opportunities, and the answer is no. Ralph Keeney suggests in Figure 5.2 how 1000 decisions that deserve some thinking should be resolved (Keeney, 2004). He suggests that a full decision analysis be done for only 5% of these decisions, and a partial decision analysis for 20% of them. The remaining 75% can be resolved simply by clear thinking that is consistent with decision analysis.

Clearly, the process that an organization uses to make a decision must be tailored to both the organization and the decision. Answers to three key questions will guide the selection of the appropriate process.

5.3.1.1 How Urgent Is the Decision? The timeframe available for making the decision may be of overriding importance in choosing how it is made, particularly if that timeframe is very short. In an emergency, for example, we simply may not have the luxury of time to give full consideration to which alternative is optimal. Even if the timeframe is measured in days rather than minutes, the press of time may curtail the process that we would ideally want to use. So, we must choose a process that results in a timely decision.

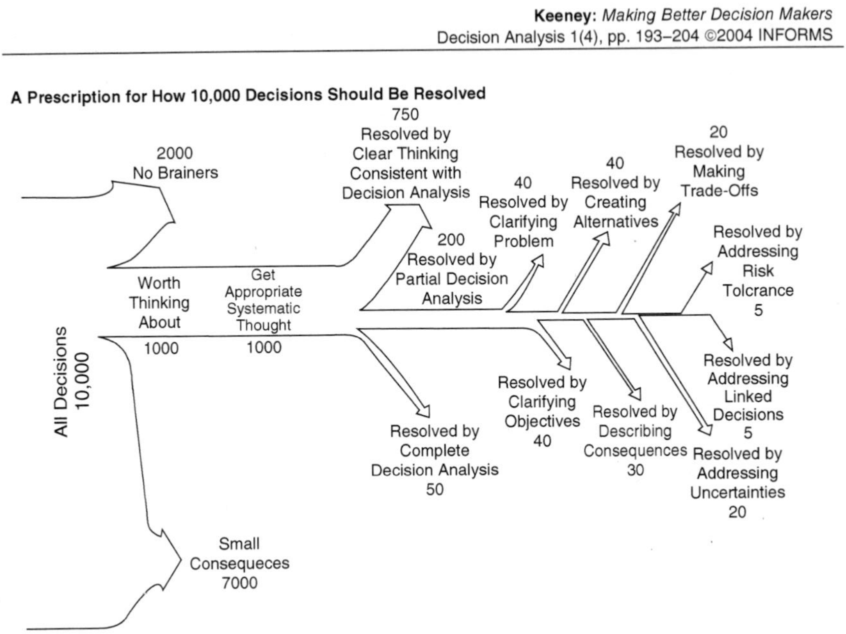

FIGURE 5.2 Suggested prescription for resolving decisions. (Reprinted by permission, Keeney RL, Making better decision makers, *Decision Analysis*, volume 1, number 4, pp. 193–204, 2004. Copyright 2004, the Institute for Operations Research and the Management Sciences (INFORMS), 7240 Parkway Drive, Suite 300, Hanover, MD 21076 USA.)

5.3.1.2 How Important Is the Decision?

The degree of importance of the decision should dictate the level of effort devoted to making it. A useful rule-of-thumb is the one percent rule, which states that one should be willing to spend 1% of the resources that will be allocated in a decision to ensure that the choice is a good one. So, for example, when deciding on the purchase of a \$1000 household appliance, one should be willing to spend \$10 to gather information that could improve the choice. By the same token, a company deciding on a \$100 million investment should be willing to spend \$1 million to ensure that the investment decision is made well.

Failures commonly occur in both directions. Some very important decisions are made with far too little effort given to making them well. And sometimes, far too much effort is inappropriately expended to make relatively minor decisions.

5.3.1.3 Why Is This Decision Difficult to Make?

As stated in Chapter 2, the degree of difficulty in a decision can be characterized in three different dimensions of complexity—content complexity, analytic complexity, and organizational complexity. A decision that is difficult in only one of these dimensions calls for an approach that is specific to that dimension. A decision that has only content complexity can be made well by collecting and analyzing the required information. A decision that has only analytic complexity can be made well through good analysis. And a decision that has only organizational complexity

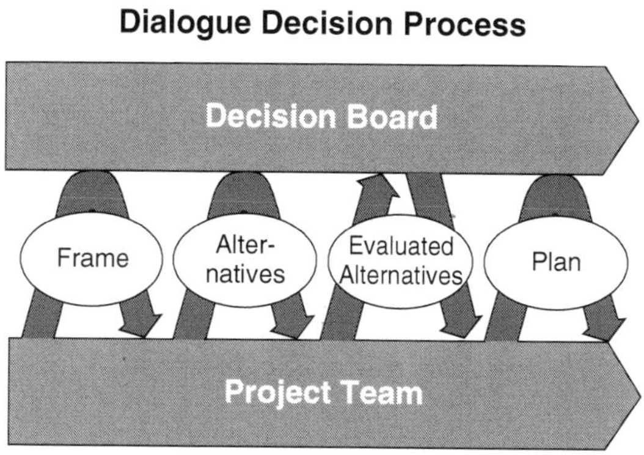

FIGURE 5.3 The dialogue decision process.

can be made well with excellent facilitation techniques (see Chapter 4). But a decision that is difficult because it is complex in two or three of the dimensions calls for a carefully designed process.

5.3.2 TWO BEST PRACTICE DECISION PROCESSES

5.3.2.1 Dialogue Decision Process. The dialogue decision process (Spetzler, 2007) has been successfully applied within hundreds of organizations to achieve good decision making in the face of content, analytic, and organizational complexities. As its name suggests, the process centers on a structured dialogue between two teams: (1) A team of decision makers, here called the decision board, and (2) a team created to assist the decision makers, here called the project team. The process is shown in Figure 5.3. Taken together, the arrows in this figure are suggestive of a snake, so the process is sometimes informally referred to as "snaking."

The decision board should comprise those people whose agreement is both necessary and sufficient for the decision to be made and successfully implemented. A failure mode to avoid is to exclude from the decision board someone whose veto could block the decision. On the other hand, the decision board should not include anyone whose agreement is not needed to make the decision. Scheduling meetings of the decision board is generally a major challenge, and the more members on the board, the more difficult the scheduling task. Typically, decision board members are asked to commit the time to attend three or four meetings, each of 2–3 hours duration, during a period of 2–4 months.

The project team consists of staff members from the organization who collectively possess the skills to conduct the dialogue decision process, have access to key information sources (including key stakeholder and subject matter experts) required to make the decision, and have the credibility and trust of the decision board members. The project team may also include decision professionals who are internal or external consultants. At least one of the internal members of the project team should commit to work full-time on the decision for the duration of the process, and the remaining members should commit to be involved at least half-time. The project team should have enough members to give it sufficient capacity to carry out its responsibilities, but not so many members that their time is used inefficiently. People who are needed for only one specific part of the process, such as some subject matter experts, need not be on the project team but rather can meet with the team as needed.

The dialogue decision process centers on a typical sequence of 3 or 4 meetings between the decision board and project team, driven by specific deliverables made up of a subset of the six elements of decision quality.

5.3.2.2 First Decision Board Meeting: Frame. At the beginning of the project, the decision board must be in agreement that the decision being addressed is appropriately framed. Sometimes, this is done by a distinct frame-review meeting; other times by one-on-one interactions. Either way, decision board members discuss the frame for the decision as proposed by the project team, amend it as appropriate, and finally approve the frame as amended. The frame should provide clear answers to the following questions: What is the problem or opportunity being addressed in this decision? What range of choices are within scope for this decision and what possible actions are outside the scope? What value metrics should be used to compare the possible alternative courses of action? What risk factors need to be accounted for in this decision? What is the timeframe for making the decision? Who should be involved? (See Chapter 6 for a full discussion of framing a decision and Chapter 7 for the development of value metrics.)

5.3.2.3 Second Decision Board Meeting: Alternatives. Following the agreement on frame, the project team works to identify and develop the full range of decision alternatives within the agreed-upon frame. Emphasis is placed on using creativity-enhancing techniques to help identify possibly great alternatives that might be difficult to see initially. Each alternative is checked to ensure that it is both feasible and attractive. (See Chapter 8 for a full discussion of generating good alternatives.) After confirming that the alternatives are consistent with the agreed-upon frame and that the set of alternatives span the decision space, the project team presents them to the decision board, who may eliminate some from further consideration and may suggest new alternatives. By the end of this meeting, the decision board and project team will have agreed on the set of alternatives to be carried forward for evaluation.

5.3.2.4 Third Decision Board Meeting: Evaluated Alternatives. The Project Team then embarks on the task of evaluating the defined set of alternatives. This task typically requires the creation of an analytic structure, including a quantitative model for comparing the alternatives and the collection of relevant information to support the assessment of inputs for the model. (See Chapters 9, 10, and 11 for full discussions on conducting the analysis to evaluate alternatives.) The *decision analysis cycle* provides a logical framework for the analysis of

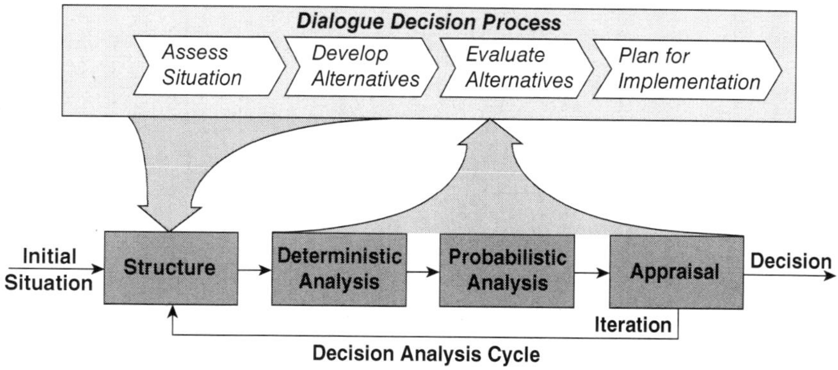

FIGURE 5.4 The decision analysis cycle.

alternatives (see Figure 5.4). The goal of the analysis is to create insight and understanding of which alternative is best and why it is better than the others. A crucial step at this point in the process is to use the insights gained from the evaluation of the initial set of alternatives to create new "hybrid" alternatives that may be superior to any of the original alternatives. Calculating the value of obtaining additional information to reduce key uncertainties may stimulate the creation of valuable hybrid alternatives.

The project team presents to the decision board the insights gained from the evaluation of the alternatives. This might be done in one meeting or it might be done in two, particularly if the analytic results are complex and require additional time to fully "digest." In this case, the first presentation of results is in the spirit of inquiry, to familiarize decision board members with its content, probe for weaknesses in the analysis, and for ways that the analysis and/or the alternatives could be improved, and reach agreement on additional work to be done. Alternatively, one-on-one briefings can be made to decision board members to allow them to preview the analytic results and to ask questions in private. These briefings also help the Project Team understand what issues are likely to come up and prepare to address them constructively prior to the final decision meeting.

Chapter 13 discusses the presentation of decision analysis results to decision makers in the spirit of enlightenment. The goal at this point of the process, whether done in one meeting or two, is to have the decision board members agree on the course of action to be undertaken—that is, to have the decision board make the decision.

5.3.2.5 Fourth Decision Board Meeting: Implementation Plan. Once the decision board has chosen a course of action, planning for its implementation should be done. Depending on the situation, the decision board may or may not meet once again to review and approve the implementation plan. Chapter 14 describes the activities in decision implementation.

In addition to having formal meetings with the decision board, it is always a good practice for project team members to have more frequent, informal

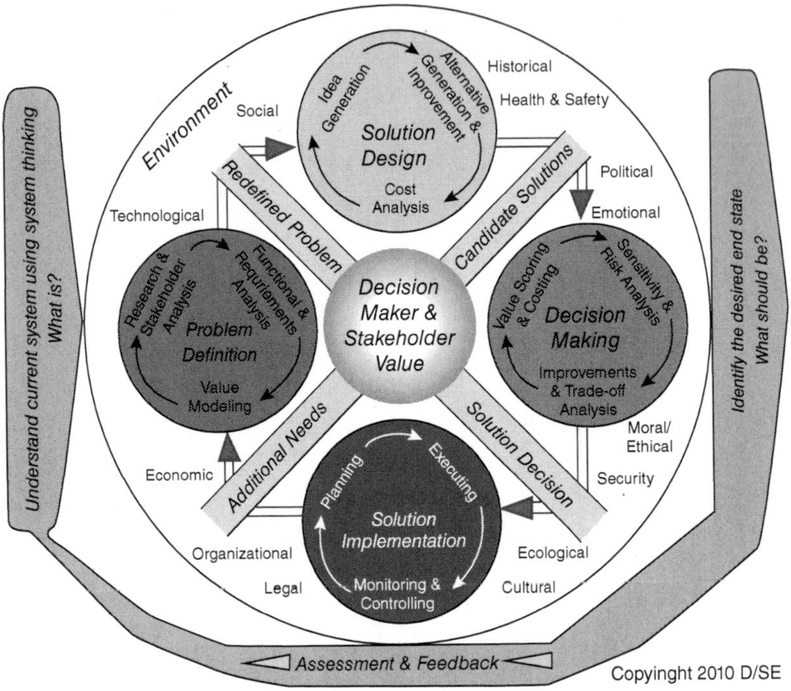

FIGURE 5.5 Systems decision process. (See insert for color representation of the figure.)

communication with the decision makers when that is possible. Such communication would improve the dialogue on the decision and provide more timely feedback on important issues to resolve.

The systems decision process (Parnell et al., 2011) is an example of the dialogue decision process tailored for system decision making performed by systems engineers and program managers responsible for large complex systems. Figure 5.5 provides a graphic of the systems decision process.

The systems decision process is designed to be used to make the decision for a system to proceed from its current stage to the next stage in the life cycle process. For example, it could be used to approve the system concept that will proceed into system design. The phases of the systems decision process map directly to the steps of the dialogue decision process: the systems decision process uses problem definition for framing, solution design to develop alternatives, decision making to evaluate alternatives, and solution implementation to plan for implementation. The project team is called the system design and analysis team, while the decision board is made up of the decision makers and key stakeholders responsible for approving the system design for the next life cycle stage. Furthermore, the spokes of the diagram define the interaction points with the decision makers and stakeholders. Also similar is the use of a decision analysis value model to determine the best design solution. The systems decision process also includes a list of the potential environmental factors that may be important to the decision.

5.3.2.6 Decision Conferencing Another best practice decision process is decision conferencing, which is described in detail in Appendix C. In a decision conference, decision makers and stakeholders work together in an intensive 2- to 3-day session under the guidance of skilled facilitators to identify key issues, evaluate alternatives, reach decisions, and plan implementation. Required information is elicited directly from the participants rather than derived from external data sources. A simple value model is developed during the session to help evaluate and compare alternatives.

Larry Phillips (Phillips, 2007), a highly experienced leader of decision conferences, identifies three reasons why decision conferencing is valuable: (1) it helps decision makers and stakeholders create a shared understanding of the issues; (2) it develops a sense of common purpose among the participants; and (3) it fosters a commitment among the participants on the way forward.

Decision conferencing is most useful for decision situations in which there is a clear sense of urgency so that all key participants are willing to devote the time and effort required and in which the chief challenge is in gaining agreement and commitment of the decision makers rather than in overcoming difficult analytic and content challenges.

Decision conferencing and the dialogue decision process are often combined in a series of intensive workshops interspersed with intensive "homework." This approach has been especially useful with teams that are geographically dispersed.

5.3.3 TWO FLAWED DECISION PROCESSES

Next, we describe two common types of decision processes that have major flaws.

5.3.3.1 Strictly Analytical Decision Processes. The focus of a strictly analytical process is on the technical complexity of the decisions and not the organizational complexity. Typically, this approach uses complex analytic models to compare the existing alternatives. Generation of additional alternatives may or may not receive appropriate attention. Interaction with decision makers and key stakeholders typically occurs only at the start and end of the process. The final presentation provides analytic insights about the alternatives. But since the decision makers and stakeholders have not participated in the process, they may raise issues or alternatives that were not considered in the analysis. Furthermore, they may not be willing to accept the results of an analysis that they consider a "black box" and do not understand. Thus, a strictly analytical process may result in "spinning wheels," a situation in which the decision point is never reached

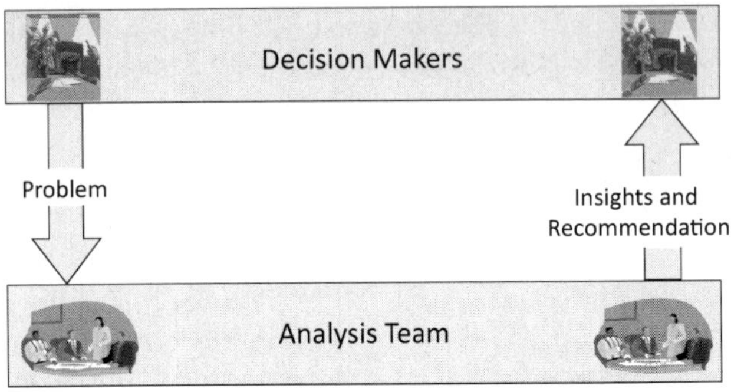

FIGURE 5.6 Strictly analytical process.

because the process is endlessly cycled. Figure 5.6 is a diagram of a strictly analytical process.

5.3.3.2 Advocacy Decision Processes. The second flawed type of decision process is the advocacy process. In this process, an advocate (who may be just one person or a group of people) prepares a presentation to the decision makers with the goal of convincing them to undertake a particular course of action. The advocate is a "champion" of the recommended course of action, either because he or she believes that it is best for the organization or for other reasons.

The advocacy process shares with the strictly analytical process the flaw that decision makers are presented with a recommendation that they were not involved in exploring and evaluating. Their role in the advocacy process is to try to identify the weaknesses of the recommended course of action and to either accept or reject the "pitch." The advocacy process has two additional unique flaws. First, possible alternatives to the recommended course of action are generally not presented for consideration to avoid unwanted distraction from the advocated course of action. Second, the analysis supporting the advocated actions tends to focus on its upside and minimize its downside, with information that is unfavorable to the recommendation being downplayed or omitted. Figure 5.7 illustrates the advocacy process.

5.4 Decision Processes in Illustrative Examples

See Table 1.3 for further information on the illustrative examples. In all three illustrative examples, the decision process had to be determined.

5.4.1 ROUGHNECK NORTH AMERICAN OIL STRATEGY

Roughneck Oil's North American division owned or controlled assets that could lend themselves to a variety of energy-related businesses, including:

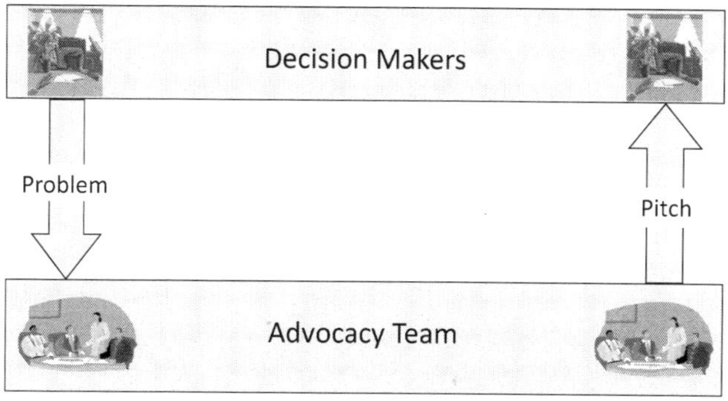

FIGURE 5.7 Advocacy process.

- Conventional oil and gas exploration and production
- Enhanced oil recovery
- Coalbed methane
- Tar sands
- Power generation.

The dialogue decision process was used for this decision. The project was initiated by the Roughneck North American VP, who identified his boss, the SVP for The Americas, as the final decision maker. The North American VP invited six additional functional area leaders (conventional exploration and production, marketing, enhanced oil recovery, facilities, business development, and strategy) under that SVP to join the decision review board.

In consultation with an external decision analysis team (consultants), the Roughneck North American VP created the core team, comprising 40 key contributors in key functional areas. He asked the core team to frame the project around their concerns, and asked them to work together throughout the project. There had been two previous decision analysis projects, so a few of the participants were familiar with the process, but it was new to many of them.

5.4.2 GENEPTIN PERSONALIZED MEDICINE

As a growing biotech company, DNA Biologics was transitioning its decision making process. Historically, it had relied on informal, experience-based decision-making process. As it matured, it started to adopt a more formal, analysis- and consensus-based decision approach—the dialogue decision process shown in Figure 5.8, which is a variation of the process described in Section 5.3.2.1. The project team worked with decision makers and other stakeholders within the organization to establish shared ownership, alignment, and fact-based decision making to achieve consensus on the best decision strategy.

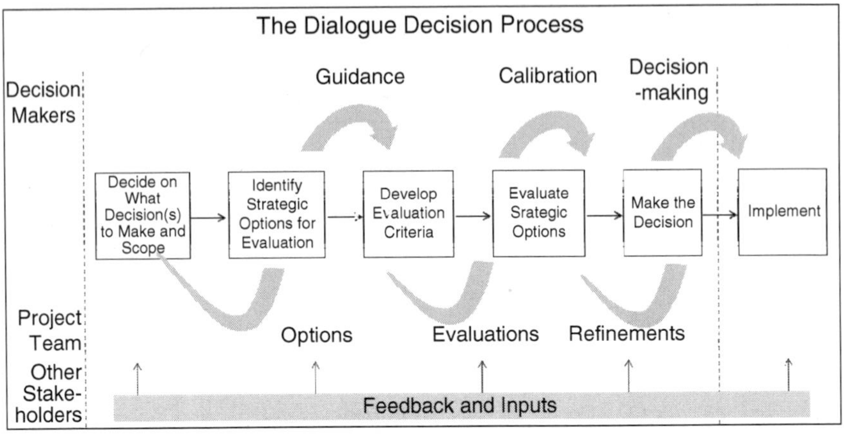

FIGURE 5.8 The Geneptin dialogue decision process.

For the Geneptin decision, the decision board comprised the two joint decision makers: the VP of Drug Development and the VP of Commercialization. The project team was headed by a program team leader, and included members from multiple functional areas, including commercial (global/strategic marketing), clinical development, pricing and market access, biomarker development, regulatory, medical affairs, and business development (for potential diagnostic partnership aspects).

5.4.3 DATA CENTER

The government agency needed to be able to justify the data center location decision to budget approvers in the executive branch and the Congress. In addition, the decision had to be defensible to explain the decision to governors, senators, congressmen, and mayors in locations *not* selected. Multiple agency directorates were involved in the decision: technology (design the mission applications), mission (operate the mission applications), information technology (operate data center and communications), and logistics (design and acquire the facilities, power, and cooling). There would be multiple approval levels to obtain the funds and approve the location decision within and outside of the agency. The agency director and deputy director were the decision makers.

The agency did not have a process to decide the location of the next data center. Based on the recommendations of agency advisors, the systems decision process was used. A senior executive was assigned to lead the project team. About 25 experts from all the key organizations were assigned to the team. The decision analysis was performed by two decision analysts on the team. With the exception of the decision analysts, the project team did not have experience with the systems decision process.

5.5 Organizational Decision Quality

This chapter discusses the choice of an appropriate process for making single decisions. Often, however, an organization that has realized the benefits of high quality in several important decisions will strive to routinely achieve that quality in *all* of its decision making. This goal is given the name *organizational decision quality* (ODQ). Carl Speltzler (Spetzler, 2007) states five criteria by which one can judge whether an organization has reached this goal:

1. The organization consistently makes high-quality decisions, correctly recognizing and declaring decisions, framing them appropriately, addressing them with a good balance of content, analytic rigor, and facilitation, and understanding and avoiding the behavioral traps that can degrade decisions.
2. The organization has a broad array of tools, techniques, and processes in place as well as decision support staff members who are skilled in using them appropriately.
3. The decision makers within the organization understand their roles in the process and have the right knowledge and skills to participate effectively.
4. The organization is well aligned around decision quality, utilizing a common language for and understanding of decision quality as well as a strong shared desire to regularly achieve it.
5. The organization continuously seeks to learn and improve its decision competency.

Achieving organizational decision quality requires the concerted efforts of many people over a period of many years (perhaps 5–10), with systematic training programs and repeated application of decision analysis processes. The path is long and the investment high, but organizations that have achieved the goal report that the rewards easily justify the investment.

5.6 Decision Maker's Bill of Rights

The Society of Decision Professionals (SDP) has published a document called "The Decision Maker's Bill of Rights"[1] that states the expectations that any decision maker should have. Among the several variations of the document is the following:

[1]The Bill of Rights was conceived by Jim Felli and Jay Anderson of Eli Lilly and Company.

The Decision Maker's

Bill of Rights

As a decision maker, I have the right to:

1. A decision **frame** that structures the decision in the context most relevant to my needs
2. Creative **alternatives** that allow me to make a selection among viable and distinct choices
3. Relevant and reliable **information** upon which to base my decision
4. A clear understanding of the consequences of each alternative based on my **values** and choice criteria
5. A logical **analysis** that allows me to draw meaningful conclusions and choose the best alternative
6. **Clarity** of action for the choice I select, and
7. A decision process that gains **alignment and commitment** to action.

The decision maker Bill of Rights is an excellent way to engage the key decision makers at the beginning of a decision analysis project to help them understand why they benefit directly from the use of the decision processes recommended in this chapter.

5.7 Summary

Our overall goal in the decision process is to help senior leaders make good decisions, where goodness is defined by the six essential elements of decision quality. We describe the dialogue decision process as a best practice that can be used to achieve consistently good decisions in the face of complexity of three different types: content complexity, analytic complexity, and organizational complexity. We also describe decision conferencing as a best practice that is useful for urgent decision situations in which decision makers and key stakeholders are willing and able to work together intensively for a few days, using only readily available information and fairly simple analytic structures.

Both the dialogue decision process and decision conferencing require a significant investment of time and effort and therefore should be used only for those decisions where such an investment is justified. Table 5.1 suggests the process that is appropriate for each type of decision.

TABLE 5.1 **Fitting the Process to the Decision**

	Nature of the Decision	Timeframe	Appropriate Process
Quick	Small everyday decisions or emergencies	Decided in the moment	Reflex Develop decision fitness and good habits
Significant	Important but easy or complex but not important	Decided in hours	Due Consideration Use DQ elements as a checklist and avoid decision traps
Strategic	Of strategic importance and complex	Decided in days, weeks, or months	Rigor Use dialogue decision process or decision conferencing

KEY TERMS

Advocacy process A decision process focused on advocacy (collecting data to support the recommendation) versus analytics to determine the best alternative.

Analytical process A decision process that focuses primarily on using analytic models of the decision content and complexity and not organizational complexity.

Decision conference A decision process that puts decision makers, stakeholders, and subject matter experts in a facilitated meeting to define the problem, develop alternatives, evaluate alternatives, and make decisions and plan implementation.

Decision Maker's Bill of Rights List of seven expectations of decision makers for a quality decision process.

Dialogue decision process A decision process that focuses on a structured dialogue between the Decision Board and the Project Team.

Decision process An organizational process that considers the technical and social processes of decision making.

Decision quality A high quality in each of six essential elements: clear frame, creative alternatives, credible information, clear values, logical reasoning, and commitment to action.

Five Rules The five axioms for a mathematically and logically sound decision.

REFERENCES

Campbell, A. & Jo Whitehead, J. (2010). How to test your decision-making instincts. *McKinsey Quarterly*, May, 1–4.

Kahneman, D., & Klein, G. (2010). Strategic decisions: When can you trust your gut? *McKinsey Quarterly, 2*, 58–67.

Keeney, R.L. (2004). Making better decision makers. *Decision Analysis, 1*(4), 193–204.

Matheson, D. & Matheson, J. (1998). *The Smart Organization*. Boston: Harvard Business School Press.

Parnell, G., Driscoll, P., & Henderson, D. (eds.). (2011). *Decision-Making in System Engineering and Management*, 2nd ed. New York: John Wiley & Sons.

Phillips, L. (2007). Decision conferencing. In W. Edwards, R. Miles, & D. von Winterfeldt (eds.), *Advances in Decision Analysis*, pp. 375–399. Cambridge: Cambridge University Press.

Spetzler, C.S. (2007). Building decision competency in organizations. In W. Edwards, R. Miles, & D. von Winterfeldt (eds.), *Advances in Decision Analysis*, pp. 451–468. Cambridge: Cambridge University Press.

CHAPTER SIX

Frame the Decision Opportunity

STEVEN N. TANI and GREGORY S. PARNELL

A pessimist is one who makes difficulties of his opportunities and an optimist is one who makes opportunities of his difficulties.

—Harry S. Truman

Opportunities multiply as they are seized.

—Sun Tzu

Handbook of Decision Analysis, First Edition. Gregory S. Parnell, Terry A. Bresnick, Steven N. Tani, and Eric R. Johnson.

6.1 Introduction

The decision frame is the lens that we use to view the decision problem or opportunity. We believe that a good decision frame is critical to decision quality (see Chapter 5). Creating a good frame is the first task that a decision practitioner should undertake when working on a decision. Almost every decision process begins with a step that focuses on describing the decision problem or the decision opportunity. As described in Chapter 5, the dialogue decision process begins with "decision framing." Decision framing is the first step in decision quality. Clemen and Reilly's decision analysis process flow chart begins with "Identify the decision situation and understand objectives" (Clemen & Reilly, 2001). The scalable decision process begins with "defining the problem or opportunity" (Skinner, 2009). The systems decision process begins with "problem definition" (Parnell et al., 2011).

Inadequate or poor framing is an all-too-common cause of failure to achieve good decision making within an organization. When postmortems are conducted on decisions that have "gone wrong," the reasons identified for failure frequently point back to low quality in the framing of the decisions. "We did a great job of solving the wrong problem." "We couldn't agree on a path forward within the short timeframe that we had." "We failed to involve a key decision maker and he vetoed the recommendation." "We didn't talk to all of the key stakeholders and didn't have a full understanding of the key issues." "We overlooked an important nonfinancial objective, so the decision makers were not willing to accept the results of our analysis." "We tried to solve all of the company's problems instead of focusing on the most important ones and we never got anywhere."

In effect, the frame defines the decision that is being made. The frame makes clear which possible courses of action may be considered as part of the decision and which may not. The frame specifies *when* the decision must be made, and it identifies the measures by which the potential outcomes of the decision should be evaluated. The frame guides us regarding who should be involved in making the decision.

Remember that there is no objectively "correct" frame for any decision situation. Rather, the right frame is one that is explicitly approved as appropriate by the decision makers. That is why the first meeting of the decision board in the dialogue decision process (see Chapter 5) is devoted to reaching agreement on the framing of the decision.

The decision frame is the critical first step in decision analysis. The decision frame helps us define the decision.

In this chapter, we begin by discussing how a decision is declared (Section 6.2). We then discuss what constitutes a good frame for a decision (Section 6.3) and describe a number of best practices for achieving good decision framing (Section 6.4). We conclude with a description of how the decisions were framed in the illustrative examples (Section 6.5).

6.2 Declaring a Decision

Before a decision can be framed, someone must declare that the decision needs to be made. All too often a worry festers, but no one takes the step to declare that alternatives should be formulated and action taken. Sometimes, it can be a great service to someone facing such a situation to initiate a conversation that leads to the idea that a decision should be made.

It is quite possible that the mere declaration of a decision will lead the person to formulate an excellent alternative, which he or she can then confidently choose, without further ado; no extensive analysis required! Sometimes, a brief discussion of objectives can clarify the issue, and, again, enable a quick decision to be taken without extensive process.

On other occasions, the decision maker may be tempted to "trust their gut," but this may lead to low quality in the decision. Several authors (Kahneman & Klein, 2010), (Campbell & Whitehead, 2010) point out that if the decision maker has not received good feedback on previous analogous decisions, or has some motivational involvement with this one, deciding without more structured thought is problematic. It can be a great service to help someone in this situation formulate the intention to articulate objectives, generate alternatives, and ascertain which alternatives meet the objectives best. This conversation is an ideal time to help the person formulate a vision statement (see Section 6.4) for a decision process, and begin thinking about the resources required to realize the vision.

6.3 What Is a Good Decision Frame?

We often use the analogy of framing a photograph when discussing decision framing. An experienced photographer carefully frames a photograph to focus the viewer's attention on those features of the subject that the photographer believes are of greatest importance. In doing this, the photographer consciously excludes some features of the situation and includes others within the frame. The framing of the photograph creates a specific perspective on the subject—long-distance view versus close-up and wide-angle versus telephoto view.

A decision frame specifies three key aspects of the decision:

1. Purpose
2. Perspective
3. Scope.

Purpose. The decision frame should make clear *why* the decision is being made. Although this may seem obvious in most situations, a carefully considered answer to this question may open up the opportunity to identify creative and valuable alternatives. For example, it might seem obvious that the purpose of a particular decision is to choose the best factory configuration to produce a new product. But further thought might lead to a broader definition of purpose—to find the most cost-effective way of producing the new product—that could lead to consideration of a wider range of alternatives and possibly to a more valuable final choice.

The purpose of the decision may also include a specification of *when* the decision must be made, particularly if the choice must be made urgently. For example, the purpose of a decision might be to choose whether or not to exercise an option before it expires 2 days from now.

And finally, the purpose of the decision should make clear which consequences of the decision are important. In other words, the frame should specify which value measures should be used to compare alternatives. For example, the purpose of a decision might be to choose the method of producing the new product that achieves the best trade-off between financial profit and environmental impact.

Perspective. The frame of a decision should give an indication of how the decision situation is to be viewed. What perspective(s) should we use to look at this situation? Is this more a production planning problem or a marketing strategy problem? Should we think of this as a strictly financial opportunity or does it have a social responsibility aspect as well? For example, the second author worked on the Army's 2005 Base Realignment and Closure (BRAC) (Ewing et al., 2006). This was the fifth round of BRAC. The perspective for the first four BRAC rounds was to reduce defense resources by moving units and closing installations. However, for the 2005 BRAC, the framing was much broader. The perspective was to transform Army missions and infrastructure to better meet future national security threats and to reduce defense resources.

With a clear perspective on the decision situation, the frame informs which issues need to be addressed and helps identify which people should be involved in making the decision.

Scope. The decision frame should clearly delineate which possible actions may be considered as part of the decision and which may not. The frame sets a boundary around those alternatives that are within scope. Once agreement on the frame is reached, alternatives that are outside the boundary can be disregarded. For example, the scope may include choices involving the reconfiguration of a production plant while explicitly excluding the choice of shutting down the plant and outsourcing production. The decision frame should make clear whether or not the "do nothing" alternative is within scope.

Importance of a Decision Frame

This is a personal story that our decision analysis colleague Carl Spetzler likes to tell to illustrate the importance of agreeing on the frame for a decision. One morning, his wife said to him, "I think it is time to repaint and carpet our house." Carl looked around and saw that she was right. His response was, "Should we consider doing some remodeling in the kitchen and playroom first? After all, we are going to become empty nesters in six months." Pretty soon they were talking about hiring an architect to redo their bedroom area. And as the ideas kept growing—along with the dollar signs—they concluded that perhaps they should consider selling the house and buying another that may already have these amenities. After a while, Carl's wife asked, "How long do you intend to work before retirement?" Soon the question had grown to "So, what are we going to do with the rest of our lives?"

Painting and carpeting would be a 2-month and $2,000 project, whereas planning the rest of their lives was a huge question that might take a couple of years to resolve in multiple rounds with large financial and quality-of-life implications. What decision should they focus on?

Clearly, before proceeding, they would need to agree on the frame of the decision to work on, particularly the scope that specifies which kind of choices are under consideration and which are not.

6.4 Achieving a Good Decision Frame

The fundamental requirement to achieve good framing of a decision is communication—well-structured and effective communication among those involved in making the decision and the key stakeholders. Three major communication techniques are interviews, surveys, and facilitated groups (See Chapter 4 for a full discussion of the advantages and disadvantages of each technique). In practice, we use a combination of these techniques. A highly recommended way to foster good communication is to conduct a framing workshop. This event is typically an all-day session, or even a multiday session, involving perhaps 15–20 people who are the decision makers, subject matter experts, and other stakeholders or their appointed representatives. If scheduling constraints make such a workshop impossible, it can be conducted in several shorter sessions or key individuals who are not able to attend may be interviewed.

The suggested agenda for the framing workshop is as follows:

1. Vision statement
2. Issue-raising
3. Categorization of the issues
4. Decision hierarchy
5. Values and trade-offs
6. Initial influence diagram
7. Decision schedule and logistics.

6.4.1 VISION STATEMENT

Creating a vision statement for the decision is an effective way to make sure that there is agreement on the purpose of the decision among the decision maker(s) and key stakeholders. The vision statement[1] answers three questions:

1. What are we going to do?
2. Why are we doing this?
3. How will we know that we have succeeded?

A good way to create the vision statement is to solicit and display answers to each question from all workshop participants. Eliminate duplications among the responses and then discuss substantial differences in the answers, trying to reach consensus on them. If necessary, appeal to the decision makers for final resolution of differences. Once consensus is reached on the key thoughts in the statement, have a volunteer work offline to find a concise but clear way to express the vision statement.

The vision statement refers to the process of making the decision, NOT the consequences of the decision. So, the third question ("How will we know that we have succeeded?") should be understood to mean success at the time the decision is taken, not when the final outcome is known. Workshop participants may be tempted to include their desired solution in the criterion of success. Highlighting this issue up front can help combat this tendency when it arises. An example of a Vision Statement is shown in Figure 6.1.

It may be tempting to skip working on the vision statement because "Everyone knows what this decision is about." If that is the case, then creating the vision statement will take only a few minutes. But the vision statement exercise may reveal unsuspected differences of opinion about the purpose at

[1]Other common names for the vision statement are the purpose statement, opportunity definition, or problem definition.

Example Vision Statement

We will decide how to manufacture the new product in the most cost-effective way. We need to do this to support the successful launch of the product next year. We will know that we have succeeded if all participants are satisfied that we have chosen the best path forward.

FIGURE 6.1 Example vision statement.

hand. These differences must be resolved so that there is full agreement on the purpose. Otherwise, the entire decision effort may be doomed to failure.

The Power of a Clear and Compelling Vision Statement

In a past consulting project, the second author was asked to review and help a decision analysis team whose results were not accepted by decision makers during the external review process. The existing vision statement was not clear or compelling. We spent the first hour of a 2-day meeting working on the vision statement. We used the exact procedure described above. At the final presentation to the senior decision maker, the new vision statement was presented first. She said, "This is the first time I have understood the purpose of this decision!"

6.4.2 ISSUE RAISING

A central step in framing a decision is issue raising. When successfully done, issue raising brings to light all of the many perspectives that can be used to view the decision situation and lays the groundwork for creating a good decision frame.

Issue raising can be enhanced by the active participation of everyone in the framing workshop under the guidance of a skilled facilitator (see Chapter 4 for discussion of facilitator skills and techniques). The facilitator makes sure that all participants understand the purpose of the decision, as declared in the vision statement. Then each participant is asked to write down (in silence) as many issues as come to mind that bear on the decision situation. An issue is any statement or question relating to the situation at hand. The facilitator should insist that each issue be expressed as a complete sentence or question. So, "market share" is not a valid issue statement, but "Our market share is falling." is a valid issue statement. If possible, each issue should be written on a large Post-it® note. After participants have had sufficient time to write their issue statements (say,

5–10 minutes), the facilitator asks one participant to read one of his or her issue statements. No discussion of the issue statement is permitted, except for questions of clarification. If Post-it notes are used, the issue is placed on the wall. Otherwise, it is recorded in a computer document that is projected and visible to all. The facilitator then selects other participants in turn to each read one issue statement. As issues are placed on the wall or in the computer document, attention is paid to grouping related issues together. After all issue statements have been read and posted, the facilitator asks the participants to review the issues and reflect on them. This may lead to the identification of additional issues, which are then added to the original set.

Working with scenarios may help to enrich the set of issues. To do this, the facilitator divides the group into small teams and assigns a scenario to each. The scenarios, which are prepared in advance, portray significantly different possible futures that could strongly impact the outcome of the decision at hand. Each team is asked to raise issues that would be relevant within their assigned scenario. These are then collected when the teams are brought back together and added to the issues previously raised.

For a major decision situation, the number of issues raised is typically more than one hundred. The facilitator may appoint a subteam to review all of the issue statements and to make sure that they are appropriately grouped by theme.[2] Duplicate issue statements may be consolidated.

The facilitator may also have participants vote on which issues they think are most important to address in reaching a good decision. This will give some sense of priority ordering of the issues, but no issue should be discarded from consideration at this stage.

The following are suggested good practices for issue raising:

- Make sure that participants representing as many diverse perspectives as possible are in the workshop.
- Make sure that everyone's issue statements are heard.
- Remind participants that issue raising is just an early step in the decision process, not a search for the solution.
- Allow plenty of time for issue raising—1–3 hours.
- Prohibit judgmental statements about issues.
- Strive for a goal of quantity, not quality.

6.4.3 CATEGORIZATION OF ISSUES

After the many issues are raised, they should be categorized into four groups

- **Decisions.** Issues suggesting choices that can be made as part of the decision.

[2]Common names for this process are affinity diagramming or binning.

- **Uncertainties.** Issues suggesting uncertainties that should be taken into account when making the decision.
- **Values.** Issues that refer to the measures with which decision alternatives should be compared.
- **Other.** Issues not belonging to the other categories, such as those referring to the decision process itself.

For many issue statements, categorization is clear. For example, the statement "We need to expand our productive capacity" is a decision issue, while the statement "When will our competitor introduce an upgraded product?" is an uncertainty issue. But some issue statements are more difficult to categorize. For example, consider the issue statement "Can we use new technology to reduce production costs?" This might be viewed as an uncertainty about the impact of new technology on production costs. Alternatively, it might be viewed as a decision about whether to install new technology. The context of the situation may make it clear that one of these interpretations is relevant and the other is not, or, it may be that both are relevant. If so, the one issue statement should be split into two statements that are placed in separate categories.

Once the issues have been categorized, they are used to inform later steps in the decision process. The decision issues provide the basis for defining the scope component of the decision frame (see decision hierarchy, below) as well as serve as raw material for the creation of alternatives (see Chapter 8). The uncertainty issues provide a checklist to be used when developing the analytic structure for evaluating alternatives (see Chapters 9 and 11). The value issues likewise provide a checklist for defining the measures used in evaluating the decision alternatives (see Chapter 7).

6.4.4 DECISION HIERARCHY

The decision hierarchy is a valuable conceptual tool for defining the scope of the decision. The decision hierarchy (see Fig. 6.2) is portrayed as a pyramid structure with three levels to which are assigned the many possible choices that might be part of the decision at hand. The topmost level is for those high-level context-setting choices that are out of scope because they are assumed to have been made already. For the purposes of the decision at hand, we "take as given" the choices in the top level of the hierarchy. For example, in a decision focused on the manufacturing strategy for a product line, the choice of whether or not to keep that product line as a company offering might be assigned to the top level of the hierarchy—we take as given that the company will continue to make and sell the product line.

The middle level of the decision hierarchy contains those choices that are inside the scope of the decision under consideration. Different combinations of these choices will be defined as alternatives for the decision.

The lowest level of the decision hierarchy has choices that are outside the scope of the decision because they can be deferred until later and/or delegated to others. These might be quite important choices that are separable from the

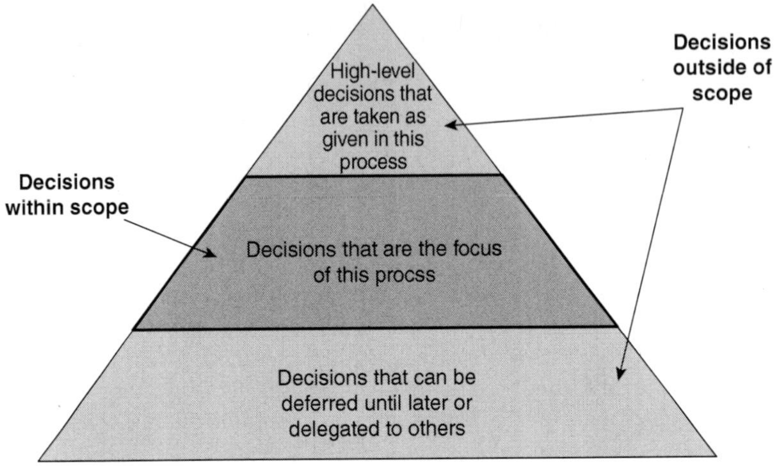

FIGURE 6.2 Format of the decision hierarchy.

decision under consideration. Or they might be low-level choices that have very little impact on the overall outcome of the decision.

A good method for creating the "first draft" decision hierarchy is to take all of the issues categorized as decisions and have the team discuss in which level of the hierarchy each one belongs. Make sure to include any decisions that were not mentioned in the issue raising but which are clearly a possible part of the decision situation. The resulting decision hierarchy is preliminary because the decision makers have the final say on where each choice should be.

The decision hierarchy is an effective vehicle for fostering a good conversation about the scope of the decision. And when completed and approved by the decision makers, the decision hierarchy is a clear statement of that scope. Only choices in the middle level of the hierarchy should be considered as part of the decision.

6.4.5 VALUES AND TRADE-OFFS

A part of the framing workshop should be devoted to reaching agreement on the value measures to be used to evaluate and compare alternatives for the decision. See Chapter 7 for a full discussion of values and trade-offs.

6.4.6 INITIAL INFLUENCE DIAGRAM

Depending on the situation, it may or may not be useful to devote time in the framing workshop to creating an initial version of the influence diagram that describes the structure of analysis to be used in evaluating the decision alternatives. Sometimes, creating the initial influence diagram helps clarify the discussion of the decision scope. Or the initial influence diagram can be used to define information-gathering tasks that need to be undertaken immediately. See Appendix B for important background information on influence diagrams. The use of influence diagrams for modeling is also presented in Chapter 9.

6.4.7 DECISION SCHEDULE AND LOGISTICS

The framing workshop, because it usually occurs at the start of the decision process, is a good occasion to establish the schedule for making the decision and to agree on the major tasks and logistical details. The schedule for the decision should be a timetable of steps that lead to making the decision. If the dialogue decision process (see Chapter 5) is to be used, then the dates of the decision board meetings should be established, at least provisionally. The composition of both the decision board and the project team should be agreed upon at this time. If possible, logistical details, such as designating a special meeting space for the project team, should be worked out in the framing workshop.

6.5 Framing the Decision Opportunities for the Illustrative Examples

See Table 1.3 for further information on the illustrative examples. In this section, we discuss the use of decision framing for the illustrative examples and some of the challenges in applying the techniques.

6.5.1 ROUGHNECK NORTH AMERICAN STRATEGY (RNAS)

With the help of an external team of decision analysis consultants, the Roughneck North American Vice President (VP) began the decision analysis project. There had been two previous DA projects, so a few of the participants were familiar with the process, but it was new to many of them. The VP identified a group of 40 key contributors in key functional areas and asked the decision analysis consultants to frame the project around the concerns of this group. The strength of this approach is that it gives voice to various constituencies who may not normally be able to raise concerns at a strategic level.

The first step for RNAS framing was an issue-raising session, in which participants raised their major issues, which the decision analysts then categorized as decisions, uncertainties, objectives, or other. This project illustrates the strengths and weaknesses of bottom-up framing of a decision project. It allowed the VP to "take the pulse" of the various business units, and gave many participants a feeling of inclusion in the process. However, upon reviewing the results, the decision analysis consultants were concerned that addressing them would call for multiple fragmentary perspectives, rather than generating and leveraging a coherent perspective on the business. The VP agreed.

Based on a review of the areas of concern and the kinds of insights desired, the VP asked that the decision analysts redefine their framing as a portfolio

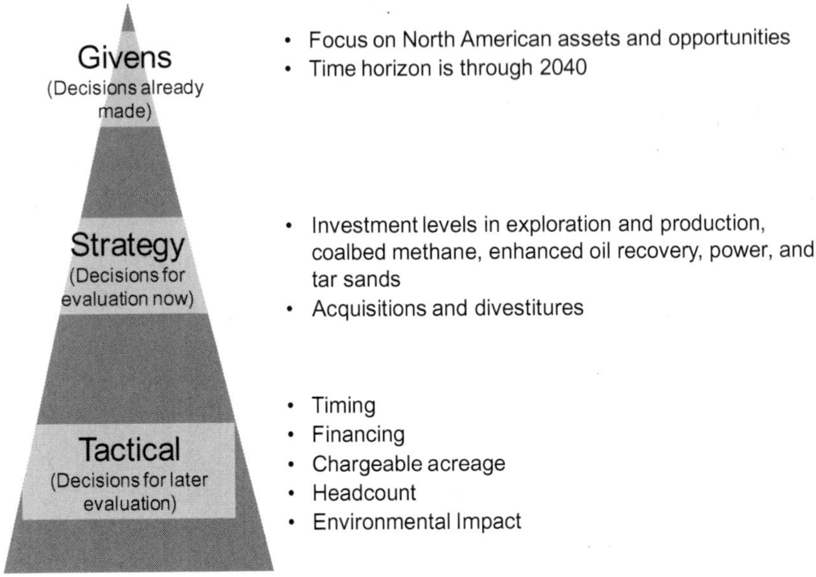

FIGURE 6.3 RNAS decision hierarchy.

resource allocation problem. The portfolio consisted of the three existing business areas (conventional E&P, coal bed methane, and enhanced oil recovery), along with two possible areas of expansion (tar sands, and electric power generation). A decision hierarchy (Fig. 6.3) was developed and used in this project.

The vision statement for the project was to "develop the RNAS portfolio investment and divestment strategy to achieve the financial objectives of the firm."

6.5.2 GENEPTIN PERSONALIZED MEDICINE

In framing the Geneptin decision, the project team developed the decision hierarchy shown in Figure 6.4. For an oncology drug candidate, Geneptin could be explored for many indications during its life cycle. Geneptin, for example, was being explored for the metastatic breast cancer indication, but could also be investigated for such other indications as adjuvant breast cancer and other tumor types. As the result of discussions within the decision team, to simplify and contain the scope of the decision analysis, the project team was instructed to focus the analysis on the metastatic breast cancer indication only. Thus, the key decisions to be addressed by the team were: (1) whether to develop Geneptin as a therapy for all patients with metastatic breast cancer or to pursue a personalized

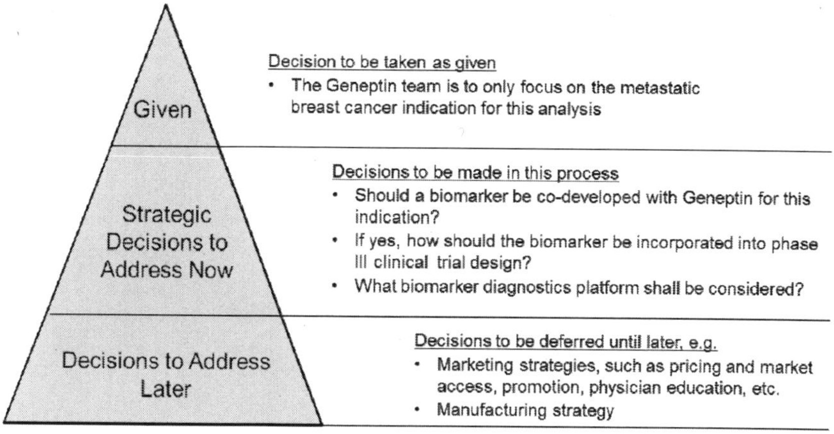

FIGURE 6.4 Geneptin decision hierarchy.

medicine approach; and (2) if Geneptin is to be developed as a personalized medicine, how should the biomarker diagnostic test be incorporated into the Phase III trials.

6.5.3 DATA CENTER DECISION

Three people from the Agency's advisory board were selected to do the opportunity framing. The decision makers believed that a new data center was needed. At that time, all of the agency's data centers were located in one area. Several senior leaders viewed this problem as an opportunity to make data center operations more secure by selecting a location outside of this area. There would be multiple approval levels to obtain the funds and approve the location decision within and outside of the agency. The agency needed to select the best data center location and justify the decision to budget approvers in the Executive Branch and Congress.

The decision framing team interviewed several key decision makers and stakeholders and identified some key assumptions. First, the full process from site selection to initial operational capability was projected to take at least 3 years. Second, it was decided that the actual IT equipment would be ordered *after* the site had been selected so the agency could order the latest IT technology. A third assumption was that after the facility became operational, it would be a "lights-out facility" remotely operated by mission managers with minimal support staff on site. Table 6.1 summarizes each stakeholder's key concerns and location preference.

After the interviews were conducted, an issue identification workshop was held with representatives of the key stakeholders. The stakeholder issue identification matrix in Table 6.2 summarizes the major issues from the interviews and the workshop. The interviews and workshop were effective techniques

TABLE 6.1 Concern List by Stakeholder

Stakeholder	Concerns	Location Preference
Agency director	Defensible decision Mission capability Cost	Out of metro DC
Director, board of advisors	Defensible decision Mission capability	Out of metro DC
Mission manager	Mission capability Usable mission floor space Loss of power or cooling Office space Travel distance from DC	Washington State (Columbia River)
IT director	Power Physical security Site accessibility for mission support personnel	Tennessee (Tennessee Valley Authority) at a National Laboratory location
Facilities director	Facilities acquisition cost Power cost	Existing agency facility in Texas
Security manager	Location from nearest road Physical security	Existing agency facility
Data communications manager	Bandwidth (from mission managers in DC to data center) Latency Information security	Facilities with high bandwidth communication links
Power manager	Primary and backup power sources Low cost power	Any location with reliable, low-cost power
Cooling manager	Primary and backup cooling sources Low cost cooling	Any location with reliable, low-cost cooling
Systems engineering director	Clear requirements Use a life cycle cost model	None
Life cycle cost manager	Use life cycle cost model	None

to get issues from the senior leaders and their key representatives involved in this important agency decision.

After the interviews and the workshop, the decision hierarchy was developed (Fig. 6.5). The vision statement for the project was to "select the most cost effective large data center location and design for future agency mission support."

TABLE 6.2 Stakeholder Issue Identification Matrix

		Decision Makers/Stakeholders					
		Decision Authority	Client (Investment Manager)	Data Center Owner (Budgeter)	User (Data Center Operator)	Consumer (Mission User)	Other: Local communities
Environment factors	Cultural						Acceptance of mission
	Economic	Budget	Cost overruns Schedule slips Skilled labor	Resource (power and water) costs	Power and water interruptions and costs	Data center downtime Travel distance	Local support Labor availability
	Emotional					Travel time	
	Historical	Past cost overruns	Past cost overruns				
	Legal	Acquisition compliance	Acquisition compliance		Supplier loses license (e.g., nuclear power)	Data center downtime	
	Moral/ethical		Acquisition ethics				
	Natural environment		Environmental requirements	Local weather Natural hazards at site location	Local weather Natural hazards at site location	Data center downtime	
	Organizational		Acquisition funding	Operations and support funding	Operations and support funding		
	Political	Defend location decision	Budget approval				Political protests due to mission
	Security			Information security Terrorist attack		Mission downtime	Local facilities
	Social				Accidents		
	Technological	Mission data requirements	Performance risk	Future mission data needs	Mechanical and electrical failures	Mission data limits	

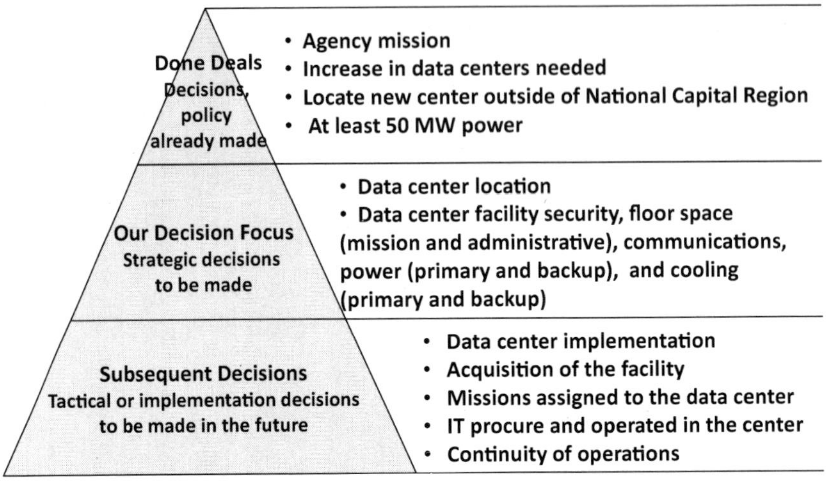

FIGURE 6.5 Data center location decision hierarchy.

6.6 Summary

Achieving a good frame is essential to making a good decision. A decision practitioner should always resist the temptation to "plunge in" and start working on a decision before a clear frame is agreed upon by the decision makers and key stakeholders.

This chapter introduces helpful tools for framing the decision opportunity. The vision statement is an effective tool to obtain agreement on the purpose of the decision analysis. Issue raising is an important technique to involve many key individuals in defining a clear decision frame. The stakeholder issue matrix is also useful for initial thinking about key stakeholders and for summarizing the key issues. The decision hierarchy is a proven technique for clarifying the scope of the decision. These tools should be a part of every decision analyst's tool kit.

Just as important, a worthy decision practitioner should never think of the decision frame as being permanent. Even if it took considerable time and effort to create the decision frame, it is always possible that the frame will need to be changed during the decision process. Unexpected external developments may occur. Or new information may come to light. Or new insights into the situation may arise. Any of these events may trigger the need to reexamine and possibly change the decision frame. It is good practice to ask periodically during the decision process whether the frame needs to be reexamined.

KEY TERMS

Decision frame The decision frame is the lens that we use to view the decision problem or opportunity.

Decision hierarchy The decision hierarchy is the primary conceptual tool for defining the scope of the decision. The decision hierarchy (see Fig. 6.2) is portrayed as a pyramid structure with three levels which summarize the decisions that have been made, the decision made in this decision, and the subsequent decisions.

Issue raising Issue raising brings to light all of the many perspectives that can be used to view the decision situation and lays the groundwork for creating a good decision frame.

Stakeholder issue identification matrix A matrix that has stakeholder groups in one dimension, environmental factors bearing on the decision in the second dimension, and key issues in the cells.

Vision statement A vision statement is an effective way to make sure that there is agreement on the purpose of the decision among the decision maker(s) and key stakeholders. The vision statement answers three questions: What are we going to do? Why are we doing this? How will we know that we have succeeded?

REFERENCES

Campbell, A. & Whitehead, J. (2010). How to test your decision-making instincts. *McKinsey Quarterly*, May, 1–4.

Clemen, R.T. & Reilly, T. (2001). *Making Hard Decisions with Decision Tools*. Belmont, CA: Duxbury Press.

Ewing, P., Tarantino, W., & Parnell, G. (2006, March). Use of decision analysis is the army base realignment and closure (BRAC) 2005 military value analysis. *Decision Analysis, 3*(1), 33–49.

Kahneman, D. & Klein, G. (2010, March). Strategic decisions: When can you trust your gut? *McKinsey Quarterly, 2*, 58–67.

Parnell, G., Driscoll, P., & Henderson, D. (eds.). (2011). *Decision-Making in System Engineering and Management*, 2nd ed. New York: John Wiley & Sons.

Skinner, D.C. (2009). *Introduction to Decision Analysis: A Practitioner's Guide to Improving Decision Quality*, 3rd ed. Sugar Land, TX: Probabalistic Publishing.

CHAPTER SEVEN

Craft the Decision Objectives and Value Measures

GREGORY S. PARNELL, TERRY A. BRESNICK, and ERIC R. JOHNSON

Management by objective works—if you know the objectives. Ninety percent of the time you don't.

—Peter Drucker

Our age is characterized by the perfection of means and the confusion of goals.

—A. Einstein

(*Continued*)

Handbook of Decision Analysis, First Edition. Gregory S. Parnell, Terry A. Bresnick, Steven N. Tani, and Eric R. Johnson.

7.1 Introduction

In the previous chapter, we discuss the critical role of the decision analyst in framing the decision. With this understanding of the decision frame, which may include a partial list of objectives, we next strive to identify the full list of objectives that the decision maker(s) seek to achieve with the decision. Of course, the identification of a significant new objective can change our frame. In addition to improving the decision frame, decision analysts use the objectives as the foundation for developing the value measure(s) used to evaluate the alternatives and select the best alternative(s).

We believe that it is good practice in any decision analysis to consider values and objectives *before* determining the full set of alternatives. Our experience is that decision analysis studies that develop alternatives first and use value measures only to distinguish those alternatives often miss several important objectives and therefore miss the opportunity to create a broader and more comprehensive set of alternatives that have the potential to create more value.

As the opening quote from Peter Drucker notes, although objectives are essential for good management decision making, they may be difficult to identify. In order to develop a complete list of decision objectives for an important decision, the decision analyst should interact with a broad and diverse group of decision makers, stakeholders, and subject matter experts.

To identify objectives, it is generally not sufficient to interact with only the decision maker(s) because, for complex decisions, they may not have a complete and well-articulated list of objectives. Instead, significant effort may be required to identify, define, and, perhaps, even carefully craft, the objectives based on many interactions with multiple decision makers, diverse stakeholders, and recognized subject matter experts. The ability to craft a comprehensive, composite set of objectives that refines the decision frame requires several of the soft skills of decision analysis (see Chapter 4).

The chapter is organized as follows. In Section 7.2, we describe shareholder and stakeholder value, which is the basis for the decision objectives. In Section 7.3, we describe why the identification of objectives is challenging. In Section 7.4, we list some of the key questions we can use to identify decision objectives and the four major techniques for identifying decision objectives: research, interviews, focus groups, and surveys. In Section 7.5, we discuss key considerations for the financial objective of private companies and cost objectives of public organizations. In Section 7.6, we discuss key principles for developing value measures to measure a priori how well an alternative could achieve an objective. In Section 7.7, we describe the structuring of multiple objectives, including objective and functional value hierarchies; describe four techniques for structuring objectives: Platinum, Gold, Silver, and Combined Standards; identify some best practices; and describe some cautions about risk and cost objectives. In Section 7.8, we describe the diverse approaches used to craft the objectives for the three illustrative problems. We conclude with a summary of the chapter in Section 7.9.

7.2 Shareholder and Stakeholder Value

Decision objectives should be based on shareholder and stakeholder value. Value can be created at multiple levels within an enterprise or organization. An organization creates value for shareholders and stakeholders by performing its mission (which usually helps to define the potential customers); providing products and services to customers; improving the effectiveness of its products and services to provide better value for customers; and improving the efficiency of its operations to reduce the resources required to provide the value to customers. Defining value is challenging for both private and public organizations. Private companies must balance shareholder value with being good corporate citizens (sometimes called stakeholder value). The management literature includes discussion of stakeholder value versus shareholder value for private companies (Charreaux & Desbrieres, 2001). Public organizations, which operate without profit incentives, focus on stakeholder value with management and employees being key stakeholders.

Next, we consider the shareholder and stakeholder objectives in a private company example and the stakeholder objectives in a public organization.

7.2.1 PRIVATE COMPANY EXAMPLE

Consider a large publicly owned communications company operating a cellular network that provides voice and data services for customers. The stakeholders include the shareholders, the board of directors, the leadership team, the employees (managers, designers, developers, operators, maintainers, and business process personnel), the customers, the cell phone manufacturers, the companies that sell the cell phone services, and the communities in which the company operates. The competition includes other cellular communications companies and companies that provide similar products and services using other technologies (satellite, cable, etc.). The environment includes natural hazards (e.g., earthquakes, hurricanes, and storms) that create operations risk.

Since the company is publically owned, the board of director's primary objective is to increase shareholder value. Stakeholders have many complementary and conflicting objectives. For example, the board may want to increase revenues and profits; the leadership may want to increase executive compensation; the sales department may want to increase the number of subscribers; the network operators want to increase availability and reduce dropped calls; the safety office may want to decrease accidents; the technologists may want to develop and deploy the latest generation of network communications; the cell phone manufacturers may want to sell improved technology cell phones; companies that sell the cell phones and services may want to increase their profit margins; operations managers want to reduce the cost of operations; the human resources department may want to increase diversity; and the employees may want to increase their pay and benefits.

7.2.2 GOVERNMENT AGENCY EXAMPLE

Next, consider a government agency that operates a large military communications network involving many organizations. There are no shareholders in this public system that provides communications to support military operations. However, like the private company example, there are many stakeholders. Some of the key stakeholders are the Department of Defense (DoD) office that establishes the communications requirements and submits the budget; Congress who approves the annual budgets for the network; the agency that acquires the network; the agency that manages the network; the contractor personnel who manufacture and assemble the network; the contractor, civilian, and/or military personnel who operate the network; the information assurance personnel who maintain the security of the network; the mission commanders who need the network to command and control their forces; and the military personnel whose lives may depend on the availability of the network during a conflict. The environment includes natural hazards (e.g., earthquakes, hurricanes, and storms) that create operations risk. Instead of business competitors, the network operators face determined adversaries who would like to penetrate the network to gain intelligence data in peacetime or to disrupt the network during a conflict.

The stakeholders have many complementary and conflicting objectives. For example, the DoD network management office wants the best network for the budget; Congress wants an affordable communications network to support national security; the acquiring agency wants to deliver a network that meets the requirements, on time and on budget; the network management agency wants to insure an adequate budget; the contractors want to maximize their profits and obtain future work; the network operators want to increase network capabilities and maximize availability of the network; the information assurance personnel want to maximize network security; the mission commanders want to maximize the probability of mission success; and the military personnel who use the network want to maximize availability and bandwidth.

7.3 Challenges in Identifying Objectives

The identification of objectives is more art than science. The four major challenges are (1) indentifying a full set of values and objectives, (2) obtaining access to key individuals, (3) differentiating fundamental and means objectives, and (4) structuring a comprehensive set of fundamental objectives for validation by the decision maker(s) and stakeholders.

In a complex decision, especially if it is a new opportunity for the organization, the identification of objectives can be challenging. In a research paper (Bond et al., 2008), Ralph Keeney and his colleagues concluded that "in three empirical studies, participants consistently omitted nearly half of the objectives that they later identified as personally important. More surprisingly, omitted objectives were as important as the objectives generated by the participants on their own. These empirical results were replicated in a real-world case study of decision making at a high-tech firm. Decision makers are considerably deficient in utilizing personal knowledge and values to form objectives for the decisions they face." To meet this challenge, we must have good techniques to obtain the decision objectives.

A second challenge is obtaining access to a diverse set of decision makers (DMs), stakeholders (SHs), and subject matter experts (SMEs). In Chapter 5, we discuss the importance of using a decision process with access to the key decision makers. However, sometimes, clients are reluctant to provide the decision analysis team access to senior decision makers and diverse stakeholders who have the responsibilities and the breadth of experience that are essential to providing a full understanding of the decision objectives. In addition, it can be difficult to obtain access to the recognized experts instead of individuals who have more limited experience. Many times, the best experts resist meetings that take their focus away from their primary area of expertise. In addition, even if we have access, we may not have the time in our analysis schedule to access all the key individuals. To be successful, the decision analysis team must obtain access to as many of these key individuals as possible in the time they have allocated for the study.

The third challenge is the differentiation of fundamental and means objectives (R.L. Keeney, 1992). Fundamental objectives are what we ultimately care about in the decision. Means objectives describe how we achieve our fundamental objectives. An automobile safety example helps to clarify the difference. The fundamental objectives may be to reduce the number of casualties due to highway accidents and to minimize cost. The means objectives may include to increase safety features in the automobile (e.g., air bags and seat belts), to improve automobile performance in adverse weather (e.g., antilock brakes), and to reduce the number of alcohol impaired drivers (e.g., stricter enforcement). The mathematical considerations of multiple objective decision analysis discussed in Chapter 3 require the use of fundamental objectives in the value model.

The fourth challenge is structuring the knowledge about fundamental objectives and value measures that we obtain in an organized manner that makes it easy for decision makers, stakeholders, and experts to validate that the objectives and value measure form a necessary and sufficient set of measures to evaluate the alternatives

7.4 Identifying the Decision Objectives

7.4.1 QUESTIONS TO HELP IDENTIFY DECISION OBJECTIVES

The key to identifying decision objectives is asking the right questions, to the right people, in the right setting. Keeney (Keeney, 1994) has identified 10 categories of questions that can be asked to help identify decision objectives. These questions should be tailored to the problem and to the individual being interviewed, the group being facilitated, or the survey being designed.[1] For example, the strategic objectives question might be posed to the senior decision maker in an interview, while the consequences question is posed to key stakeholders in a facilitated group.

1. **Strategic objectives.** What are your ultimate or long range objectives? What are your values that are fundamental? What is your strategy to achieve these objectives?
2. **A wish list.** What do you want? What do you value? What should you want? What are you trying to achieve? If money was not an obstacle, what would you do?
3. **Alternatives.** What is a perfect alternative, a terrible alternative, a reasonable alternative? What is good or bad about each?
4. **Problems and shortcomings.** What is wrong or right with your organization or enterprise? What needs fixing? What are the capability, product, or service gaps that exist?

[1]Our focus with these questions is the decision objectives. However, the answers may provide valuable insights on issues, alternatives, uncertainties, and constraints that can be used in later phases of the decision analysis.

5. **Consequences.** What has occurred that was good or bad? What might occur that you care about? What are the potential risks you face? What are the best or worst consequences that could occur? What could cause these?
6. **Goals, constraints, and guidelines.** What are your goals or aspirations? What limitations are placed upon you? Are there any legal, organizational, technological, social, or political constraints?
7. **Different perspectives.** What would your competitor or your constituency be concerned about? At some time in the future, what would concern you? What do your stakeholders want? What do your customers want? What do your adversaries want?
8. **Generic fundamental objectives.** What objectives do you have for your customers, your employees, your shareholders, yourself? What environmental, social, economic, or health and safety objectives are important?
9. **Structuring objectives.** Follow means–ends relationships: why is that objective important, how can you achieve it? Use specification: what do you mean by this objective?
10. **Quantifying objectives.** How do you measure achievement of this objective? If not, would you measure achievement of this objective? Which objective is the most important? Why is objective A three times as important as objective B?

7.4.2 HOW TO GET ANSWERS TO THE QUESTIONS

The four techniques to obtain answers to these questions and help identify objectives and value measures are research, interviews, surveys, and facilitated group meetings. Chapter 4 describes the key features of interviews, focus groups, and surveys in decision analysis and the best practices for using each technique. In this section, we discuss the use of these techniques to identify objectives.[2] The amount of research, the number of interviews, the number and size of focus groups, and the number of surveys we use depends on the scope of the problem, the number of decision levels, the diversity of the stakeholders, the number of experts, and the time allocated to defining objectives and identifying value measures.

7.4.2.1 Research. Research is an important technique to understand the problem domain; to identify potential objectives to discuss with decision makers, stakeholders, and experts; and to understand suggested objectives. The amount of research depends on the decision analysts' prior understanding of the problem domain, knowledge of key terminology, and amount of domain knowledge expected of the decision analyst in the decision process. The primary research sources include the problem domain and the decision analysis literature. Research

[2]The techniques can also be used to obtain the functions and the value measures.

should be done throughout the objective identification process. Many times, information obtained with the other three techniques requires research to fully understand or to validate the objective recommendation.

7.4.2.2 **Interviews.** Senior leaders and "world-class" experts can do the best job of articulating their important value and objectives. Interviews are the best technique for obtaining objectives from senior decision maker(s), senior stakeholders, and "world-class" experts since they typically do not have the time to attend a longer focus group or the interest in completing a survey. However, interviews are time consuming for the interviewer due to the preparation, execution, and analysis time. Since interviews are very important and take time, it is important to use the interview best practices we describe in Chapter 4.

7.4.2.3 **Focus Groups.** Focus groups are another useful technique for identifying decision objectives. We usually think of focus groups for decision framing and product market research; however, they can also be useful for identifying decision objectives. While interviews typically generate a two-way flow of information, focus groups create information through a discussion and interaction between all the group members. As a general rule, focus groups should comprise between 6 and 12 individuals. Too few may lead to too narrow a perspective, while too many may not allow all attendees the opportunity to provide meaningful input. Chapter 4 provides best practices for group facilitation.

7.4.2.4 **Surveys.** In our experience, surveys are not used as frequently to identify potential decision objectives as interviews and focus groups. However, surveys are a useful technique for collecting decision objectives from a large group of individuals in different locations. Surveys are especially good for obtaining general public values. Surveys are more appropriate for junior to mid-level stakeholders and dispersed experts. We can use surveys to gather qualitative and quantitative data on the decision objectives. A great deal of research exists on techniques and best practices for designing effective surveys. Chapter 4 provides a summary of the best practices.

7.5 The Financial or Cost Objective

The financial or cost objective may be the only objective, or it may be one of the multiple objectives. Shareholder value is an important objective in any firm. In many private decisions, the financial objective may be the only objective or the primary objective. Firms employ three fundamental financial statements to track value: balance sheet, income statement, and cash flow statement. In addition, firms commonly use discounted cash flow (net present value [NPV]) to analyze the potential financial benefits of their alternatives. For public decisions, the cost objective is usually a major consideration. The cost can be the full life cycle cost or only a portion of the costs of the alternative.

7.5.1 FINANCIAL OBJECTIVES FOR PRIVATE COMPANIES

In order to understand the financial objectives for private companies, we begin with the three financial statements. Next, we consider the conversion of cash flows to NPV.

7.5.1.1 Balance Sheet Statement. A balance sheet is developed using standard accounting procedures to report an approximation of the value of a firm (called the net book value) at a point in time. The approximation comes from adding up the assets and then subtracting out the liabilities. We must bear in mind that the valuation in a balance sheet is calculated using generally accepted accounting principles, which make it reproducible and verifiable, but it will differ from the valuation we would develop taking account of the future prospects of the firm.

7.5.1.2 Income Statement. The income statement describes the changes to the net book value through time. As such, it shares the strengths (use of generally accepted accounting procedures and widespread acceptance) and weaknesses (inability to address future prospects) of the balance sheet.

7.5.1.3 Cash Flow Statement. A cash flow statement describes the changes in the net cash position through time. One of the value measures reported in a cash flow statement is "free cash flow" (which considers investment and operating costs and revenues, but not financial actions like issuing or retiring debt or equity). Modigliani and Miller (Modigliani & Miller, 1958) argue that the anticipated pattern of free cash flow in the future is a complete determinant of the economic value of an enterprise. Many clients are comfortable with this viewpoint; hence, a projected cash flow statement is the cornerstone of many financial decision analyses.

7.5.1.4 Net Present Value. To boil a cash flow time pattern down to a one-dimensional value measure, companies usually discount the free cash flow to a "present value" using a discount rate. The result is called the NPV of the cash flow.

We usually contemplate various possible alternatives, each with its own cash flow stream and NPV cash flow. In many decision situations, clients are comfortable using this as their fundamental value measure. Chapter 9 provides information on building deterministic NPV models. Chapter 11 provides more information on analyzing the impact of uncertainty and risk preference. See Chapter 12 for a discussion of the choice of discount rate.

7.5.2 COST OBJECTIVE FOR PUBLIC ORGANIZATIONS

For many public organizations, minimizing cost is a major decision objective. Depending on the decision, different cost objectives may be appropriate. The most general cost objective is, subject to acceptable performance levels, to minimize life cycle cost, the full cost over all stages of the life cycle: concept development, design, production, operations, and retirement. However, costs that have already been spent (sunk costs) should never be included in the analysis. Since sunk costs should not be considered and some costs may be approximately the same across alternatives, in practice, many decision analysts consider the delta life cycle costs among the alternatives. In public organizations, the budget is specified by year and may or may not be fungible between years. When multiple years are analyzed, a government inflation rate is used to calculate net present cost.

7.6 Developing Value Measures

In order to quantitatively use the decision objectives in the evaluation of the alternatives, we must develop value measures for each objective that measure the *a priori* potential value, that is, before the alternative is selected. The identification of the value measures can be as challenging as the identification of the decision objectives. We can identify value measures by research, interviews, and group meetings with decision makers, stakeholders, and subject–matter experts. Access to stakeholders and experts with detailed knowledge of the problem domain is the key to developing good value measures.

Kirkwood (Kirkwood, 1997) identifies two useful dimensions for value measures: alignment with the objective and type of measure. Alignment with the objective can be direct or proxy. A direct measure focuses on attaining the full objective, such as NPV for shareholder value. A proxy measure focuses on attaining an associated objective that is only partially related to the objective (e.g., reduce production costs for shareholder value). The type of measure can be natural or constructed. A natural measure is in general use and commonly interpreted, such as dollars. We have to develop a constructed measure, such as a five-star scale for automobile safety. Constructed measures are very useful but require careful definition of the measurement scales. In our view, the use of an undefined scale, for example, 1–7, is not appropriate for decision analysis since the measures do not define value and scoring is not repeatable.

Table 7.1 reflects our preferences for types of value measures. Priorities 1 and 4 are obvious. We prefer direct and constructed to proxy and natural for two reasons. First, alignment with the objective is more important than the type of scale. Second, one direct, constructed measure can replace many natural and proxy measures. When value models grow too large, the source is usually the overuse of natural, proxy measures.

TABLE 7.1 **Preference for Types of Value Measure**

Type	Direct Alignment	Proxy Alignment
Natural	1	3
Constructed	2	4

7.7 Structuring Multiple Objectives

Not all problems have only the financial or the cost objective. In many public and business decisions, there are multiple objectives and many value measures. Once we have a list of the preliminary objectives and value measures, our next step is to organize the objectives and value measures (typically called structuring) to remove overlaps and identify gaps. For complex decisions, structuring objectives can be quite challenging. In this section, we introduce the techniques for identifying and structuring, using hierarchies. The decision analysis literature uses several names: value hierarchies, objectives hierarchies, value trees, objective trees, functional value hierarchy, and qualitative value model.

7.7.1 VALUE HIERARCHIES

The primary purpose of the objectives hierarchy is to identify the objectives and the value measures so we understand what is important in the problem and can do a much better job of qualitatively and quantitatively evaluating alternatives (see Chapter 9). Most decision analysis books recommend beginning with identifying the objectives and using the objectives to develop the value measures. For complex decisions, we have found that it is very useful to first identify the functions that create value that the solution must perform (Parnell et al., 2011). For each function, we then identify the objectives we want to achieve for that function. For each objective, we identify the value measures that can be used to assess the potential to achieve the objectives. In each application, we use the client's preferred terminology. For example, functions can be called missions, capabilities, activities, services, tasks, or other terms. Likewise, objectives can be called criteria, evaluation considerations, or other terms. Value measures can be called any of the previously mentioned terms (see Chapter 3).

The terms objectives hierarchy and functional value hierarchy are used in this book to make a distinction between the two approaches. The functional value hierarchy is a combination of the functional hierarchy from systems engineering and the value hierarchy from decision analysis (Parnell et al., 2011). In decisions where the functions of the alternatives are the same, or are not relevant, it may be useful to group the objectives by categories to help in structuring the objectives.

Both hierarchies begin with a statement of the primary decision objective as the first node in the hierarchy. An objectives hierarchy begins with the objectives in the first tier of the hierarchy, (sometimes) subobjectives as the second tier and value measures as the final tier of the hierarchy (see Fig. 7.1). A functional value hierarchy uses functions as the first tier, (sometimes) subfunctions as the second tier, objectives as the next tier, and value measures as the final tier of the hierarchy (see Fig. 7.2).

In the car purchase case illustrated in Figures 7.1 and 7.2, the objectives and value measures are the same in both hierarchies. However, in our experience, this is seldom the case. A more typical case is shown in Figure 7.3. When the randomly ordered objectives hierarchy is logically organized by functions, the

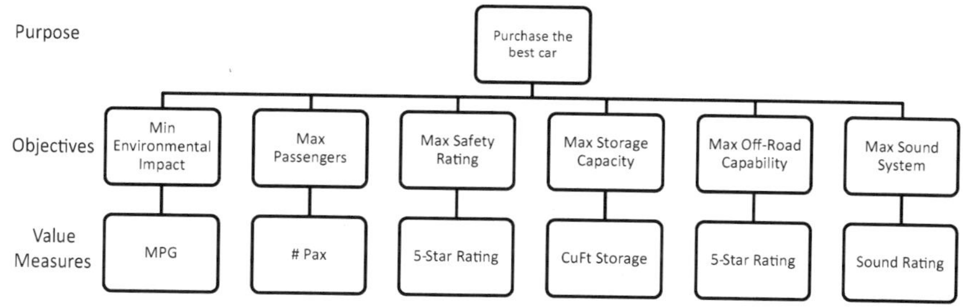

FIGURE 7.1 Objectives hierarchy for car purchase.

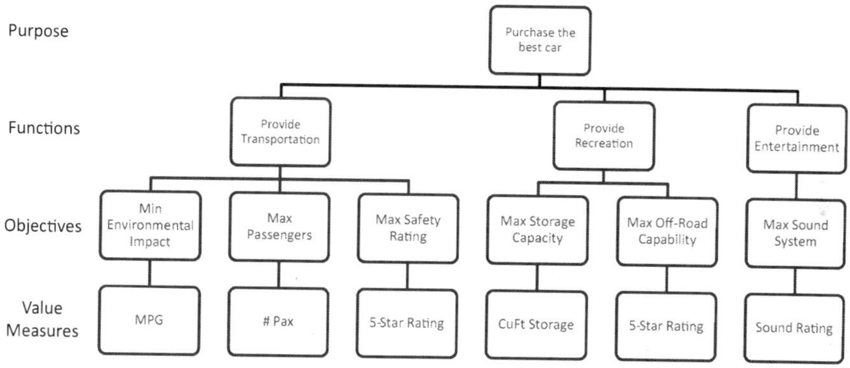

FIGURE 7.2 Functional value hierarchy for car purchase.

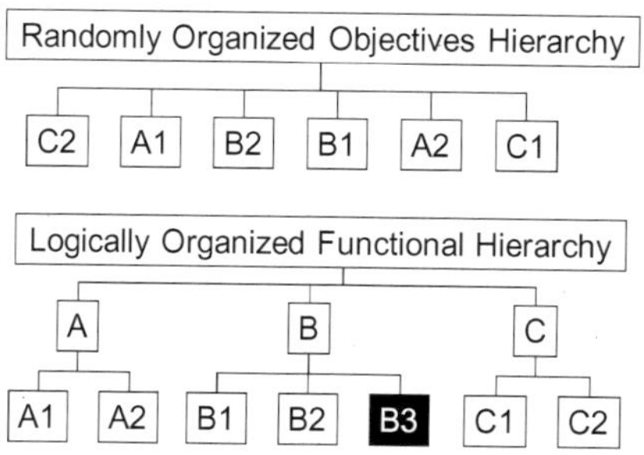

FIGURE 7.3 Comparison of objectives and functional objectives hierarchy.

objectives and measures make more sense to the decision makers, and, many times, we identify missing objectives (and value measures). The benefit of identifying the functions is threefold. First, a logical structure of the functional objectives hierarchy is easier for the decision analyst to develop. Second, it may help identify additional objectives (and value measures) that might be missed in the objectives hierarchy. Third, the logical order helps the decision makers and stakeholders understand the hierarchy and provide suggestions for improvement.

In either the value hierarchy or functional hierarchy, we can create the structure from the top down or from the bottom up. Top-down structuring starts with listing the fundamental objectives on top, and then "decomposing" into subobjectives until we are at a point where value measures can be defined. It has the advantage on being more closely focused on the fundamental objectives, but often, we initially overlook important subobjectives. Bottom-up structuring starts by discussing at the subobjective or subfunctional level, grouping similar things together, and defining the titles at a higher level for the grouped categories. Value measures are then added at the bottom of the hierarchy. It has the advantage of discussing issues at a more concrete and understandable level, but if we are not careful, we may drift from the fundamental objectives in an attempt to be comprehensive. In theory, both approaches should produce the same hierarchy. In practice, this rarely happens. Our experience has shown that top-down structuring is easier for the novice, while bottom-up structuring, with the guidance of an experienced decision analyst, provides a better understanding of the levels where trade-offs are actually being made.

7.7.2 TECHNIQUES FOR DEVELOPING VALUE HIERARCHIES

The credibility of the qualitative value model is very important in decision analysis since there are typically multiple decision-maker and stakeholder reviews. If the decision makers do not accept the qualitative value model, they will not (and should not!) accept the quantitative analysis. We discuss here four techniques for developing objectives that were developed by the first author and his colleagues: the platinum, gold, silver, and combined standards.

7.7.2.1 Platinum Standard. A platinum standard value model is based primarily on information from interviews with senior decision makers and key stakeholders. Decision analysts should always strive to interview the senior leaders (decision makers and stakeholders) who make and influence the decisions. As preparation for these interviews, they should research potential key problem domain documents and talk to decision-maker and stakeholder representatives. Affinity diagrams (Parnell et al., 2011) can be used to group similar functions and objectives into logical, mutually exclusive, and collectively exhaustive categories. For example, interviews with senior decision makers and stakeholders were used to develop a value model for the Army's 2005 BRAC value model (Ewing et al., 2006).

7.7.2.2 Gold Standard. When we cannot get direct access to senior decision makers and stakeholders, we look for other approaches. One approach is to use a "gold standard" document approved by senior decision makers. A gold standard value model is developed based on an approved policy, strategy, or planning document. For example, in the SPACECAST 2020 study (Burk & Parnell, 1997), the current U.S. Space Doctrine document served as the model's foundation. In addition, environmental value models have been directly developed from the Comprehensive Environmental Restoration and Liability Act (Grelk et al., 1998). Many military acquisition programs use capability documents as a gold standard since the documents define system missions, functions, and key performance parameters. Many times, the gold standard document has many of the functions, objectives, and some of the value measures. If the value measures are missing, we work with stakeholder representatives to identify appropriate value measures for each objective. It is important to remember that changes in the environment and leadership may cause a gold standard document to no longer reflect leadership values. Before using a gold standard document, confirm that the document still reflects leadership values.

7.7.2.3 Silver Standard. Sometimes, the gold standard documents are not adequate (not current or not complete) and we are not able to interview a significant number of senior decision makers and key stakeholders. As an alternative, the silver standard value model uses data from the many stakeholder representatives. Again, we use affinity diagrams to group the functions and objectives into mutually exclusive and collectively exhaustive categories. For example, inputs from about 200 stakeholders' representatives were used to develop the Air Force 2025 value model (Parnell et al., 1998). This technique has the advantage of developing new functions and objectives that are not included in the existing gold standard documents. For example, at the time of the study, the Air Force Vision was Global Reach, Global Power. The Air Force 2025 value model identified the function, Global Awareness (later changed to Global Vigilance), which was subsequently added to the Air Force Vision of Global Vigilance, Reach, and Power.

7.7.2.4 Combined Standard. Since it is sometimes difficult to obtain access to interview senior leaders, and many times key documents are not sufficient to completely specify a value model, the most common technique is the combined standard. First, we research the key gold standard documents. Second, we conduct as many interviews with senior leaders as we can. Third, we meet with stakeholder representatives, in groups or individually, to obtain additional perspectives. Finally, we combine the results of our review of several documents with findings from interviews with some senior decision makers and key stakeholders and data from multiple meetings with stakeholder representatives. This technique was used to develop a space technology value model for the Air Force Research Laboratory Space Technology R&D Portfolio (Parnell et al., 2004), the data center location decision we use for the data center illustrative example, and many other value models.

7.7.3 VALUE HIERARCHY BEST PRACTICES

The following are some recommended best practices for developing value hierarchies using any of the above four techniques:

- Put the problem statement at the top of the hierarchy in the language of the decision maker and key stakeholders. A clear problem statement is a very important tool to communicate the purpose of the decision to the decision maker(s), senior stakeholders, and the decision analyst team.
- Use fundamental objectives and not means objectives in the hierarchies.
- Select terms (e.g., functions, objectives, and value measures) used in the problem domain. This improves understanding by the users of the model.
- Develop functions, objectives, and value measures from research and stakeholder analysis.
- Carefully consider the use of constraints (screening criteria) in development of your objectives and value measures. Constraints can create objectives and value measures. However, as we discuss in the next chapter, the overuse of constraints can reduce our decision opportunities.
- Logically sequence the functions (e.g., temporally). This provides a framework for helping decision makers and stakeholders understand the value hierarchy.
- Define functions and objectives with verbs and objects. This improves the understanding of function or objective.
- Identify value measures that are direct measures of the objectives and not proxy measures. Proxy measures result in more measures and increased data collection for measures that are only partially related to the objectives.
- Vet the value hierarchy with decision makers and stakeholders.

7.7.4 CAUTIONS ABOUT COST AND RISK OBJECTIVES

Two commonly used objectives require special consideration: the cost and risk objectives.

7.7.4.1 Cost Objective. Mathematically, minimizing cost can be one of the objectives in the value hierarchy and cost can be a value measure in the value model. However, for many multiple objective decisions (especially portfolio decision analysis in Chapter 12), it is useful to treat cost separately and show the amount of value per unit cost. In our experience, this is the approach that is the most useful for decision makers who have a budget that they might be able to increase or they might have to accept a decrease.

7.7.4.2 Risk Objective. Risk is a common decision making concern, and it is tempting to add minimization of risk to the set of objectives in the value hierarchy. However, this not a sound practice. A common example is helpful to further explain why we do not recommend this approach. Suppose there are three objectives: maximize performance, minimize cost, and minimize time to complete the schedule. It may be tempting to add minimize risk as a forth objective. But what type of risk are we minimizing and what is causing the risk? The risk could be performance risk; cost risk; schedule risk, performance and cost risk; cost and schedule risk; performance and schedule risk; or performance, cost, and schedule risk. In addition, there could be one or more uncertainties that drive these risks. In Chapter 11, we introduce probabilistic modeling to model the sources of risk and their impact on the value measures and the objectives. We believe this is a much sounder approach than the use of a vague risk objective in the value hierarchy.

There is also a more basic reason why adding a risk objective is not sound. To make decisions that are consistent with the Five Rules of decision analysis (see Chapter 3), risk-taking preferences should be handled via a utility function that maps value outcomes to the utility metric. Therefore, making the level of risk one of many components in the calculation of overall value is not a "best practices" approach. See Chapter 11 for a full discussion of how risk should be treated in decision analysis.

That said, if circumstances dictate that risk must be part of the value hierarchy, Phillips proposes an interesting approach for representing probabilities as preferences such that they can used in an additive multicriteria model (Phillips, 2009). He suggests including a "confidence" criterion, assessing probabilities, and converting them to preference scores (or "expected values") using a "proper scoring rule." A proper scoring rule, due to its formulation, encourages an expert to report states of uncertainty accurately in order to avoid severe penalties for misassessments. If the expert, such as a weatherman, judges a probability to be high and this turns out to be correct, he is given a low penalty score. But whatever his state of uncertainty, he can also minimize the penalty score by expressing his uncertainty accurately when reporting it as a probability. One such proper scoring rule uses a logarithmic rule in which the score for a probability p is proportional to $\log_{10}p$. The logarithmic nature of the scale ensures that scores for compound events can be derived by summing scores associated with the component events. The expected score associated with an inaccurate report of uncertainty is always worse than the expected score for the expert's real belief. Surprisingly, this statement is true whatever the expert's true belief. In the long run, the penalty will be minimized by reporting accurately. It also provides an audit trail enabling different assessors to be compared after many assessments have been made. The person with the lowest total penalty is both knowledgeable about the events in question, and is good at reporting their uncertainty accurately. The major purpose of using the proper scoring rule is to make it easier to obtain unbiased assessments. While the authors believe that there can be value in using this approach under certain circumstances, we believe that the best practice is to directly assess probabilities and use them in an expected value or expected utility calculation rather than converting probabilities to preferences.

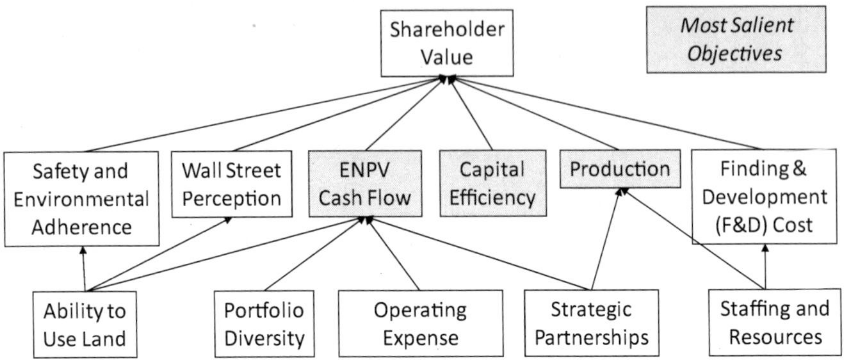

FIGURE 7.4 RNAS means–ends objectives hierarchy.

7.8 Illustrative Examples

See Table 1.3 for further information on the illustrative examples. In this section, we review three very different approaches used to identify and structure the objectives used in the three illustrative examples.

7.8.1 ROUGHNECK NORTH AMERICAN STRATEGY (by Eric R. Johnson)

The decision frame for the Roughneck North American Strategy (RNAS) project is described in Chapter 6. The RNAS project used a means-ends objectives hierarchy (Fig. 7.4) to structure the objectives, which were identified using interviews and team meetings. The primary fundamental objective is shareholder value. The means–objectives hierarchy shows how the means objectives contribute to each other and the fundamental objective. The shaded means objectives were salient.

7.8.2 GENEPTIN (by Sean Xinghua Hu)

The decision frame for Geneptin is described in Chapter 6. To make the best decision for the development strategy of Geneptin, the product team used a dialogue decision process to involve senior decision makers and key stakeholders. As in most of pharmaceutical development and commercialization strategy decision making, the key metric the product team and company senior management used was economic value, as measured by the expected net present value (ENPV) of cash flow. Other metrics that are commonly of interest in the pharmaceutical industry, such as the probability of technical and regulatory success (PTRS) and investment level (i.e., drug development and launch costs), are used in the calculation of ENPV.

7.8.3 DATA CENTER LOCATION (by Gregory S. Parnell)

The decision frame for the data center problem is defined in Chapter 6. For the data center location decision analysis, we used the Combined Standard approach. We began by researching commercial data center technologies and trends. Second, we interviewed a large number of senior leaders and stakeholders. Third, we used a focus group meeting of stakeholder representatives to develop the functional value hierarchy.

The initial functional value hierarchy was developed in a 1-day focus group meeting of about 15 participants facilitated by a senior decision analyst. The following organizations were represented: mission; IT; data communications; facilities; security; power and cooling; program manager; systems engineering; and life cycle cost. The actual value model used about 40 value measures. For the illustrative example in this book, we use the most important 10 value measures (Fig. 7.5). We used the following approach.

1. **Brainstorm screening criteria, functions, objectives and measures.** The facilitator requested that the participants write down on sticky notes any screening criteria that all sites must satisfy, the functions that data centers would perform, the objectives for the data center, and the value measures we should use.
2. **Label screening criteria, functions, objectives, and measures.** The facilitator, with the group's help, labeled each note as a screening criteria (S), function (F), objective (O), or value measure (M).[3] This process resulted in

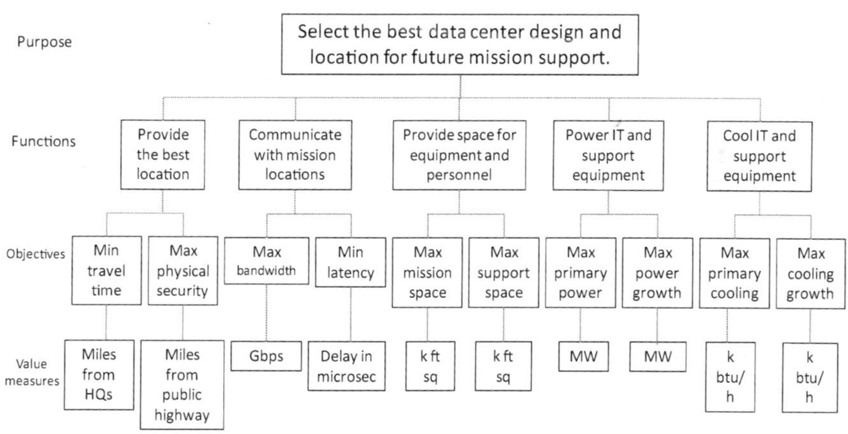

FIGURE 7.5 Data center functional value hierarchy.

[3]Sometimes we do the sticky notes in four steps; screening criteria, functions, objectives, and value measures. However, our experience is that many participants do not clearly understand the distinctions between the four concepts at this stage of the analysis, so we get all their ideas and then label them.

adding additional notes during the discussion. For example, screening criteria (expandable power) may also require an objective (maximize growth power) since we value the potential to expand the power.

3. **Group the screening criteria.** The screening criteria were grouped into seven categories. The screening criteria were security (minimum distance from a highway), power (minimum power, reliable power, and expandable power), floor space (minimum floor space, floor loading, ceiling height, and expandable), cooling (minimum cooling, redundancy, and expandable), communications (minimum bandwidth, reliable, and redundant lines), environmental (no environmental hazards, e.g., earthquake or hurricane areas), and schedule (meet expected completion date).
4. **Group and aggregate functions.** The facilitator asked three individuals to group all the function notes into categories by affinity (hence the name affinity diagramming). Next, they aggregated the categories into five sequential functions. Three to five functions are easy to remember. The five resulting data center functions are shown in Figure 7.5.
5. **Structure the objectives with the functions.** Next, we aligned the objectives with the most appropriate function. For each function, we grouped the objectives and, sometimes, rewrote an objective to capture several initial objectives or added new objectives. The 10 objectives for our illustrative example are shown in Figure 7.5.
6. **Develop a value measure for each objective.** Next, we aligned the original value measures with the appropriate objective. We also had to add new value measures when objectives did not have a value measure. We attempted to use one direct, natural measure for each objective (Kirkwood, 1997). The 10 value measures for our illustrative example are shown in Figure 7.5.

In Chapter 9, we describe the quantitative value model that was developed to evaluate the data center location alternatives.

An additional important objective was to minimize the life cycle cost. As part of our analysis, we developed a life cycle cost model that we used to evaluate each alterative data center location. However, we did not put life cycle cost in the functional value hierarchy. Instead, we plotted value versus cost to identify the nondominated alternatives (see Chapter 9).

7.9 Summary

Decision objectives are based on shareholder and stakeholder value. The crafting of the objectives and the value measures is a critical step in the decision analyst's support to the decision maker and helps qualitatively define the value we hope to achieve with the decision. It is not easy to identify a comprehensive set of objectives and value measures for a complex decision. We describe the four techniques for identifying decision objectives: research, interviews, focus groups, and surveys. Research is essential to understand the problem domain and the decision analysis modeling techniques. Interviews are especially useful for senior leaders. Focus groups work well for stakeholder representatives. Surveys are especially useful to obtain public opinion. We note that the financial or cost objective is almost always an important objective. Next, we describe the important role of hierarchies in structuring objectives and providing a format that is easy for decision makers and stakeholders to review and provide feedback. We present objectives and functional hierarchies. We recommend functional value hierarchies for complex system decisions. Four techniques are useful for structuring objectives: Platinum (senior leader interviews), Gold (documents), Silver (meetings with stakeholder representatives), and Combined (using all three standards) Standards. The combined standard is the most common. The illustrative examples provide three very different approaches to identifying objectives.

KEY TERMS

Combined standard An objective identification process that uses the Platinum (senior leader interviews), Gold (documents), and Silver (meetings with stakeholder representatives) standards.

Focus groups Decision objectives are obtained through a facilitated discussion and interaction between 6 and 12 group members.

Functional value hierarchy A hierarchy displaying the decision purpose, functions to be met by the decision, the decision objectives for each function, and the value measure(s) for each objective that will be used to evaluate the alternatives.

Fundamental objective The most basic objective we're trying to achieve. For example, reduce driving deaths.

Gold standard An approved document is used as the foundation for identifying and structuring the decision objectives.

Interviews An interview is a meeting with a leader or SME using a series of planned questions and follow-on discussion to identify the decision objectives.

Life cycle cost The total cost of an alternative across all stages of the alternative life cycle. Sunk costs should not be included when evaluating alternatives.

Means objective An objective that describes how we can achieve a fundamental objective. For example, "increase use of safety belts" is a means objective for "reduce driving deaths."

Net present value The calculation of the present worth from a sequence of cash flows using a discount rate.

Objective identification The process of identifying potential decision objectives.

Objective structuring The process of organizing decision objectives and value measures into a hierarchical structure.

Objectives hierarchy A hierarchy displaying the decision purpose, the decision objectives, and the value measures that will be used to evaluate the alternatives.
Platinum standard The primary source of decision objectives are interviews with senior leaders.
Proper scoring rule An approach that allows an assessor to convert probabilities to preference scores for use in a MODA model. A rule is "proper" only if it encourages reporting of probabilities that are accurate representations of the state of uncertainty.
Risk The likelihood of a bad outcome.
Shareholder value Value is defined in financial terms understandable to shareholders.
Silver standard The primary sources of decision objectives are meetings with stakeholder representatives.
Stakeholder value Value is defined by meeting the objectives of the key stakeholders.
Survey A survey is a prepared list of questions that is sent to stakeholders to help identity the functions, objectives, and/or value measures.

REFERENCES

Bond, S., Carlson, D., & Keeney, R.L. (2008). Generating objectives: Can decision makers articulate what they want? *Management Science, 54*(1), 56–70.

Burk, R.C. & Parnell, G.S. (1997). Evaluating future space systems and technologies. *Interfaces, 27*(3), 60–73.

Charreaux, G. & Desbrieres, P. (2001). Corporate goverance: Stakeholder value versus shareholder value. *Journal of Management and Governance, 5*(7), 107–128.

Ewing, P., Tarantino, W., & Parnell, G. (2006). Use of decision analysis is the army base realignment and closure (BRAC) 2005 military value analysis. *Decision Analysis, 3*(1), 33–49.

Grelk, B.J., Kloeber, J.M., Jackson, J.A., Deckro, R.F., & Parnell, G.S. (1998). Quantifying CERCLA usiing site decision maker values. *Remediation, 8*, 87–105.

Keeney, R. (1994). Creativity in decision making with value-focused thinking. *Sloan Management Review, 35*, 33–34.

Keeney, R.L. (1992). *Value-Focused Thinking: A Path to Creative Decision Making*. Cambridge, MA: Harvard University Press.

Kirkwood, C. (1997). *Strategoc Multiple Objective Decision Analysis with Spreadsheets.* Belmont, CA: Duxbury Press.

Modigliani, F. & Miller, M. (1958). The cost of capital, corporation finance and the theory of investment. *American Economic Review, 48*(3), 261–297.

Parnell, G.S., Conley, H., Jackson, J., Lehmkuhl, L., & Andrew, J. (1998). Foundations 2025: A framework for evaluating air and space forces. *Management Science, 44*(10), 1336–1350.

Parnell, G.S., Burk, R.C., Schulman, A., Kwan, L., Blackhurst, J., Verret, P., et al. (2004). Air force research laboratory space technology value model: Creating capabilities for future customers. *Military Operations Reseach*, *9*(1), 5–17.

Parnell, G.S., Driscoll, P.J., & Henderson, D.L. (Trans.). (2011). *Decision Making for Systems Engineering and Management*, 2nd ed. Hoboken, NJ: John Wiley & Sons.

Phillips, L. (2009). *Multi-criteria Analysis: A Manual.* Wetherby: Department for Communities and Local Government.

CHAPTER EIGHT

Design Creative Alternatives

STEVEN N. TANI and GREGORY S. PARNELL

Alternatives are the raw material of decision making. They represent the range of potential choices you have for pursuing your objectives. Because of the their central importance, you need to establish and maintain a high standard for generating alternatives

—Hammond et al. (1999)

Nothing is more dangerous than an idea when it is the only one you have

—Emile Chartier, French philosopher

Handbook of Decision Analysis, First Edition. Gregory S. Parnell, Terry A. Bresnick, Steven N. Tani, and Eric R. Johnson.

8.1 Introduction

Identifying a good set of alternatives is an important part of decision making. In most decision situations, it is quite easy to identify an initial set of alternatives. Should we close the factory or invest in upgrading its technology? Should we proceed with development of Product A or Product B or both?

The challenge for any decision practitioner is to help the client push beyond the initial set of alternatives and to lead a search for additional, possibly much better, alternatives. This is an opportunity to create great value. Doing an excellent job of analyzing an existing set of alternatives cannot create more value than that of the best alternative in the set. But taking the time and effort to search diligently for new alternatives might develop an alternative that has much greater value than any in the initial set. This has been our experience as decision analysts.

There are several excellent references for creating good alternatives. Keller and Ho (Keller & Ho, 1988) identified five categories of alternative generation techniques including attribute-based, state-based, composite, option-based, and creativity techniques. Keeney has authored two books (Keeney, 1992; Hammond et al., 1999) that include chapters on generating alternatives, including one titled "How to Make Smarter Choices by Creating Better Alternatives." The textbook by Clemen and Reilly (Clemen & Reilly, 2001) includes a chapter titled "Creativity and Decision Making."

This chapter is organized as follows. In Section 8.2, we identify the characteristics of a good set of alternatives. In Section 8.3, we discuss the obstacles to creating a good set of alternatives. Next, we discuss the two phases of generating a good set of alternatives—the expansive phase in Section 8.4 and the reductive phase in Section 8.5. In Section 8.6, we discuss suggestions for improving the alternative set. In Section 8.7, we discuss alternative generation for our illustrative examples.

8.2 Characteristics of a Good Set of Alternatives

The decision analyst's goal is to create a *small* set of alternatives that are *feasible*, *complete*, *compelling*, and *diverse*.

Small set. Considerable effort is usually required to fully define and evaluate each alternative. A quantitative model must be created, and inputs for the model must be assessed from subject matter experts (SMEs). Therefore, it is usually good practice to limit the number of alternatives to be designed and evaluated to just a handful, say three to five. It is often useful for the initial set of alternatives to be a "spanning set" of choices that cover the range of feasibility, and after an initial pass at evaluation, to develop additional alternatives in the part of the set that seems most interesting. Of course, this good practice may not apply to decision making for portfolios where the number of alternatives may be much greater. For example, in a military base realignment and closure (BRAC) decision, all 99 installations needed to be evaluated (Ewing et al., 2006). See Chapter 12 for a discussion of portfolio decision analysis.

Feasible. Each alternative to be evaluated must be a course of action that can actually be taken. There must not be any barrier (e.g., physical, legal, financial, cultural, social, or ethical) that would make the alternative impossible to pursue. While the final set of alternatives must be feasible, we should not be too restrictive in early phases. Particularly in the brainstorming and creative phase of alternative generation, it is useful to suspend constraints to allow thinking to diverge. We should examine the initial constraints and see if, in fact, they really are constraints. Many times we can relax the initial constraints after creative thinking about the alternatives. This is particularly true of financial constraints. Initially, we may be given guidance that the budget is fixed. But when we identify some really creative and innovative ideas, the decision makers may be able to obtain the additional funds from other programs. However, once we settle on the final constraints, we use screening criteria (West, 2011) with thresholds to eliminate infeasible alternatives.[1] An example of a threshold for the data center location problem would be alternative locations with at least 50 MW of primary power.

Complete. Each alternative should be a full description of the course of action to be taken. A setting for each of the "levers" in the decision should be specified. For example, if a decision is to be made on how to deploy a military battalion in an exercise, each alternative should specify mission and tasking for every unit within the battalion.

Compelling. Each alternative should have something in it that makes it appealing. Perhaps an alternative offers a chance to capture a major share of a valuable new market segment. Another might be a path to greatly reduce costs. If it is difficult to see why it might be a good idea to pursue a particular alternative, that alternative probably should not be in the set to be evaluated. Of course, many times, the baseline[2] alternative and/or the "do nothing" alternative are also included for comparison.

---

[1]Common names are screening criteria or go-no go table.

[2]Other names for the baseline alternative are the current, status quo, or momentum alternative.

Diverse. The alternatives in the set to be evaluated should be significantly different from each other. Collectively, they should embody all of the ideas that have been put forward. A useful analogy here is exploring for oil. The set of alternatives to be evaluated are like exploratory wells—they should be spaced widely throughout the potential oil field to generate as much information as possible about the location of oil deposits. The production well that is ultimately sunk may or may not be located exactly where an exploratory well is sited, depending on what is learned from the exploration. So too, the decision alternative that is ultimately undertaken may or may not be one of those in the set that are evaluated (see discussion of hybrid alternatives in the next section).

8.3 Obstacles to Creating a Good Set of Alternatives

Conceptually, an alternative is easy to create. It is just a matter of laying out a clear plan of action. But creating a good set of alternatives can be difficult. Alternatives are created in the minds of people. and it is well established that our human minds are subject to a number of cognitive biases (see Chapter 2) that can interfere with the process of creating alternatives (as well as with assessing probability ranges for uncertainties).

For example, the cognitive bias called "anchoring" refers to the strong tendency for a person's judgments to be influenced by an existing thought. This bias can make it difficult for someone to think of an alternative that differs radically from one already in his mind. And the bias called "availability" refers to the fact that thoughts that are easily recalled tend to overshadow those that are more difficult to recall. This bias can make it difficult for someone to go beyond the most obvious and prominent alternatives, being swayed too strongly by recent successes, for example, compared with earlier failures.

Another bias that can interfere with creating good alternatives is called the "comfort zone bias." This bias refers to the tendency for people to be more willing to contemplate actions that are familiar rather than those that are unfamiliar. So, an alternative that would require actions for which the person feels ill-prepared or unskilled would tend to be overlooked, even if that alternative might offer great value.

Motivational bias can also limit the number of alternatives. Some individuals like to provide only one alternative, the one they recommend. Other individuals like to provide three alternatives to decision makers: one that has obviously low value, one that has significant value but is unaffordable, and an affordable alternative that is being recommended. We do not advocate such approaches.

James Adams posits a series of conceptual blocks that he defines as "mental walls that block the problem-solver from correctly perceiving a problem or conceiving its solution" (Adams, 1974). These blocks as adapted from Adams can be summarized as follows:

Perceptual Blocks

- Stereotyping—seeing what you expect to see.
- Difficulty in isolating the problem—solving the wrong problem.
- Delimiting the problem too closely—imposing too many constraints upon the problem and its solutions.
- Inability to examine the problem from multiple perspectives.
- Saturation—inability to perceive and retain all the information around us.
- Failure to use all of our senses—we should not neglect any sensory inputs.

Emotional Blocks

- Fear of taking a risk—we tend to be afraid of making a mistake or failing.
- No appetite for chaos—we struggle to deal with ambiguity and uncertainty.
- Judging rather than generating ideas—we tend to judge and analyze too early in the decision-making process.
- Inability to incubate—failing to give ideas time to mature before eliminating them prematurely.
- Lack of challenge or excessive zeal—lack of motivation will inhibit creativity as will excessive motivation to succeed quickly.
- Reality and fantasy—ignoring one of these critical resources to creativity in problem solving.

Cultural and Environmental Blocks

- Taboos—we eliminate alternatives too quickly in our thought process because they seem culturally incorrect.
- Lack of humor in problem solving—we may take problem solving too seriously and leave no room for humor, which can be inspiring.
- Reason and intuition—many decision analysts tend to think that reason, logic, and numbers are good, while feeling intuition, and qualitative assessments are bad; both are essential for creativity.
- Left-handed and right-handed thinking—we tend to favor one side of the brain over the other; both sides are essential for creativity—left side (right hand, primary creativity) for order, reason, logic and mathematics, right side (left hand, secondary creativity) for imagination, artistry, and intuition.
- Tradition and change—it is hard to overcome the inertia of tradition that can impede the ability to see the need for change in developing alternatives.
- Supportive versus nonsupportive environments—physical, economic, and organizational support are often needed to bring ideas into action.
- Autocratic bosses—may make it difficult to push new ideas through.

Intellectual and Expressive Blocks

- Choosing the "correct" problem solving language—we need the ability to go from one problem solving language to another to include analytical/mathematical, visual/special, analogy, and so on; some work for one problem, but not for others.
- Flexibility versus fluency—we need both, flexibility to generate many alternatives that may be similar, fluency to generate alternatives that are different in nature.
- Incorrect information—lack of, or bad, information expands rapidly and makes problem solving difficult; this ties in closely with "value of information" concepts of decision analysis.
- Inadequate language skills and imprecision in verbal expression—poor communication inhibits "selling" good alternatives to others.

8.4 The Expansive Phase of Creating Alternatives

The process of generating a good set of alternatives has two phases, the first expansive (or divergent), and the second reductive (or convergent).

The goal of the expansive phase is to generate as many ideas as possible that might lead to the definition of creative alternatives. In this phase, divergent and creative thinking comes to the fore and analytic thinking takes a back seat. Typically, this phase consists of one or more facilitated "brainstorming" sessions ideally involving participants having a wide range of perspectives on the situation.

The rules of brainstorming should be strictly enforced in these sessions. Dare to think "crazy" and "wild" ideas. Build on ideas already contributed. Do not criticize any idea. A good brainstorming session will generate a large collection of ideas.

A variety of facilitation techniques (see Chapter 4) can be used in these sessions to stimulate creative thinking and to help people break through the mental barriers that constrain them. One is to have participants make a list of challenges. A challenge is a statement that begins, "We would achieve great success if only we could. . . ." The list of challenges can then be used to stimulate thinking about actions that would overcome them. These actions may form the basis of good alternatives. For example, consider a company trying to decide how to expand into an overseas market. They might make the following challenge statement: "We would be in great shape if only we did not have to invest so much in a distribution network there." That challenge statement might stimulate creative thinking about new alternatives in which the company could sell in the foreign country without having its own distribution system there.

Another facilitation technique is to have participants imagine how various other parties would respond to the situation at hand. For example, if a struggling industrial company is trying to find a new business strategy, one might ask participants to answer questions, such as these: "What would Microsoft do if it owned this business?" "How would Procter and Gamble act in this situation?" "How would Samsung run this business?" The idea of this facilitation technique is to have participants consider how companies that are successful in other industries would respond to the situation at hand. It is also often useful to have participants consider how actual or potential competitors would act in the situation

A third facilitation technique is called "backcasting" (or "crystal ball"). In this technique, participants are asked to imagine that they are transported via a time machine 5 or 10 years into the future (or that they look that far into the future via an infallible crystal ball). There, they learn that the decision they are considering has had a highly successful outcome and they are asked to state reasons why that success had occurred. The exercise is then repeated except that this time they learn that the outcome was very bad. This exercise is designed to help participants change their perspective on the situation by looking back at it from the future. The ideas generated in this exercise may form the basis of good alternatives.

A fourth facilitation technique makes use of scenarios. A scenario is a brief description of the future state of the world. An example of a scenario might be "very high energy costs, stagnant world economy." An alternative-generating exercise based on scenarios would be to pose a number of widely divergent scenarios and have participants describe courses of action that would be successful within each scenario. A good alternative might then be created by combining actions that seem to be robust across a range of the scenarios. Scenario analysis, combined with value-focused thinking, was used extensively in the Air Force 2025 (Parnell et al., 1998) and Blue Horizons (Geis et al., 2011) studies.

A fifth facilitation technique is to refer to the means–objectives at the bottom of the objective hierarchy, if one was constructed (see Chapter 7). These can often suggest doable constructive alternatives. Keeney (Keeney, 1992) provides example of means–ends networks.

8.5 The Reductive Phase of Creating Alternatives

The goal of the reductive phase is to take the large number of ideas generated in the expansive phase and winnow down (converge) to a small number of well-crafted alternatives for evaluation. Unlike the previous phase, this one calls for highly organized, analytic thinking.

A good way to start the reductive phase is to categorize the many ideas and group them by similarity. Then, create one or more alternative "themes" to represent the ideas in each group. A theme is a short expression of the key idea of the alternative. For examples, the alternative themes might include the following: "Lower costs," "Increase share of market," and "Improve customer service."

Sometimes, it is useful to prioritize the themes via voting among the participants to focus attention on just a handful. However, care must be taken not to prematurely exclude from consideration "wild" ideas that may lead to potentially valuable alternatives. Keep in mind that the overall goal is to generate a set of alternatives that have the potential to create value and are significantly different from each other.

A very useful conceptual tool to use in the reductive phase is called a "strategy table" (Howard, 1988). A strategy table contains a number of columns, each of which represents a particular choice that is part of the overall decision.[3] Figure 8.1 shows an example strategy table that describes a new product development decision. The four columns represent the components of the decision: (1) Which technology to employ in the product, (2) the timetable for development, (3) the target geographic markets for the product, and (4) whether or not to seek a development partner. Within each column are listed the possible options for that choice. Technically, these options should be defined so that for every alternative, there is one and only one option selected in each column. However, it may be permissible to allow overlapping options in a column if doing so will spur clear creative thinking about alternatives. The first column in the example strategy table in Figure 8.1 shows a choice of three options for product technology: current, in-licensed, or experimental technology.

Once all of the columns are structured, an alternative is defined by selecting one option in each column. Figure 8.2 shows the example strategy table with two alternatives defined, "Go Slow" and "Aggressive."

As alternatives are defined in the strategy table, it may be necessary to add new columns or to add options within existing columns to accommodate new alternatives. The strategy table can help refine the set of alternatives to be evaluated. For example, if two alternatives share the same option in almost every column, they might be deemed as being too similar and merged into one alternative. Or, if an option in a column is not selected by any of the alternatives, it might be possible to create an interesting new alternative starting with that option.

	Product Technology	Target Development Time	Target Market	Seek Development Partner?
	Current technology In-licensed technology Experimental technology	24 months 30 months 36 months 42 months	No. Am. No. Am and EU No. Am, EU, and Asia	No Yes

FIGURE 8.1 Example strategy table.

[3]The strategy table is similar to Zwicky's morphological box, a common tool used in systems engineering to generate alternative system designs.

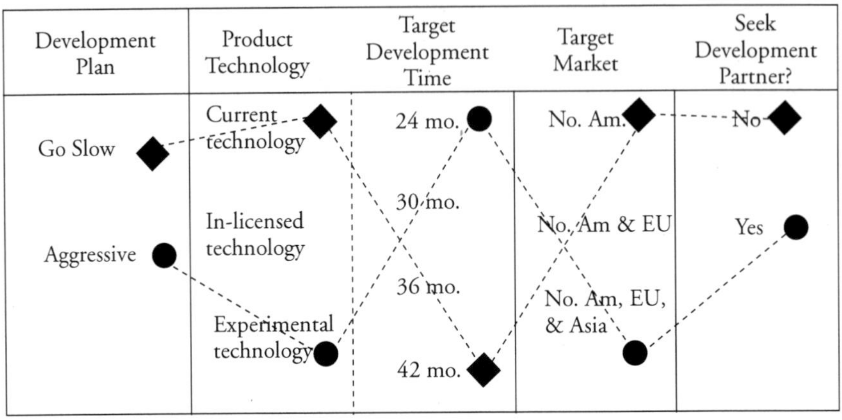

FIGURE 8.2 Defining two alternatives in a strategy table.

TABLE 8.1 Strategy Table in Matrix Format

Development Plan	Product Technology	Target Development Time	Target Market	Seek Development Partner
Go slow	Current technology	42 months	No. Am	No
Aggressive	Experimental technology	24 months	No. Am, EU, and Asia	Yes

It is useful to note that while the strategy table is a useful tool for generating alternatives, it can pose difficulties in communicating alternatives to decision makers and stakeholders. A better tool for communication is a simple matrix in which each row represents an alternative, and the options defining that alternative can be read by scanning across the matrix columns for that row. Table 8.1 displays the information in Figure 8.2 using this matrix format.

For some, it is helpful to think of a strategy table as a control panel of levers. Each column in the table is a control lever that can be set in any of a number of pre-defined positions (the options). A complete alternative is then defined by a specific combination of settings of all levers on the panel.

> The strategy table is an important technique for generating and communicating a creative and diverse set of alternatives.

For complex decisions, it may be helpful to use *nested* strategy tables, in which a lower-level strategy table defines a column in the top-level strategy table. Figure 8.3 illustrates an example of nested strategy tables. In this example, the top-level strategy table defines an overall business strategy decision. One of the columns

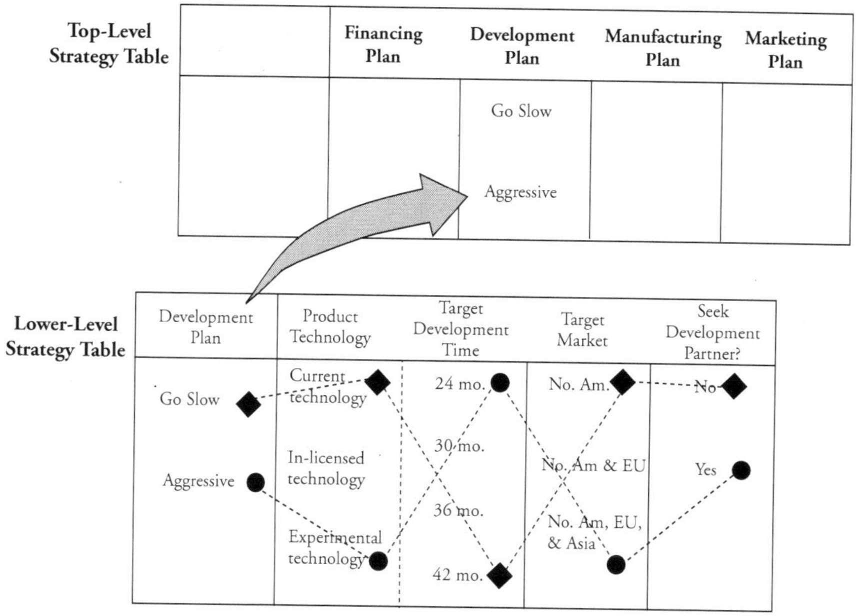

FIGURE 8.3 Nested strategy tables.

of the top-level strategy table is the choice of product development plan, which is defined by the lower-level strategy table that we have seen in Figure 8.2. The options in the column of the top-level strategy table are the decision alternatives defined in the lower-level table. The use of nested strategy tables enables a natural delegation of responsibilities because a separate subteam can develop each lower-level table. It also makes the top-level strategy table more concise, which facilitates the conversation about high-level alternatives.

It is good practice to check that the model that is used to evaluate the alternatives is consistent with the strategy table. That is, the calculations of the model should be responsive to a change of option in every column of the strategy table. If not, the column may not be necessary or the model may need to be changed.

In situations in which "downstream" decisions (i.e., decisions that will be made in the future after some current uncertainty is resolved) are important, it may be useful to employ a decision tree structure to think about and to communicate clearly the sequential nature of alternatives. See Section 11.2.1 for a discussion of decision trees.

8.6 Improving the Set of Alternatives

The set of alternatives can often be further improved after they are evaluated. The results of the analysis that is conducted to compare the alternatives can many times provide insights on how to create even better alternatives. This idea is presented briefly here and discussed in more detail in Chapter 11.

One common technique for finding better alternatives is to look at the sources of value in each of the original alternatives. How much of the overall value is coming from each market segment, or product line, or geographic region? With this information, it may be possible to design a new alternative that emphasizes high-value features and de-emphasizes low-value features. Such an alternative is called a "hybrid" because it combines the good features of several of the original alternatives.

Another technique is to examine the top bars in the tornado diagram (see Chapter 9) for each alternative and to ask the question, "Is there any way that we can make it more likely that this important value driver turns out in our favor?" For example, consider a new product introduction decision situation. Imagine that the top bar in the tornado diagram for one alternative is the time to market given the planned budget—the sooner the new product is launched, the greater the value. Seeing this, we might ask if the launch timing could be accelerated by increasing the development budget. If so, we could redesign the alternative to achieve greater value with a higher budget.

A third way to improve alternatives after evaluation is to examine the value of perfect information (see Chapter 11) on key uncertainties. It might be possible to design a new alternative that exploits high value of information. For example, a new alternative that includes a period of initial information-gathering or experimentation could be created. Or an alternative that includes built-in flexibility, such as a real-option, might be quite valuable. A real option is not a derivative financial instrument, but an actual future option that a business may enable by making a decision now. For example, by investing in R&D, the company creates future options for current and future projects.

It is important to remember that the time to search for good hybrid alternatives is after the evaluation of the original alternatives has been completed (see Chapters 9 and 11).

8.7 Illustrative Examples

See Table 1.3 for further information on the illustrative examples. The strategy generation table was useful in all three examples, but it was displayed using three different formats.

8.7.1 ROUGHNECK NORTH AMERICAN STRATEGY (by Eric R. Johnson)

As discussed in Chapter 6, the project was reframed as a portfolio resource allocation problem among five business areas: conventional E&P, coalbed methane, enhanced oil recovery, tar sands, and electric power generation. All North American opportunities were to be considered, including acquisition or divestiture of assets. The project team set up a strategy table whose five columns corresponded to the five business areas. They defined four levels of investment for each business area, and elicited from the functional experts what would be an approach to each business area consistent with each of the four levels of investment. The four levels of investment were named divest, maintain, growth, and leader (i.e., nonorganic growth by acquisition). These strategies are shown in the first four rows of Figure 8.4.

In order to ensure that all good ideas found a home in the strategy table, they defined strategies in which a given level of investment was deployed across all business areas. By doing so, they were not suggesting that anyone in Roughneck leadership intended to fully divest North American assets, or make a string of large acquisitions in each of the business areas. By defining these as "strategies," they required only that the strategies be taken seriously enough to define a coherent point of view that could be evaluated in reasonable fashion. The project team had every expectation that an intermediate course of action would be chosen, but they knew that choice of an optimal hybrid strategy would be better done once they had completed and reviewed the initial evaluation phase.

After reviewing initial evaluation results, the team formulated two hybrid strategies, which are shown in the last two rows of Figure 8.4.

8.7.2 GENEPTIN PERSONALIZED MEDICINE (by Sean Xinghua Hu)

The situation with Geneptin was relatively simple, and the number of strategic alternatives was few, with the primary decision elements being whether to employ a biomarker, which diagnostic platform to use, and how to incorporate the biomarker into Phase III trial design. The Geneptin team generated two main strategic alternatives to advance into further analysis: all comers and a prospective HER2-overexpression biomarker-stratified personalized medicine approach. Figure 8.5 shows the Geneptin strategy table.

IHC (immunohistochemistry) and FISH (fluorescence *in situ* hybridization) were two possible technologies for diagnostics tests. In addition to the prospective use of the biomarker to identify patients to be included in the trial, options could be considered to include an all comers arm (a trial patient "arm" is a group of patients who receive the same treatment during a clinical trial. A clinical trial can have one or more "arms." When there is more than one arm, the trial could be designed to be "double blinded," in which neither the physicians nor the patients know during the trial which patient is receiving which treatment) in addition to the biomarker positive one, or retrospectively analyze results to derive information on biomarker positive patients. These options were not included in either of the strategies that were evaluated.

In other personalized medicine situations, other decisions might need to be considered:

Strategy Themes	E & P	CBM	EOR	Power	Tar Sands
Divest	Divest assets	Divest assets	Divest assets	None	None
Maintain	Develop current assets	Develop current assets	Develop current assets	None	None
Growth	The above plus: Additional exploration Downspacing	The above plus: Two acquisitions. Aggressive exploration	The above plus: Multiple small acquisitions. New pipeline	Build cogen	One 50k barrel/d plant
Leader	The above plus: Major acquisition	The above plus: Major acquisition	The above plus: Two large acquisitions	Build cogen. Build coal-fired plant	Two 50k plants
Divest Hybrid	Divest assets	As leader	As leader	Build cogen	One 50k barrel/d plant
Team Hybrid	As leader	As leader	As leader	Build cogen	One 50k barrel/d plant

FIGURE 8.4 RNAS strategy table.

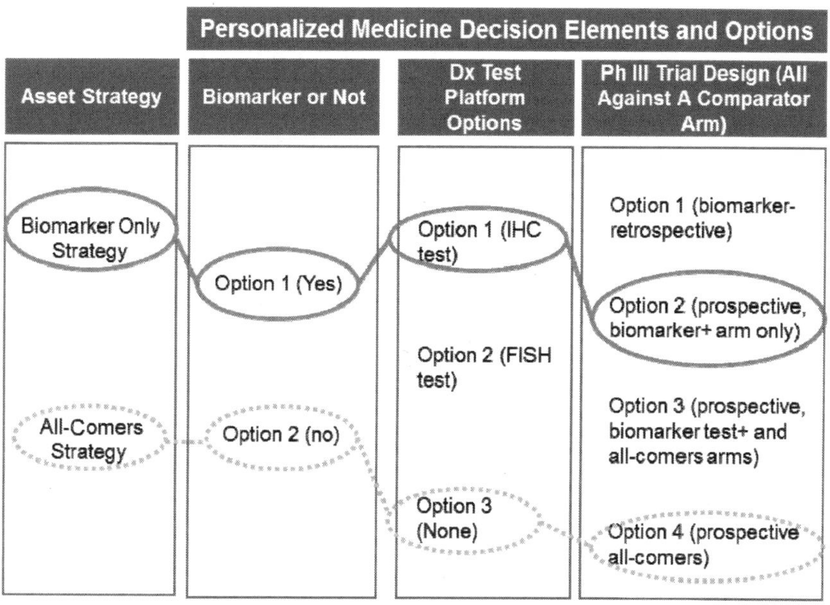

FIGURE 8.5 Geneptin strategy table.

- **Choice of biomarkers.** There can be different types of biomarkers (e.g., class biomarker: a biomarker that has general applicability to a class of drugs; drug-specific biomarker: a biomarker specific for a drug and not applicable to other drugs in the same class; indirect biomarker: a biomarker used to stratify patients based on suitability to another drug, for example, a drug can target the patient segment identified by a biomarker to be resistant to another drug); or multiple biomarker choices for the same type of biomarker (e.g., KRAS or EGFR mutations for lung cancer).
- **Types of** diagnostic (Dx) **tests.** *In vitro* diagnostics kits versus "home brew" or "laboratory-developed tests (LDTs)."
- **Dx capabilities.** Partner with a Dx company versus build Dx capabilities internally.
- **Dx company choices.** Who to partner with, if a Dx partner is desired.
- **Dx pricing.** Value pricing versus commodity/cost plus pricing.

8.7.3 DATA CENTER LOCATION (by Gregory S. Parnell)

Developing creative alternatives in the data center location problem was a challenge due to the organizational culture of the agency, the involvement of state

TABLE 8.2 **Data Center Strategy Generation Table**

	Location	Comm	Floor Space	Power	Cooling
Tennessee	Secure	Existing	Existing	Unlimited	Unlimited
Texas	Expand security	Expand	Expand	Primary and Backup	Primary and Backup
Washington State	New security	New	New	Primaryand expand backup	Primaryand expand backup
				Expand primary and backup	Expand primary and backup

political leaders who advocated the location of the data center in their states, and the time urgency of the needed capability. As the initial need for the data center was defined, key agency leaders became strong advocates for existing alternatives that would build on past decisions the agency had made. It took significant leadership by the senior decision makers to expand the list of alternatives beyond the initial two alternatives to a broader set of alternatives. Table 8.2 show how the strategy generation table was used to generate alternatives that could perform each of the five functions: provide a secure location; communicate with mission users; provide space for mission equipment and personnel; power equipment; and cool equipment and personnel.

8.8 Summary

The development of creative alternatives that have the potential to create high value for the organization is one of the most important tasks of the decision analyst. If we do not have good alternatives, our opportunity to create high value is significantly diminished. In this chapter, we review the best practices for generating high quality alternatives. In addition, we discuss the cognitive biases and motivational biases that can impact the quality of our alternatives. We introduce and demonstrate the strategy table that is a very effective technique to develop a wide range of alternatives and communicate the scope of the alternatives to senior leadership. We emphasize the importance of developing hybrid alternatives that contain the best features of the initial set of alternatives after the initial decision analysis. The hybrid alternatives can offer significant potential value.

KEY WORDS

Alternative generation Identifying a good set of alternatives is an important part of decision making. The decision can be only as good as the best alternative.

Expansive phase of alternative generation The goal of the expansive phase is to generate as many ideas as possible that might lead to the definition of creative alternatives. This phase uses creativity techniques to expand ideas for divergent alternative generation.

Hybrid alternative A new alternative is called a "hybrid" because it combines the good features of several of the original alternatives.

Nested strategy tables For complex decisions, it may be helpful to use *nested* strategy tables, in which a lower-level strategy table defines a column in the top-level strategy table.

Real option A real option is a future option that a business may enable by making a decision now. For example, by doing R&D on a project, it enables future development of the product.

Reductive phase of alternative generation The phase during which a large set of alternatives is pared down (converged) to a smaller set that will be evaluated

Screening criteria Criteria used to determine the feasibility of an alternative.

Strategy table A strategy table contains a number of columns, each of which represents a particular choice that is part of the overall decision. Selecting one option from each column defines a strategy.

REFERENCES

Adams, J.L. (1974). *Conceptual Blockbusting: A Guide to Better Ideas.* Stanford, CA: Stanford University Press.

Clemen, R.T. & Reilly, T. (2001). *Making Hard Decisions with DecisionTools.* Pacific Grove, CA: Duxbury Thomson Learning.

Ewing, P., Tarantino, W., & Parnell, G. (2006). Use of decision analysis is the army Base Realignment and Closure (BRAC) 2005 military value analysis. *Decision Analysis, 3*(1), 33–49.

Geis, J.P., Parnell, G.S., Newton, H., & Bresnick, T. (2011). Blue horizons study assesses future capabilities and technologies for the United States Air Force. *Interfaces, 41*(4), 338–353.

Hammond, J.S., Keeney, R.L., & Raiffa, H. (1999). *Smart Choices: A Practical Guide to Making Better Decisions.* Boston: Harvard Business School Press.

Howard, R. (1988). Decision analysis: Practice and promise. *Management Science, 34*(6), 679–695.

Keeney, R.L. (1992). *Value-Focused Thinking: A Path to Creative Decision Making.* Cambridge, MA: Harvard University Press.

Keller, L.R. & Ho, J.L. (1988). Decision problem structuring: Generating options. *IEEE Transactions on Systems, Man, and Cybernetics, 18*(5), 715–728.

Parnell, G.S., Conley, H., Jackson, J., Lehmkuhl, L., & Andrew, J. (1998). Foundations 2025: A framework for evaluating air and space forces. *Management Science, 44*(10), 1336–1350.

West, P. (2011). Solution design. In G.S. Parnell, P.J. Driscoll, & D.L. Henderson (eds.), *Decision Making for Systems Engineering and Management*, 2nd ed., pp. 366–368. Hoboken, NJ: John Wiley & Sons.

CHAPTER NINE

Perform Deterministic Analysis and Develop Insights

ERIC R. JOHNSON, GREGORY S. PARNELL, and STEVEN N. TANI

All models are wrong, but some are useful.

—George E P Box

Handbook of Decision Analysis, First Edition. Gregory S. Parnell, Terry A. Bresnick, Steven N. Tani, and Eric R. Johnson.

9.1 Introduction

Our decision analysis approach is to develop and refine a model of how value is created, and to use this to identify even better ways to create value. The model embodies what we call a *composite perspective* because it incorporates the combined views of many participants. How the model responds to changes in inputs representing decision choices and uncertain factors may not be obvious from the individuals' points of view. In a word, the behavior of value in the composite perspective is an *emergent* phenomenon.

We use a "value dialogue" to compare the emergent behavior of the composite perspective to the individual perspectives and, where they differ, improve one or the other to bring them into better alignment. Having done so, we can use the improved composite perspective to understand how much value each alternative creates, and to create even better alternatives. Doing this requires these steps:

- **Modeling.** Build representations of experts' perspectives about important events, stitch them together to create a composite perspective on value creation, and represent this in a model.
- **Exploration.** Explore possible value outcomes in the model in response to different decision choices and uncertainty outcomes.
- **Analysis.** Characterize the emergent behavior in terms that allow direct comparison with individuals' perspectives and generation of improved strategies.

Quantitative modeling is an essential step when applying decision analysis to most important organizational decisions, because such decisions are made in the face of considerable complexity and uncertainty. It is well established that the human brain cannot simultaneously process more than a few pieces of information without error. Modeling and analysis allow us to overcome this limitation through a "divide and conquer" approach. A complex problem is decomposed into smaller, simpler pieces that the human brain can comprehend and then put back together in an integrated analytic structure.

Every model is built for a specific purpose. When a new jetliner is being designed, for example, a model is built to test its aerodynamics in a wind tunnel. This model should be quite accurate regarding the aerodynamic surfaces of the aircraft, but need not be accurate regarding its interior structure. The purpose of a decision model is to give insight into the reasons why one possible course of action is better than the others. We seek to develop a *requisite decision model.* Phillips (Phillips, 1984) defines a requisite decision model as "a model whose form and content are sufficient to solve a particular problem." It should be accurate enough to differentiate the various alternatives under consideration. It can be quite inaccurate regarding details that do not contribute to a comparison of the alternatives.

A best practice is to follow the dictum, "Be clear about the purpose of the model." Do not waste time and effort working on features of the model that do not serve its purpose of distinguishing between alternatives. A good strategy for building a decision model is to start simple and to add complexity only where it is needed to improve the insights produced by the model. Also, resist the temptation to use a completed decision model for other purposes. For example, a model that was built for comparing high-level strategies for a business unit would likely be ill-suited for use in optimizing the operations of that business.

A *good* decision model is one that is both useful and tractable. A useful decision model is one that generates clear insights for decision making. It is both error-free and readily understood by the decision analyst. A tractable model is one that can be built and used within the constraints of time and level of effort for the decision. There is often a trade-off to be made between usefulness and tractability; good engineering judgment is needed to make this trade-off appropriately.

When creating a decision model, always be conscious of two different audiences. The first audience is the computer. The model, of course, must make sense to the computer. It should be free of errors and do what is intended. The second audience is the team of people who look at the model to try to understand how it works. It is certainly possible to create a model that the computer understands and runs flawlessly but which is impenetrable for the analysts and/or decision maker(s). Such a model is not good because if the team cannot understand how it works, they will have little credibility in its results and will be unwilling to make decisions based on them. Also, if it is difficult for the analyst to understand a model, it increases the risk of logical errors being introduced when the model is revised. Therefore, a good decision model is one that can be easily understood by the users.

In decision analysis, we can use models deterministically, probabilistically, or both. We discuss deterministic analysis in this chapter and probabilistic simulation and analysis in Chapter 11.

This chapter is organized as follows. In Section 9.2, we introduce the influence diagram as a tool for planning the model. In Section 9.3, we discuss the advantages of spreadsheet software as a modeling platform. In Section 9.4, we provide our guidelines for building a spreadsheet decision model. In Section 9.5, we describe how we organize a spreadsheet model. For complex models, we need to verify that they are correct, so in Section 9.6, we describe debugging a spreadsheet model. In Section 9.7, we present techniques for deterministic analysis. In Section 9.8, we perform a deterministic analysis of RNAS, our illustrative singe-objective problem. In Section 9.9, we present a simple example to illustrate multiple objective decision analysis having a monetary value metric. In Section 9.10, we present deterministic multiple objective decision analysis with a non-monetary value metric, using the most common model, the additive model. In Section 9.11, we apply the methods presented in Section 9.10 to the data center problem. In Section 9.12, we summarize the chapter.

9.2 Planning the Model: Influence Diagrams

Just as it is an excellent idea to draw up a set of plans before starting the construction of a building, it is wise to make a plan for a decision model before creating it. An especially useful tool for planning a decision model is the influence diagram (Howard & Matheson, 2005), sometimes also called a "value map," or "decision diagram." An influence diagram is a graphical representation of the calculation of the value metrics, as they depend on the choice of decision alternative and on external factors, which may be uncertain. Appendix B provides an introduction to influence diagrams. Chapter 10 gives a detailed discussion of how to structure the influence diagram for a decision problem, and how to populate it with pertinent expertise.

An influence diagram is a useful high-level "blue print" for the model. It identifies the key external factors that must be in the model to calculate the value metrics and shows the basic structure of those calculations. It also shows the interdependencies among the uncertain external factors, the decisions, and the values.

However, the influence diagram does not contain all of the information needed to build the model. For example, it does not specify the sequencing or dynamics of the external factors (i.e., how they change over time). Nor does it generally specify the formulas to be used in the calculations.

As an example, Figure 9.1 displays the influence diagram for a decision on the launch strategy for a new product (i.e., what features to put in the product and how much to spend on marketing it) and on the selling price of the product. Uncertainty in the product's market share is affected by its features, by its price, and by the size of the marketing budget. Unit sales are the product of market share and market size, which is also uncertain. Revenue is then calculated as unit sales times unit price. Expenses are equal to the variable manufacturing cost (unit sales times unit cost) plus development expenses, marketing expenses, and other annual fixed expenses.

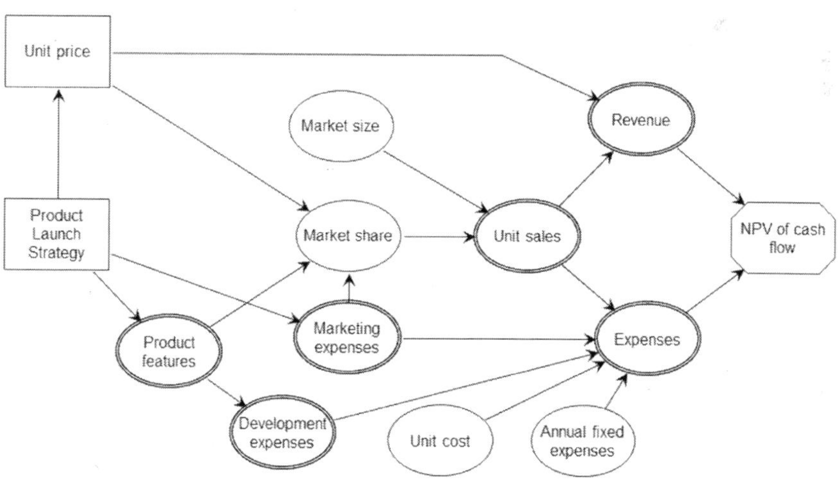

FIGURE 9.1 Example influence diagram.

9.3 Spreadsheet Software as the Modeling Platform

Spreadsheet software, such as Microsoft® Excel, offers a number of advantages that make it a good choice as the platform for a decision model. The art of modeling in Excel is recognized as a key analyst skill and is taught in many undergraduate and graduate programs (Powell & Baker, 2011).

Instant feedback. Calculations in a spreadsheet are seemingly instantaneous, with both inputs and outputs clearly visible. That makes it easy to check the logic of the model as it is being built. It also makes it easy to do "what-if" calculations with the model.

Ubiquity and familiarity. Spreadsheet software is ubiquitous, particularly in the world of large organizations where decision analysis practitioners typically work. It is generally safe to assume that everyone on a team assembled to work on a major decision has ready access to such software and is at least somewhat familiar with it. This leads to a major advantage of building decision models in spreadsheets—it is very easy to share the models among the team members. All team members are able to participate more fully in analyzing the decision and generating valuable insights for the decision makers.

Graphical representations. Many decision makers and stakeholders find it easier to interpret and remember graphical representations instead of tables of data. Spreadsheet software includes charting capabilities that make it easy to create graphical representations of results from decision models to aid in communicating important insights.

Computational power. Although spreadsheet software does not offer as much computational speed as, for example, compiled computer languages, most decision models written in spreadsheets run sufficiently fast to be useful. Built-in macros can be used to increase the speed of the analysis.

Third-party add-in software. Many third-party software packages are available that extend the capabilities of spreadsheet software. For the decision practitioner, notable among these are packages that perform probabilistic analysis on decision models (see Chapter 11), either in the form of decision tree analysis or as Monte Carlo simulation analysis.

Macro language. Spreadsheet software includes a macro programming language feature (e.g., VBA for Excel) that enables a user to write routines that perform operations that would be difficult to accomplish via spreadsheet formulas. Although using macro code is not required to build a good decision model, it can sometimes make a big improvement in both the model and the life of the analyst!

Documentation. Documenting the formulas, names, variables, and so on is an imperative for quality model building. There are two forms of documentation appropriate for spreadsheet modeling. The simplest form is an inserted comment in a particular spreadsheet cell that describes its function or purpose. The second form is the documentation worksheet and is described in Section 9.5.7.2.

Spreadsheet software does have some limitations when it comes to decision modeling. Working with arrays having four or more dimensions is difficult. It is difficult to build models with complex dynamic interactions in spreadsheets. And working with very large spreadsheet models tends to be cumbersome and slow. But overall, except for large, complex models, the advantages of spreadsheets compensate for the disadvantages. The remaining discussion assumes that modeling and analysis is done in a spreadsheet environment.

9.4 Guidelines for Building a Spreadsheet Decision Model

We present several guidelines for developing a spreadsheet decision model that are especially useful if many members of the decision team participate in developing and using the model.

9.4.1 KEEP INPUTS SEPARATED FROM CALCULATIONS

All inputs to the model should be located together in one place, clearly separated from the calculations. Think of this as the "control panel" for the model—everything needed to operate the model is there.

A corollary of this guideline is that the calculation formulas in the model should not contain any "hard-wired" numbers. The quantity represented by a hard-wired number in the formula should instead be expressed as an input variable. Even conversion constants should follow this rule. The only numbers that should appear in calculation formulas are 0 and 1. If this guideline is followed, calculation formulas will contain only inputs and the results of other calculations.

Many decision analysts use cell coloring to support this rule. For example, yellow might be the inputs, green the calculations, and blue the decisions.

9.4.2 PARAMETERIZE EVERYTHING

To describe how a quantity (such as the size of a market segment) changes over time, it is tempting to simply type into the spreadsheet a different number for each year. However, this is a bad practice. It is much better to use parameters to describe the dynamics of the quantity. For example, one way to parameterize the dynamics is to create two input variables, one for the initial amount and the other for the annual growth rate. Parameterizing the model dynamics makes it much easier to include the effects of uncertainty in the model.

9.4.3 USE RANGE NAMES FOR READABILITY

Spreadsheet software allows the user to define names to refer to individual worksheet cells or ranges of cells. Using names rather than cell addresses in calculation formulas makes it much easier for a human to understand those formulas, especially when the reference is nonlocal (i.e., from one sheet to another).

For example, the formula for unit sales using named ranges might be:

=MktSize * MktShare

This is much easier for a human reader to understand than a formula using cell addresses:

=C210 * Inputs!G75

Every input variable and calculated intermediate result in the model should be given a name. When creating names, use a good naming convention. For example, use distinctive suffixes to designate attributes, such as geographies or product line. Using a well-designed naming convention has several advantages. It enhances the ability of humans to understand the model. It reduces the chance of typographical errors being introduced. And it facilitates creating formulas via copy and paste for additional geographies or product lines.

9.4.4 USE UNIFORM INDEXING FOR ROWS AND COLUMNS OF A SHEET

Uniform indexing means that the row header on the left applies to the entire row, and the column header at the top applies to the entire column. This way, if a row or column is added or deleted, the rest of the sheet retains its validity. Excel's FreezePanes facility encourages and supports this practice by allowing common row and column indexes at the left and top to be visible regardless of what portion of the sheet is currently displayed. By contrast, putting tables with different indexing structure below or beside the main table makes spreadsheets brittle—the modeler may see a way to improve a structure, but be afraid to implement it for fear of damaging something unrelated below or beside it.

9.4.5 MANAGE THE MODEL CONFIGURATIONS

Configuration management is necessary regardless of how many modelers actually touch the spreadsheet. Configuration management can be as simple as following a naming convention for the spreadsheet file as it is updated and enhanced; for example, *decisionmodel_YYYYMMDD_initials.xlsx*.

9.5 Organization of a Spreadsheet Decision Model

The model we build must represent up to five dimensions of complexity often found in decision situations:

1. Multiple decision alternatives
2. Ranges of uncertainty in key factors
3. Multiple components of value (objectives)
4. Multiple business units
5. Multiple time periods.

Three of these dimensions comprise what is called the *decision basis*: alternatives (decisions and strategies), information (uncertainties), and preferences (values components) (Howard, 1983). We add two more dimensions to the conversation (business units and time) because these are frequently considered explicitly, and the techniques for managing these dimensions can often be useful for facilitating the value dialogue discussed in Chapter 11.

We must think carefully at the beginning about how much granularity to model, and how to represent these dimensions of complexity in the model to ensure that we can deliver insightful analysis, while allowing tractable modeling. More fine-grained information requires more effort to elicit, model, simulate, and analyze, but may allow for expertise to be captured more authentically, and may give more opportunities to improve strategies in the value dialogue.

For some kinds of granularity, it is feasible to begin with less detail, and to elaborate only if initial analysis suggests that the elaboration could be fruitful. However, if we are deciding whether to disaggregate a dimension, this often needs to be decided upon at the beginning of the project.

There are two basic ways to represent any given dimension:

- **Intensively.** Represent all possible cases but only one at a time.
- **Extensively.** Simultaneously represent all possible cases explicitly.

While extensive representation of all dimensions simultaneously would make all our desired analyses straightforward, this can become intractable in a spreadsheet, which has only three natural dimensions (sheets, rows, and columns). We usually choose intensive representation for decisions and uncertainties because each of these "dimensions" is actually a multidimensional space. In some cases, a dimension may not be modeled at all. For example, time is not explicitly modeled in the data center case. The important point is to size up what needs to be generated, stored, and understood and to design the spreadsheet accordingly.

The structure and controls of the model must support analysis, and a key aspect of analysis is to summarize, or collapse, a dimension when it is not currently of interest, to allow others to be reviewed conveniently.

This section discusses the pluses and minuses of representing each dimension explicitly, and offers thoughts on how each of the dimensions can be analyzed, and what the implications are for spreadsheet structure and analysis controls.

9.5.1 VALUE COMPONENTS

A decision analysis ultimately produces a utility metric, whose expected value (EV) is optimized for decision making under uncertainty. Sometimes it is generated by aggregating scores for objectives; sometimes by aggregating the present value (PV) of line items in a P&L such as revenue, costs, taxes, and investment; and sometimes by aggregating net present values (NPVs) of business units. Sometimes there is an explicit adjustment reflecting risk attitude, sometimes not. We refer to the items directly used in the calculation of utility as value components.

Considering value components explicitly supports the following lines of thought for adding value in the value dialogue:

- Enhance a favorable item.
- Minimize or mitigate an unfavorable item.
- Optimize for a different value measure.
- Find a better tradeoff between objectives

We almost always want to consider this level of granularity explicitly. To support this, there should be an analysis control in the model that allows the analyst to select which objective or value component is being reported, and one of the options should be for the utility.

9.5.2 DECISIONS

There can be many decisions to address in a given decision situation. If so, we define strategies, which are fixed combinations of choices, one for each decision (as discussed in Chapter 8). We usually build the model to evaluate all of the strategies, but only one at a time, so there must be a strategy selector in the model to change the evaluation from one strategy to another.

Once we have developed insights into the creation of value by analyzing the initially formulated strategies, we may find it useful to explore the consequences of novel combinations of choices. While this requires experts to consider more cases, the additional work may be justified by the potential value added.

There are times when we want to use the valuation results of two strategies to calculate a value delta that highlights how the two strategies differ. For this, the model must have indicators of which strategies are involved in the subtraction. The calculation of the value delta results is normally accomplished via caching (see Chapter 11) or via data table functionality in the spreadsheet (see Section 9.8.2). An explicit strategy table data structure with codes to indicate the choice for each decision under each strategy can be useful if we want to explore hybrid strategies or search for the optimum choice on a decision that allows for a continuum of choices (e.g., in RNAS, the decision was at what oil price level to go forward with full-scale development).

9.5.3 UNCERTAINTIES

Our experts often think in terms of various possible outcomes of multiple uncertainties, but it is not always necessary to represent this explicitly. Sometimes, expertise about the phenomena in question is good enough that uncertainty does not have a material impact. If uncertainties all add as much upside as downside, ignoring them may not bias rank ordering of strategies. Many decision analyses, both MODA and financial, have been conducted in deterministic fashion, with no explicit representation of uncertainty.

Explicit consideration of uncertainties can sometimes increase the authenticity of our composite perspective, especially in cases where the outcome of an uncertainty changes the way the rest of the system behaves. For instance, the price of oil can change whether the market for drilling rigs is constrained or in surplus, which, in turn, can substantially affect the impact of other uncertainties that apply only to one of these conditions. In addition, explicit consideration of uncertainties gives us guidance on two kinds of improvements to strategies: those that aim to influence an important uncertainty, and those that learn more about it, aiming to make a more suitable decision for its likely outcome.

We use two different methods to summarize the uncertainties and their impact. In the first, we approximate the full range of uncertainty in outcome by a single deterministic calculation. In the second, we iterate through the many possible outcomes to calculate an EV and the total variability in value. The model should be designed to facilitate both of these methods.

One common approach to deterministic evaluation of uncertainties is to set all inputs to the "base case," wherein each continuous input is set to its median, and each discrete variable is set to its most likely outcome. For success or failure events, which are common in stage-gated decision making, such as oil and gas and pharmaceutical development, the base case is usually defined as success. The base case is a case that could possibly occur.

Another approach to deterministic evaluation of uncertainties is what we call *pseudo EV*.[1] The pseudo EV is the value resulting from all inputs being set at their EVs. Inputs representing binary events, such as success and failure, are assigned values of 1 or 0 (so that their EV is the probability of success), and the binary input is treated as a multiplicative factor in the calculation of value wherever appropriate. This approach gives a deterministic result that is often numerically similar to the true EV. Using the pseudo EV rather than the base case in deterministic analysis has the advantage that what-if questions can be answered with immediately calculated results that are likely to be closer to the true EV.

[1]The name pseudo EV is new in this text, but the idea itself is not. We give this slightly pejorative name to the notion out of respect for the discussion in chapter 11 of *Flaw of Averages* (Savage, 2009). In practice, we find that Savage-style counterexamples are not so common, and we can set up safeguards to ensure that we notice and address them when they do occur. The most important safeguard is never to form a decision based on pseudo EVs unless the insight can be verified using true EV analysis.

Working with pseudo EVs helps the analysis team develop improved intuitions about the behavior of EV in the model, enabling the delivery of clear succinct explanations of model behavior. However, the results of the pseudo EV approach must be interpreted carefully because it generates P&Ls that are between those of success and failure, and therefore may not represent a case that could possibly occur.

It is useful to design the model to permit the user to switch any input to an "override" setting that is distinct from its range of uncertainty, to answer "what-if" questions.

For the calculation of the full EV, the model should be structured to facilitate the iteration through all possible combinations of important uncertainties, either within a decision tree or in a Monte Carlo simulation.

9.5.4 BUSINESS UNITS

We use the term business units to refer to segments of the entire enterprise that the decision team may want to view in isolation. This notion can apply not only to business units that are reported distinctly in an organization's financials, but also to opportunities or projects that the team wants to consider on their own. These can be oil or gas fields; in pharmaceuticals, compounds or indications; and in military/government settings, projects or initiatives. Firms may treat regions or functional areas as distinct business units.

It is not necessary to consider multiple distinct business units in a decision analysis. Experts can be asked to assess results at the consolidated enterprise level, thereby reducing the amount of information that they must provide and we must manage. MODA analyses often do this, and there are also many financial decisions where the choice is made not to explicitly represent distinct business units.

In some cases, it can be helpful to consider distinct business units. Experts may feel more comfortable giving business-unit level assessments of uncertainty, especially if an expert in one business unit knows little about other business units. Key stakeholders may wish to have an explicit understanding of how each business unit fares under the strategies being considered. Having this level of granularity helps us to explore whether reallocation of resources from one to another could create value.

A corporate or enterprise analysis tool should have a selector indicating the business unit whose value should be reported, and aggregating them all should be one of the options.

9.5.5 TIME

Consequences of decisions unfold over time, but it is not always necessary to represent this explicitly. MODA analyses often call for assessments of choices or attributes that do not explicitly refer to time.

Explicit consideration of time can be helpful for investments, where we trade short-term loss for long-term benefit. Explicit consideration of the time dimension can make assessments for any given time period more reliable, insofar as expertise about levels of investment and amount of benefit may well reside in different experts' heads. Explicit analysis also may help create or reinforce useful insight about time dynamics in key stakeholders. In addition to supporting correct decision making among existing options, it can spur generation of improved strategies, by asking participants to think of ways to accelerate benefits or delay costs.

If we model financial results through time explicitly, we normally want to calculate the PV over time (as discussed in Chapter 7). Sometimes, it is of interest to review the value in all time periods at once (a time series result). In industries where sales stay near a peak level for a period of time (e.g., pharmaceuticals), experts are accustomed to considering sales in the year of peak sales; hence peak sales can sometimes be a useful summary for a time-profile of revenue.

When choosing the time horizon to model, the cleanest approach is to project the consequences far enough into the future that the years beyond the horizon will not have a material impact on the choice, either because time discounting makes them immaterial, or because we have no reason to think they would be different from one strategy to the next. If this is not done, we must estimate the impact of outcomes beyond the study horizon. The most common approach is to assume that the results of the final year are repeated into the future *ad infinitum*. This is done by adding a cash flow item to the final year equal to the previously calculated final year's cash flow divided by the discount rate being used.

9.5.6 REPRESENTATION OF BUSINESS UNITS, VALUE COMPONENTS, AND TIME: P&L CALCULATION SHEET(S)

When designing a spreadsheet decision model, it is good practice to separate model inputs and calculations, placing them in separate worksheets.

Three of the dimensions—business units, time periods, and value components—are not usually subdivided into subdimensions. This makes extensive representation tractable. In addition, we often need to aggregate across all of them, so we need all items in each dimension (and all combinations across other dimensions) to be present simultaneously. For this reason, these three dimensions are usually represented extensively in a sheet of calculations. In a private sector application, this sheet usually takes the form of a profit and loss (P&L) financial statement, so we refer to this as the P&L sheet. Models developed for public sector applications usually have similarly structured sheets for the calculation of value (though generally not expressed as profit and loss).

Depending on the size of the model and the degree of its segmentation, it may make sense to have more than one sheet of P&L calculations. For example, if the value metric is the sum of values across several business units, it may make sense to have one sheet of calculations for each.

The P&L calculations sheets contain all of the formulas needed to calculate the value metrics from the inputs on the Inputs sheet for each alternative. It is quite common for a decision model to calculate results for each of a number of future time periods, usually years. In this case, it is common practice to lay out the Calculations sheet with time periods going across the columns. Each row of the sheet then represents how a calculated quantity changes over time. Each of these time series may be given a range name that can be used in the formulas of subsequent calculations. It is also good practice to use the same formula in every cell in a time series. That is, for any row of calculation, the cell in each column contains the same formula. This makes it much easier for a reader to understand the model logic by looking at the formulas in just one column of the Calculations sheet. To adhere to this good practice, it may be necessary to use "IF" statements in the formulas for situations in which something special happens in a particular time period, such as a capital investment.

When we map out which line items to include in a P&L, there is usually some latitude to merge some line items or split some into components. Here are things to strive for as we make those choices:

1. Make it look familiar to those who routinely review P&Ls, with similar line items in similar order.
2. Avoid displaying rows that provide immaterial detail.
3. Reflect salient differences among strategies.

It can be helpful to arrange a P&L sheet with important indexing and summary information positioned at the left and top so that it is always visible. A good practice is to organize the calculations so that they "flow" in a consistent direction on the worksheet, either from top to bottom, or in drill-down fashion with results at the top and supporting details below.

9.5.7 INPUTS SHEET(S)

Often, the most challenging dimension to manage is uncertainty, because we potentially need to explore the impact of each individual uncertainty on value, and there can be dozens of uncertainties. Just as we define a strategy as a fixed set of choices, we define a scenario as a fixed combination of outcomes of uncertainties.[2] Due to the large number and complexity of scenarios, we almost always use intensive representation of uncertainty, iterating through various cases, and caching key results for aggregation.

A good practice is to create a section in the spreadsheet model called the Input Table to contain all uncertain inputs. Figure 9.2 shows part of an example Input Table. Each input variable occupies one row of the Input Table. A description of the input variable is given, along with the units of measurement. The range of uncertainty in the input variable is specified by entries in the three columns labeled "Low," "Base," and "High," usually meaning 10th, 50th, and 90th percentiles. A fourth column provides space for an "override" input setting to allow easy exploration of "what-if" cases. The setting of the input variable

[2]In "scenario planning," a scenario also carries with it a story about how the ultimate state of affairs came about, and extreme outcomes are of more interest than routine ones. The definition used here is less rich than this view, but not inconsistent with it.

Input Table								
Parameters	Units	Name	In Use	Index	Low	Base	High	Override
Peak sales	000 Units	VolPeak	900	2	500	900	1600	2500

FIGURE 9.2 Example of an Input Table.

actually used in the model calculations, which is in the column labeled "In Use," can be selected from the Low, Base, High, and Override columns by the integer 1, 2, 3, or 4 in the column labeled "Index." The cell containing the active setting of the input is given a range name, which is shown for the convenience of the analyst in the adjacent column labeled "Name." In the example shown in Figure 9.2, the range of uncertainty in the input for peak sales is specified by a low of 500, a base of 900, and a high of 1600. The override setting is shown as 2500. The input has the name "VolPeak" for peak volume and is currently set to the base amount of 900.

For an input variable that is not uncertain, such as a conversion factor, the input is entered in the Base column, and the Low and High columns are left blank. For an input representing a binary event, the Low and Base columns can be used for 0 and 1 to mean Failure and Success, or No and Yes, or False and True.

Structuring the Input Table in this way allows the user to do "what-if" calculations in the model easily by changing the index number of an input variable. The user can see whether or not the model is set to calculate the base case by observing if all the index numbers are set to 2. Finally, this structure of inputs facilitates the use of add-in software to perform sensitivity and probabilistic analysis. Decision tree software can change the entries in the "Index" column to specify a particular path through the tree, or Monte Carlo software can put into the "Base" column inputs that are randomly selected from the specified input probability distribution, while the "Index" is kept set to 2.

If the influence diagram specifies conditioning of one uncertainty upon another, or conditioning of an uncertainty on the strategy chosen, the Inputs sheet must contain all the conditional elementary distributions, as well as the logic to select the appropriate one for any given scenario.

9.5.7.1 Strategies. Representing decisions is challenging, because there can be multiple decisions, and we may want to understand the impact of each alternative for each decision on value, across time, business units, and value components. The first step usually taken is to define a handful of strategies, each of which specifies a set of choices across the decisions. This would seem to make it possible to employ an extensive treatment of strategies in decision models, maintaining a representation of all strategies simultaneously to support the identification of value difference between strategies, which is important in the value dialogue. However, this would make the model, which often has three dimensions of complexity already, even more cumbersome—for example, if there are six strategies, we would need to maintain "six copies of the model."

A better approach is to represent strategies intensively (i.e., only one at a time) and to iterate through them, caching key results as necessary for subsequent analysis. This is done by placing on the Inputs sheet an index variable specifying which strategy is currently evaluated in the model. All calculations in the model that differ by strategy must then have formulas that make use of the strategy index. When strategies are defined as combinations of choices for elemental decisions, it is useful to create in the model a *strategy table* that specifies for the currently selected strategy the choice for each elemental decision. The calculation formulas can then use as inputs the decision variables in the strategy table rather than the overall strategy index.

One can also display on the Inputs sheet the currently calculated utility or value. This makes it convenient for the user to see immediately the result of changes made to input variables.

9.5.7.2 Documentation Worksheet. Unlike the Inputs and Calculations sheets, the Documentation sheet is created exclusively for human eyes. It contains information that helps human readers more easily understand how the model works. It might contain any of the following:

- Instructions for use of the model
- Limitations of the model
- Explanations of model logic
- Assumptions underlying the model logic
- Formatting conventions.

It is good practice for the Documentation sheet to contain a history of the versions of the model. When the model is updated and a new version created, the history should record the date of the new version and what changes have been made to the model since the previous version was created. With such a history in place, an analyst can easily see what changes have been made to the model and when.

A related good practice is to increment the version number of the model frequently, particularly during the period when the model is undergoing initial development. Creating many versions of a model with relatively little change between successive pairs reduces the cost and pain if it is necessary to abandon the current version because of corruption or errors and fall back to a previous version. Also, whenever a model under development is shared with someone else, the person developing the model should immediately afterward increment the version number of the model. This is to avoid having two different versions of the model in existence with the same version number. And, of course, only one person at a time should have version control of the model. Remember that a decision practitioner's time is expensive and computer storage is very cheap.

9.6 Spreadsheet Model for the RNAS Illustrative Example

See Table 1.3 for further information on the illustrative examples. The model used in the analysis of the Roughneck North American Strategy (RNAS) example is distributed with this Handbook. The model uses Excel, VBA, and @Risk.

9.6.1 SELECTORS

The RNAS model has selector controls for four of the five dimensions of complexity. There is one selector for business units and also a selector that addresses both business units and objectives, because some objectives do not make sense for individual business units. The model allows specification of six strategies intensively in a strategy table. There are two strategy selector menus, which allow the analyst to select pairs of strategies among those six strategies for delta analysis. The names of strategies in the baseline selector are preceded by a minus sign, as a reminder that their value is to be subtracted out. This menu also includes an option for "none," which indicates that a direct tornado is desired. For uncertain inputs, the model allows overrides and has a selector that supports pseudo EV and the current simulated random scenario, but not base case. There is no selector for time because peak revenue is not of interest, and PVs are calculated and visible for every monetary time series.

9.6.2 INPUTS AND STRATEGY TABLE SHEETS

The RNAS Inputs sheet has the following characteristics:

- Instead of a selector index, there is a column that indicates which fractile of the input distribution is to be sampled.
- There is an "Override" column whose value is used if it is nonblank.
- In addition to Low P10, Base P50, and High P90 inputs, there are columns for P0 and P100 for some variables, to support sampling from a continuous distribution.
- There is a Pseudo EV column, which displays the EV of the input distribution.

The RNAS model represents strategies on the Strategy Table sheet. The strategies are in rows, decisions are in columns, and decision-specific codes in each cell indicate the choice to be instantiated in the simulation of that strategy. It is convenient to keep specifications of other possible strategies in rows below the ones being used for simulation, so that they can be copy-pasted into the active area easily, when desired.

9.6.3 CALCULATIONS SHEETS

The RNAS model has five P&L calculations sheets, one for each business unit (EnP, CBM, EOR, TarSands, and Power). On each P&L sheet, time periods are in columns, and value components are in rows. The model is laid out in drill-down fashion, with net cash flow at the top. The PV of each time series row is shown at the left. As we discuss in Chapter 11, EV results of these PVs under each of the recently simulated strategies are cached in rows or columns adjacent to the live PV (i.e., the one for the current strategy and uncertainty scenario). The Excel "Freeze Panes" feature is used to ensure that all this summary information is always visible.

Each P&L avoids the display of immaterial line items (by merging away detailed tax and depreciation calculations and intermediate sums and ratios) and highlights differences among the strategies (by splitting out business units' contributions). Because the resulting P&L structure was unfamiliar to the key stakeholders, we vetted this with them early in the process to ensure that it would be acceptable.

The RNAS model has a sheet named "PnLBrowserLive," which aggregates across business units or displays a selected one, for a selected strategy.

9.7 Debugging the Model

In a perfect world, decision models would be created without any faults. Unfortunately, of course, we live in a world that is far from perfect and decision models are often created with many unintended faults. So, an important step in the creation of any decision model is to find and correct those faults, a process called debugging.

We can define three levels of quality in a decision model

Level 1—The model calculates results (without crashing).

- The model is syntactically correct.
- The model is well-behaved over the full domain of inputs.

Level 2—The model calculations do what is intended.

- The model is logically correct.
- The model is a faithful translation of the planned structure.

Level 3—The model is well-constructed.

- The model conforms to accepted standards and conventions.
- The model is readable and easily transferable to someone else.

It is essential that any decision model achieve at least the first two levels of quality. Otherwise, it would not produce results that are useful for decision making. It is a best practice to achieve the third level of quality as well. A decision model that is easily understood by its users contributes much more value to decision making than one that is incomprehensible or not reliable.

Debugging a model is the repetition of two basic steps:

Diagnosis. Determining whether or not a bug exists and, if so, where it is and why it is a problem.

Treatment. Correcting the identified bug without creating a new bug somewhere else in the model.

Finding and correcting Level 1 bugs is generally straightforward. The spreadsheet software indicates clearly with an error message when it cannot calculate a result. Finding the bug is usually a matter of working "upstream" from the calculated value metric to the first instance of the error message. If the model is well laid out, that means working in a uniform direction on the sheet of calculations, moving from problematic results to the inputs creating them. The error message gives an indication of the type of error, so fixing the problem once it is located is usually easy.

A type of Level 1 bug that is harder to find is one that does not affect the calculation of base case results but causes an error when one or more of the inputs is set to a Low or High amount.

Finding Level 2 bugs (logic errors) takes sharp eyes and a keen mind. A good practice is to manually trace the calculations of the model from start to finish for one representative time period (i.e., going along one column of the Calculations sheet) to make sure that they do what is intended. For a business decision model, create and examine a financial summary chart showing revenue, gross profit, net income, and net cash flow to see if the model's behavior over time is satisfactory. Calculate and examine key financial ratios, such as net income per unit of sales volume, to see if these are sensible. This can be viewed as the first step of the value dialogue.

When the tornado diagram (see Section 9.8.2) is calculated, it can be used as a diagnostic tool to check on the model logic. Are the tornado bars correct in direction? For example, does value increase or does it decrease when a cost input is increased? Are the tornado bars sensible in magnitude? Is it OK if a tornado bar has zero width (i.e., changing the amount of the input variable does not affect the output value)?

Achieving Level 3 quality in a model (i.e., making it well-constructed) is much more a matter of observing good practices during the process of building the model than it is of fixing things after the model is built.

Note that debugging is the process of ensuring that the model is a faithful and useful representation of the planned analytic structure. It is different from the process of judging whether or not the analytic structure itself is appropriate.

9.8 Deterministic Analysis

Having developed a model that embodies the composite perspective of value creation, we can undertake a value dialogue in which we debug the model, develop insights, and find ways to improve upon our strategies. Analysis of the model's behavior facilitates this dialogue. The choice among ways to analyze should be governed by whether the resulting analysis will be:

- easy to implement
- understandable to the decision team
- clean, that is, not misleading in a material way
- likely to provide insights, and
- helpful for identifying constructive improvements to strategy.

This section discusses many types of analysis. Some of them maybe uninteresting in a specific decision situation, and others may not be applicable at all. Models in multiple-objective decision analysis (MODA) typically do not map out all dimensions explicitly, but rather require experts' judgments to take them into account implicitly. For instance, the time dimension is not explicitly treated in our data center case. Hence, there can be less need for collapsing dimensions of a MODA model. However, the critical comparison and value improvement steps are just as valuable in MODA as in financial analyses. The prudent analyst runs as many of these analyses as feasible, but shows only those that are insightful to the decision team.

If a promising new strategy is developed, we evaluate it alongside the others, giving a richer set of options among which to choose. We sometimes need to elicit additional expertise to do this.

Once the decision model has been fully developed, debugged, and tested, it should be used to generate insights about the decision. Four key questions in the value dialogue are:

1. What is the best course of action?
2. Why is it the best course of action?
3. How can we develop an even better course of action?
4. How sensitive is the best course of action to our assumptions?

We can understand where the value of a strategy comes from by scrutinizing its direct P&L, by sources-of-value analysis, and by direct sensitivity analysis of the specified strategy.

We can explore the sensitivity of optimal choice to assumptions via delta P&Ls, delta sources-of-value analyses, and delta sensitivity analysis, where "delta" refers to the difference of value between two specified strategies.

Although these questions can be answered completely only after a full probabilistic analysis is done (Chapter 11), a preliminary answer can be made based on the results of a deterministic analysis, which looks closely at one scenario. One form of deterministic analysis of alternatives is to look at their base case values. In a nonstage-gate decision situation, if one alternative has a base case value that is greatly inferior to other alternatives, a preliminary insight can be drawn that it is unlikely to be the best alternative.

In stage-gate situations, where failure is possible, base case value is not numerically similar to true EV because it does not reflect the probability of success, so these conclusions may not hold. In such cases, the rank ordering of strategies from pseudo EV is more likely to coincide with true EV results than is base case. Both base case and pseudo EV undervalue optionality, if there are future decisions conditioned on events whose probability cannot be calculated *a priori*.

9.8.1 SOURCES OF VALUE

The simplest step once a P&L is available is to set up a browser to review its results when one or more dimensions is collapsed, for example, when uncertainty is collapsed to a single deterministic scenario, or time series is summarized as a PV, or business units are aggregated. A P&L browser is a data structure that allows this to be done for some or all combinations of possible analytic simplifications.

One important analysis that can be drawn directly from a P&L is cash flow through time. While senior executives normally accept PV discounting, the results of time discounting and the rank ordering this gives to the strategies may not be immediately obvious or compelling to them. If strategies have distinctly different time profiles of cash flow, it can be helpful to show the cash flow profiles of the strategies together on one chart. This way we can ask the decision team:

- Do you believe the time profile for each strategy?
- Do you believe how the profiles differ from one strategy to the next?
- Do you affirm the value ordering implied by your stated time discount rate?

Another breakdown that is helpful is to see cash flow contributions by business unit. Figure 9.3 shows pseudo EV cash flow components for RNAS, showing that E&P gives near-term value, CBM medium-term value, and EOR long term value, after an investment. Power and Tar Sands are not major contributors.

Further insight into the decision can be created at this stage by doing an analysis of the sources of value. One form of this type of analysis is to show total PV for a specified strategy, disaggregated into value components in a waterfall chart. For example, suppose that total value is the sum of values contributed by different product lines. A waterfall showing the components of total value might reveal that some product lines contribute much less value than others, suggesting that a strategy that shifts resources from low-value product lines to high-value product lines might be a good choice. Figure 9.4 shows an example waterfall chart that displays the value contributions of five product lines. We can see from the waterfall that Product Line D contributes negative value, leading to the insight that we may be able to increase total value by deleting or modifying Product Line D. Sources of value analysis, of course, can be based on any other attributes, such as geographic region or market segment.

Another form of sources of value analysis that can lead to useful insights is to show the components of the *difference* in value between two alternatives. For example, suppose that a company is considering upgrading the technology of its production line to improve the yield rate and to lower energy costs. Three alternatives have been identified: (1) do nothing, (2) moderate investment, (3) high

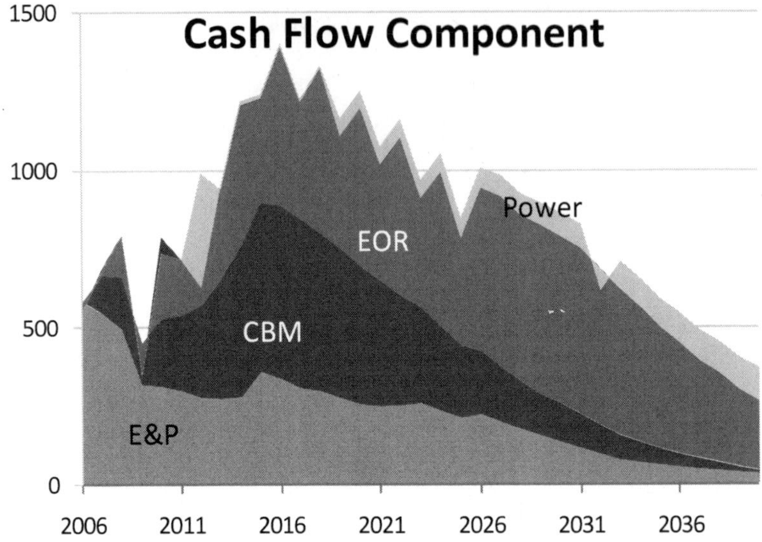

FIGURE 9.3 RNAS cash flow by business unit. (See insert for color representation of the figure.)

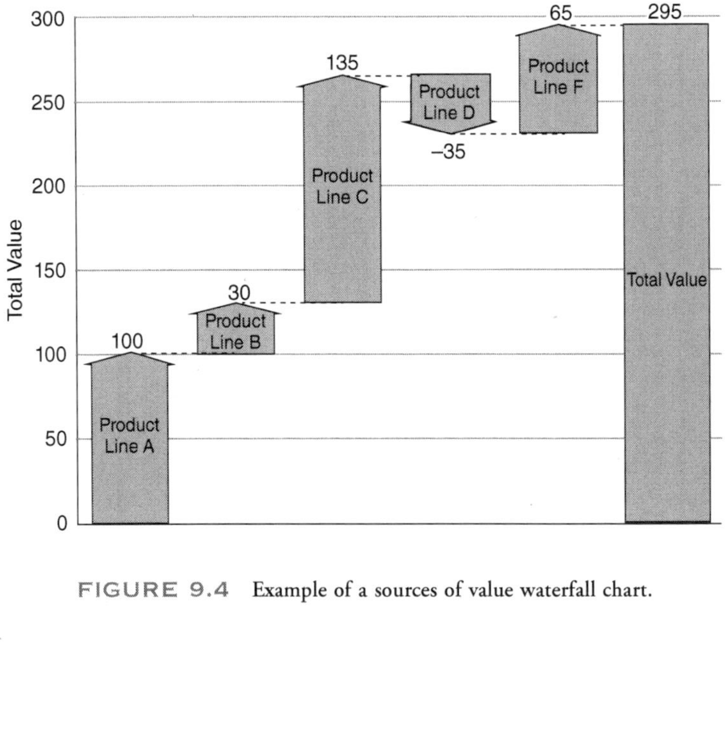

FIGURE 9.4 Example of a sources of value waterfall chart.

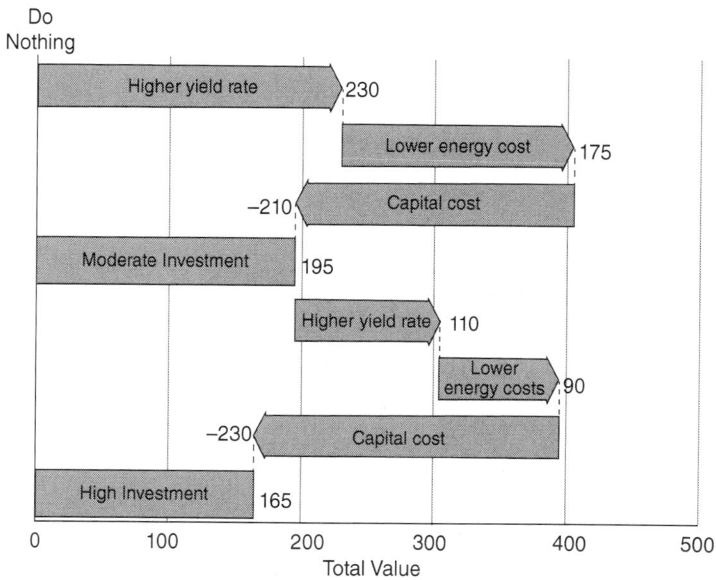

FIGURE 9.5 Waterfall chart of difference in value between alternatives.

investment. Figure 9.5 shows a waterfall chart that depicts how the value changes when going from one alternative to the next. The insight for the decision maker is that in going from Do Nothing (which has zero value) to Moderate Investment, the benefit in higher yield rate and lower energy costs easily outweighs the cost. But in going from Moderate to High Investment, the incremental benefit is not enough to justify the incremental cost, so the net incremental value is negative.

A waterfall chart displays value components for a selected strategy, or for the difference between two strategies, very clearly. Sometimes it can be useful to juxtapose value components for all strategies. A value components chart does this (see Fig. 9.6). In it, the value components of each strategy are shown in a pair of vertical stacked bars, where the negatives (investment, opex, taxes, etc.) start at zero in the left bar, and stack downward; and then the positives (revenues, etc.) are in the right bar, starting at the bottom of the negatives and working upward. The net height of the right bar shows the NPV value of the strategy. By juxtaposing all the value components, we enable comparisons across strategies. When these charts have large value components that are similar across strategies, a delta value components chart, which shows the delta of each strategy from a specified reference strategy, can sometimes show the unique characteristics of each strategy more clearly. This gives the equivalent of a waterfall difference chart for all strategies.

The first thing to do with the value components chart is to test the composite perspective it portrays:

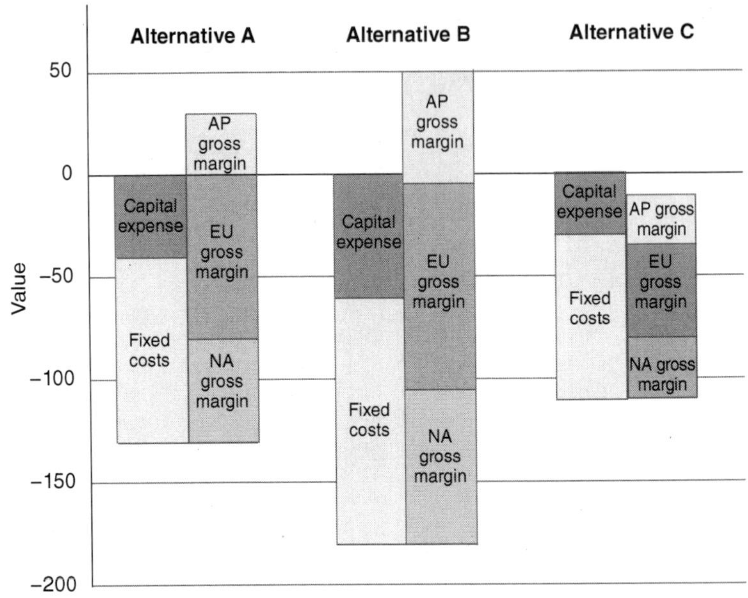

FIGURE 9.6 Example value components chart.

- Do we believe the value components for each strategy?
- Do we believe the relative value components across strategies?
- Do we believe the value ranking that emerges from the assessed trade-offs of the objectives?

Next, if value trade-offs are noted, here are some ways to use them to develop an improved strategy:

- Enhance a favorable item.
- Minimize or mitigate an unfavorable item.
- Optimize for a different value measure.
- Find a better compromise among the objectives implicated in the trade-off.

9.8.2 DETERMINISTIC SENSITIVITY ANALYSIS

The purpose of sensitivity analysis is to assess the impact on value of changing each uncertain input across its range of uncertainty. To calculate it, we must collapse all dimensions to a scalar value, normally by using a deterministic value (base case or pseudo EV), taking NPV across time, aggregating business units, and analyzing utility, or financial or MODA value. But it can also sometimes be helpful to look at a specific business unit or objective. The results of sensitivity analysis are displayed in a tornado diagram.

To conduct sensitivity analysis, we need the completed decision model and a set of preliminary ranges of uncertainty on all input variables. Getting the ranges of uncertainty for the sensitivity analysis calls for a careful balancing act. On the one hand, the ranges should represent the considered judgment of the decision-making team. On the other, it would be inappropriate to devote an enormous amount of time and effort to obtain high-quality ranges of uncertainty on the many input factors in a typical decision situation. Assuming the variable definitions have been clearly defined, we spend just a few minutes on assessing the range of uncertainty on each input factor for use in sensitivity analysis. One of the results of the sensitivity analysis is identifying which of the uncertainties really matter to the decision so that much greater attention can then be devoted to refining the ranges of uncertainty on them. In effect, sensitivity analysis is a form of *triage* in which we use results based on quickly assessed ranges to focus attention on the ranges that merit more careful assessment.

It is crucially important that the ranges of uncertainty used in the sensitivity analysis be defined consistently. The range of uncertainty for each input factor is generally specified by three numbers—low, base, and high. By convention, these are defined as the 10th, 50th, and 90th percentiles, respectively (see Appendix A, Probability). For binary uncertainties, such as events that may or may not occur, the base setting is defined as the more likely state, or in pharmaceuticals or oil and gas, as success. When obtaining these ranges of uncertainty, it is very important to avoid the uncertainty biases identified in Chapter 2. In particular, the anchoring bias should be avoided; appropriately wide ranges of uncertainty should be assessed.

Deterministic sensitivity consists of testing how a value metric varies when each input factor is varied across its range of uncertainty while all other input variables are held at their base or EV settings. Many software packages exist that automate these calculations. The result is a table of sensitivity analysis results and a corresponding tornado diagram. Figure 9.7 shows a typical tornado diagram for a decision alternative. Each tornado bar represents the amount of change in the value metric for an alternative as the input is varied across its range of uncertainty. The input variables are sorted in descending order of the width of the tornado bars so that the most important uncertainties are at the top of the tornado. A tornado diagram can be produced for each value metric of interest for each decision alternative. For example, we see in Figure 9.7 that as we vary the Peak sales input from the low of 500 to the high of 1600, total value varies from −2 to 67. Uncertainty in the Product life input has less of an impact on total value—the range for total value is smaller, going from 15 to 39.

To use a tornado diagram in the value dialogue, compare the large bars to the uncertainties that the decision team thinks should be important, and identify anything additional or missing in the tornado. Check the direction of change in value indicated by each bar against the team's intuitions. Should value increase or decrease as this input increases? Each point where intuitions differ from the result of the analysis constitutes a "tornado puzzle." Tornado puzzles are a great way to develop insights or to identify model errors. If a model result seems odd, investigate how it came about and determine whether this makes sense. If it

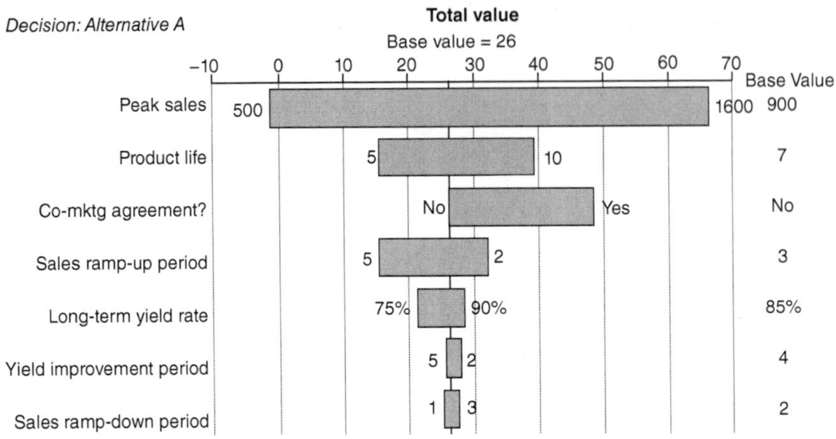

FIGURE 9.7 Example of a tornado diagram.

seems wrong, figure out how to improve the model so that it reflects the team's expertise more closely. If the result holds up, this is an insight for the team. Understand it clearly and explain it to them.

9.8.2.1 Difference Tornado. One form of tornado diagram that often provides great insight into the decision is the difference (or delta) tornado diagram. This is a tornado diagram showing the impact that uncertainties have on the difference of a value metric between two decision alternatives. If the value metric is the primary (or only) decision criterion, the difference tornado highlights which uncertainties have the power to switch the optimal choice depending on how they turn out. Typically, it is most constructive to investigate the delta between two leading strategies. Figure 9.8 shows an example of a difference tornado diagram.

Practitioners accustomed to using only direct tornados for a class of similar problems find them to be of limited use, because "the usual suspects" (e.g., the price of oil in an oil and gas project) are always at the top of the chart. It is frequently the case that these variables affect all strategies roughly equally. In such cases, the impact of the "usual suspects" largely cancels out in a delta tornado, allowing the variables that are important in the decision at hand to rise to the top.

A large bar in a delta tornado diagram that crosses zero (or comes close) indicates that additional information about this variable might cause us to change our decision. Accordingly, it can be worthwhile to convene experts to consider whether there is some way to gather information about any of the leading variables in the delta tornado diagram before the decision must be made.

Some spreadsheet software, such as Microsoft Excel, facilitates the calculation of difference tornados by providing a way to calculate an output seemingly

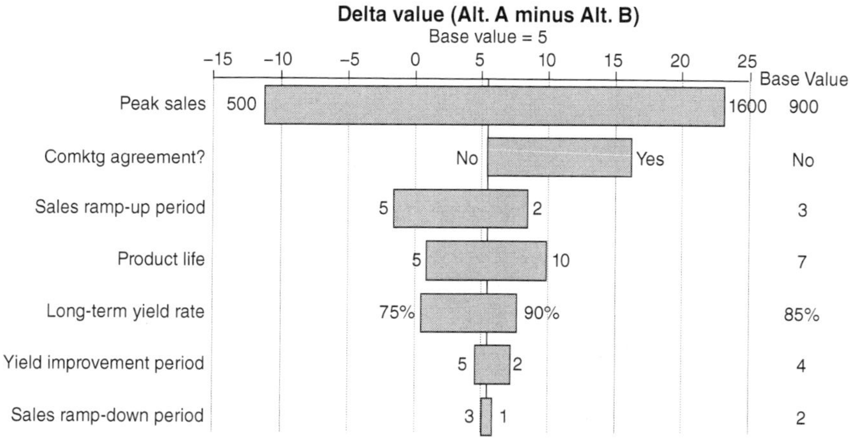

FIGURE 9.8 Example of a difference tornado diagram.

simultaneously for several different settings of the strategy selector input. (In Excel, this feature is called a Data Table.)

Another sort of deterministic sensitivity analysis is to test the impact of two key variables (e.g., the ones from the top of a difference tornado) on the optimal choice. To do this, we set up a table with one variable in the rows and the other in the columns, identify a range of plausible values for each variable, and test which strategy has the best deterministic (base case or pseudo EV) value in each cell of the table. If the strategies are color-coded, this table is called a rainbow chart. This analysis can also be done probabilistically, based on EV values, as illustrated in Section 11.3.5.

9.8.3 SCENARIO ANALYSIS

Just as a strategy is a fixed combination of decision choices, in this handbook, a "scenario" is a fixed combination of uncertainty outcomes. "Scenario Planning" (Kirkwood, 1996) investigates the behavior of a set of strategies under each of a small number of distinct scenarios, where scenarios are also understood to include a chronological story that gives coherence to the set of uncertainty outcomes, and where the scenarios are required to span the plausible range of outcomes for the most important variables. This approach lies on a continuum from no analysis at all to an analysis that considers multiple distinct uncertain factors. Its advantage over analysis with multiple distinct uncertainties is that fewer cases need to be considered. Its disadvantage is that the insights it generates are necessarily less fine-grained, and sometimes less actionable than insights emerging from evaluation with distinct uncertain factors. Accordingly, scenario analysis may help to identify one strategy that seems good, but it is less likely to suggest specific improvements to this strategy. However, scenario analysis can be used to provide insights to help design alternatives or portfolios of alternatives

(See Chapter 12) that are robust across all scenarios. The process of analysis is not fundamentally different—just compare cases to identify where value is created, and use this to understand strategies better, and to refine or improve them.

9.9 Deterministic Modeling Using Monetary Multidimensional Value Functions (Approach 1B)

We present in this section a simple illustrative example of a multiple-objective decision analysis that uses a value metric expressed in monetary terms. This is called Approach 1B in the taxonomy of decision analysis practice presented in Chapter 3.

Suppose that an industrial company has a decision to make regarding the level of technology to install in its new manufacturing facility. It has identified three possible alternatives: (1) current technology, (2) state-of-the-art technology, and (3) experimental advanced technology. The company prides itself on being both a financially successful business and a role model in socially responsible behavior. Accordingly, the two objectives for this decision are to maximize the impact on shareholder value (as measured by incremental NPV) and to minimize the environmental damage caused by the manufacturing operations (which in this case is largely in the form of carbon emissions). Deterministic evaluation of the alternatives has produced the results shown in Table 9.1.

Unfortunately, there is a direct trade-off in these alternatives between incremental NPV and the amount of carbon released, so the best choice is not clear from the results shown in the table. What is needed is a quantitative assessment of the value trade-off between NPV and carbon emissions. After much research, thought, and discussion, the decision makers agree that they are indifferent between increasing NPV by \$20 and reducing carbon emissions by 1 ton (and that this trade-off of \$20 per ton of carbon is valid for the full range of NPVs and levels of emissions relevant to this decision). This quantified trade-off allows the calculation of an overall value metric, expressed in monetary terms, for comparing the alternatives, as shown in Table 9.2.

The analysis now leads to the insight that the current technology alternative, which is best in purely financial terms, produces too much carbon to be optimal

TABLE 9.1 **Deterministic Results for Manufacturing Technology Example**

Alternative	NPV (\$ Million)	10-Year Carbon Emissions (Million Tons)
1. Current technology	\$1500	25
2. State-of-the-art technology	\$1300	10
3. Experimental technology	\$1000	2.5

TABLE 9.2 **Overall Value Metric for Manufacturing Technology Example**

Alternative	NPV ($ Million)	10-year Carbon Emissions (Million Tons)	Carbon Value ($ Million)	Total Value ($ Million)
1. Current technology	$1500	25	−$500	$1000
2. State-of-the-art technology	$1300	10	−$200	$1100
3. Experimental technology	$1000	2.5	−$50	$950

overall. And the experimental advanced technology alternative, although excellent in reducing carbon emissions, is not quite strong enough financially to be the best choice. So, at least based on deterministic analysis, the alternative with the highest overall value is the State-of-the-art technology. Of course, probabilistic analysis (see Chapter 11) may generate additional insights that could change the ranking of the alternatives.

9.10 Deterministic Modeling Using Nonmonetary Multidimensional Value Functions (Approach 1A)

The preceding section discusses the creation of a monetary value function for multiple objectives (Approach 1B in the taxonomy in Chapter 3). In some private applications and many public applications, it may not be possible or desirable to express all the value measures in monetary terms. In these applications, a nonmonetary value function (Approach 1A) may be used (Keeney & Raiffa, 1976).

For the problem of multiple and, usually, conflicting objectives, value-focused thinking (R. L. Keeney, 1992) recommends focusing first on the values or objectives that the decision is supposed to fulfill, rather than on the alternatives. Using multiple objective decision analysis, we develop a value model, which provides an unbiased, transparent, logical structure to give a numerical overall value for each alternative (see Chapter 3).[3] The model is made up of five parts: (1) an objectives or a functional value hierarchy that describes and organizes the objectives (see Chapter 7); (2) value measures that quantify each objective; (3) ranges for each of the value measures, from minimum acceptable (or

[3]This section draws on material in Parnell (2007) and Parnell et al. (2011).

available) to best possible (or achievable); (4) value functions that describe how value accumulates as one goes from low to high levels in each value measures; and (5) swing weights that specify the relative value of full-range swings in each of the different value measures. The value model must be based on preferences carefully elicited from the decision maker(s) and stakeholders. Value measures can be direct (best) or proxy, and they can be natural (best) or constructed, depending on the time and data available.

9.10.1 THE ADDITIVE VALUE FUNCTION

Multiple objective decision analysis can use any of several mathematical functions to evaluate alternatives. The simplest and most commonly used is the additive value function, which assumes mutual preferential independence (Kirkwood, 1997), which means that the assessment of the value function on one value measure does not depend on the level of the other value measures. For further detail, see Keeney & Raiffa (1976) or Kirkwood (1997). The additive value function uses the following equation to calculate the value of any combination of value measure levels

$$v(x)=\sum_{i=1}^{n} w_i v_i(x_i), \tag{9.1}$$

where for a set of value measure levels given by vector x,

$v(x)$ is the alternative's value of x

$i = 1$ to n is the index of the value measure

x_i is the alternative's score of the ith value measure

$v_i(x_i)$ = is the single-dimensional y-axis value of an x-axis score of x_i

w_i is the swing weight of the ith value measure

and

$$\sum_{i=1}^{n} w_i = 1 \tag{9.2}$$

(all weights sum to one)

When developing value functions, there are a variety of words used to define the "x-axis" and "y-axis" that describe the value curve.[4] In particular, some references use the word "score" to represent the x-axis measure, while others use it to represent the value on the y-axis. To standardize our terminology, we use the term "score" to represent the level of the value measure portrayed on the x-axis

[4]Other terms for $v(x)$ are value score, score, utility, and benefit. Other terms for the score on x_i are level, value, attribute, performance, and data.

(e.g., Car *X* gets 49 MPG fuel economy) and "value" to represent our strength of preference for that score on the *y*-axis (e.g., on a scale of 0–100, fuel economy of 49 MPG is valued at 90).

MODA quantitatively assesses the trade-offs between conflicting objectives by evaluating an alternative's contribution to the value measures (a score converted to value by single-dimensional value functions) and the importance of each value measure (swing weight). As an important technical note, the swing weights must be on a ratio scale (with an absolute 0), but the value can be on an interval scale or a ratio scale. When interval scales are used for the value functions, 0 value does not necessarily mean no value. Instead, it means the minimum acceptable value on each value measure. Because the same equation in the additive value function applies to all alternatives, no index is required for the alternatives.

9.10.2 SINGLE-DIMENSIONAL VALUE FUNCTIONS

Value functions measure returns to scale on the value measures. They are usually monotonically increasing (decreasing) for value measures aligned with a maximizing (minimizing) objective. The value functions can be discrete or continuous and can have any shape. However, in practice there are four basic shapes: linear, concave, convex, and an S-curve (Fig. 9.9 for increasing value). The linear value function has constant returns to scale: each increment of the measure score is equally valuable. For increasing value measures, the concave value function has

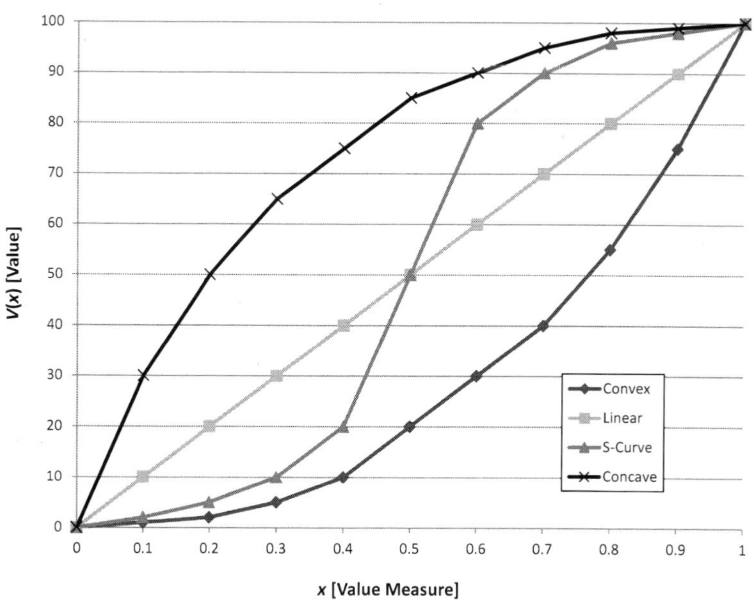

FIGURE 9.9 Four types of value functions for increasing value.

decreasing returns to scale: each increment is worth less than the preceding increment. For increasing value measures, the convex value function has increasing returns to scale: each increment of the measure is worth more than the preceding increment. For increasing value measures, the S-curve has increasing, then decreasing, returns to scale on the measure. The S-curve is sometimes used to model leadership goals.

We occasionally see value functions that first rise and then fall (nonmonotonic parabolic shape). This often happens when assessors combine two measures rather than keeping them separate and independent. For example, when assessing the value of the number of bedrooms in a house, we may hear that value increases up to 5 bedrooms, and then starts to decrease. When we ask why, we hear that there are more rooms to clean, air conditioning bills will be higher, and so on. These are legitimate trade-offs, but a better practice is to keep the benefits and costs of the number of rooms as separate value measures.

We have several techniques to assess the value functions using the preferences of experts (the decision makers and/or stakeholders) (Kirkwood, 1997). It is important to note that the experts may or may not be able to directly provide the value functions. In fact, they may not even understand the concept of returns to scale without some discussion and a couple of examples. Being able to explain a value function and help experts assess credible value functions is an important interviewing soft skill. The following are some useful approaches based the authors' experience.

1. The first, and most important, step is to carefully define the value measure whose value is being assessed. If we think of a value curve being expressed on a value measure on the *x*-axis and how we value that measure on the *y*-axis, it is critical that the *x*-axis be carefully defined. For example, suppose we want to develop a value curve for memory capacity of a laptop computer. When we say memory capacity, do we mean RAM or ROM? Does memory have to be internal, or is external memory acceptable? Is there a minimum acceptable level below which we will not go? If we do not answer questions such as these carefully, our value functions will not reflect what we really value. In addition, if we are going to do uncertainty analysis (see Chapter 11), we need to make sure that the range of the *x*-axis includes the range of uncertainty.
2. Once the measure is defined, we must then decide upon the units of value. Although the units can be any range, the three most common ranges are 0.0–1.0, 0–10, and 0–100. Since people assess value, a reasonable level of precision in assessment may be one significant digit, for example, 0.5, 5, or 50, but it is common to use two significant digits. Beyond two digits is precision without accuracy.
3. The range of each value measure must be specified. The range of the value measure will impact the swing weights and the shape of the curve (a value function may be convex in a small range and an S-curve in a larger range). There are two approaches to setting the ranges that depend

on the focus of the study. The alternative-focused thinking (AFT) approach recommends narrow ranges on value measure scores to use the full value range to help distinguish the alternatives. Often, the value measure score of the best alternative is made the top of the value scale, and the value measure score of the worst alternative is made the bottom of the value scale. The value-focused thinking (VFT) approach recommends a wide range that encourages the development of new alternatives to achieve the ideal score of the value measure.[5] The disadvantage of the AFT approach is that it does not encourage the development of better alternatives. The disadvantage of the VFT approach is that it usually results in unused value that does not differentiate the alternatives. With the introduction of the decision-focused transformation (DFT) (Dees et al., 2010), the analyst can have the best of both techniques.[6] The analysis can begin with VFT and then use DFT to transform the value space (similar to AFT) to distinguish the alternatives.

4. The two most general approaches to assessing value functions are (1) to assume a curve shape (e.g., the exponential or the S-curve) and assess the defining parameter(s) such as inflection points, or (2) to assess points on the curve and fit a curve through the points.
5. An important best practice is to get the expert(s) to agree on the shape of the single-dimensional value function and the rationale for the returns to scale shown in the curve *before* any points are assessed. This is useful because stakeholders or more senior decision makers may challenge the preference judgment of the expert(s), and the expert(s) may not be present for this discussion. A wise decision analyst records the rationale given by the expert(s) for the shape of the value function.
6. Two ways of assessing points on the value curve are absolute and relative assessments. One absolute assessment technique asks the question, "What score on the value measure provides *Y*% of the value?" Or the question, "What percent of the value do I get for *Z*% of the value measure score on the *x*-axis?" We can also divide the value measure range into several increments using a technique called bisection. A second absolute technique would be to make ratio judgments (increment 1 is *Z* times as valuable as increment 2). A relative assessment technique[7] would make relative value judgments, for example, increment 1 has greater value than increment 2 and increment 3. With enough preference assessments, and knowing that the value increments must sum to the total value, we can define a value curve. This technique is used in the balance beam approach method (Buede, 2000).

[5]The AFT approach is sometimes called "local," and the VFT approach is sometimes called "global."

[6]Approaches similar to DFT are incorporated into several MODA software packages.

[7]There is some behavioral decision analysis research that demonstrates that relative judgments are easier for experts than absolute judgments.

An Example of a Constructed Scale Assessment

When natural measures do not exist, we often use constructed scales. Frequently, the x-axis on such scales is adjectival, and the "distance" between points has no meaning. One approach that has been used very successfully is as follows. Assume we are trying to develop a 0-to-100 value curve for the water-crossing ability of a light armored vehicle (LAV).

1. Define the minimal acceptable threshold below which we would not consider an alternative no matter how well it did on other measures. That minimal threshold could be zero, the current capability, or anything else. For the LAV, we set the minimal acceptable level at "no water crossing capability" and assigned a y-axis value of 0 to that capability.
2. Define the ideal capability that sets an upper bound and assign it a value of 100. It is fine to use "stretch goals" as the ideal, but it is not helpful to set it at a point that is clearly far beyond that which is feasible. For the LAV, we set the value of 100 at "ability to swim unassisted across deep (>6′) rivers with rapid currents (>5 mph)." Note the importance of clearly specifying the definitions of x-axis points.
3. Since it is often difficult to achieve the ideal, we define a point that "backs off" a bit from the ideal by asking for the definition of the capability that might achieve a 90 or 95 value. This has the added benefit of not forcing the user to select the extreme point of the scale, yet still meet most of the need. For the LAV, we assigned a value of 90 to a capability that allows for the LAV to swim across deep (>6 ft) rivers with rapid currents (>5 mph) with the assistance of a "swim kit" that must be attached and activated before use. Note that a value of 90 does not imply that the capability satisfies 90% of what is needed, but rather, it achieves 90% of the value from the bottom to the top of the scale. This allows for the bottom of the scale to be greater than a "true zero."
4. We then define several intermediate points to complete the value curve. Experience has shown that fewer than three to five total points do not provide sufficient discrimination, while more than nine points is overkill. For the LAV, we established two additional scale points:
    - **a.** The ability to ford small streams (<1 ft depth, 15 ft across) was valued at 10.
    - **b.** The ability to ford large streams (1 ft ≤ depth < 3 ft, 25 ft across) was 40.
    - **c.** The ability to swim with a swim kit across small rivers (3 ft ≤ depth ≤ 6, >25 ft across) was 80.

9.10.3 SWING WEIGHTS

Swing weights play a key role in the additive value model. The most common mistake in MODA is assessing weights without taking into account the specific ranges of value measure scores under consideration. Kirkwood (Kirkwood, 1997) provides a mathematical proof of this statement. The following story has helped many people understand swing weights.

Explaining Swing Weights: Parnell's Car Buying Example

Recently my wife and I decided to consider buying a car. I wanted to buy an SUV with awesome off-road capability. My wife wanted to buy a minivan to transport our children and grandchildren. Once we agreed that we wanted to buy a minivan, we talked about what criteria we should use. The criteria we selected were cost, safety, performance, and comfort.

Before we assign swing weights we have to define the range of the value measure scores for each criterion. The swing weight we assign to each measure depends on the importance (an intuitive assessment) of the "swing in range" of the measure (a factual assessment).

Let us consider safety. Suppose we measure safety using a 5-star scale and we assign a value of 0 to a safety score of 1 star and a value of 100 to 5 stars. The variation in this measure, from 1 to 5 stars, represents a significant difference in the likelihood of personal injury in an accident (a factual judgment). Given this variation in safety, we would say that safety is very important to us because our family will be in the vehicle (intuitive importance assessment). Therefore, we would assign a high weight to safety since the measure has high importance given the significant "swing" (1 to 5 stars means bottom 20% to top 20%).

Suppose we think about it some more and decide to eliminate from consideration 1- and 2-star vehicles (the bottom 40% by safety rating). Clearly, our intuitive assessment of the importance of safety has not changed but the range of the measure has been reduced from 1 to 5 stars to 3 to 5 stars. So if we now assign a value of 0 to 3 stars and keep a value of 100 for 5 stars, we would then assign less weight to safety than before since we are guaranteed to buy at least a 3-star vehicle.

Finally, suppose we think some more and decide to consider only vehicles with 5-star safety ratings. Our importance assessment has not changed but now there is no variation in safety rating because we have made the 5-star safety rating a screening criterion. Therefore, we would assign a swing weight of 0 to safety since there is no longer any "swing" in safety in the decision.

In conclusion, we always assess weights based on the swings in the measure range.

Kirkwood (1997) and Clemen and Reilly (2001) describe swing weight-assessment techniques for individuals. One common way to assess weights from a group of experts is to use voting to obtain ordinal and then cardinal weights:

1. Vote. (Have each individual spread 100 points over the value measures based on the measures' importance and range.)
2. Discuss significant differences. Have the "outliers" discuss their rationales.
3. Revote until the group agrees on the ordinal ranking of the value measures.
4. Vote again requiring each person's weights to follow the group's ordinal ranking of the value measures.
5. Average the weights (cardinal ranking of weights) and normalize so they sum to one.
6. Discuss significant differences. Have the "outliers" discuss their rationales.
7. Repeat steps 4–6 until the group agrees.

If disagreements about the weights cannot be resolved, record them. Then do a sensitivity analysis during evaluation to determine if the disagreements are significant. Often, the preferred alternatives are not sensitive to the evaluated weight range. Unfortunately, this weighting technique is not useful for explaining the rationale for the weights assigned. The technique we recommend is the swing weight matrix.

9.10.4 SWING WEIGHT MATRIX

The swing weight matrix[8] was designed to help decision makers and stakeholders understand the importance and the impact of the value measure range on the decision. The swing weight matrix defines importance and impact of the range of the value measures in the decision context. The idea of the swing weight matrix is straightforward. A measure that is very important to the decision should be weighted higher than a measure that is less important. A measure that differentiates between alternatives, that is, a measure in which value measure ranges vary widely, is weighted more than a measure that does not differentiate between alternatives. The first step is to create a matrix (Table 9.3) in which the top row defines the value measure importance scale and the left side defines the impact of the range of value measure.[9] The levels of importance and variation should be thought of as constructed scales that have sufficient clarity to allow the analyst to uniquely place every value measure in one of the cells. In this example, mission

[8]The swing weight matrix was developed by Parnell in 2003 and refined by colleagues in the U.S. Army and Innovative Decisions, Inc.

[9]A 3-by-3 matrix is the most common of a larger (or smaller) number of columns and/or rows may be used for decision with a large number (or small) of value measures. Also, some decision analysts like to put A in the upper right-hand corner.

TABLE 9.3 **The Elements of the Swing Weight Matrix**

		Importance of the Value Measure to the Decision (Intuitive Judgment)		
		Critical (High)	Important (Medium)	Nice to Have (Low)
Impact of the range of the value measure on the decision (factual judgment)	Large impact	A	B2	C3
	Medium impact	B1	C2	D2
	Small impact	C1	D1	E

critical is the highest importance, mission enabling is the middle level of importance, and mission enhancing is the lowest level. A measure that is very important to the decision and has a large measure range would go in the upper left of the matrix (cell labeled A). A value measure that has low importance and has small variation in its scale goes in the lower right of the matrix (cell labeled E).

9.10.4.1 Consistency Rules. Since many individuals may participate in the assessment of weights, it is important to insure consistency of the weights assigned. It is easy to understand that a very important measure with a high variation in its range (A) should be weighted more than a very important measure with a medium variation in its range (B1). It is harder to trade off the weights between a very important measure with a low variation in its range (C1) and an important measure with a high variation in its range (B2). Weights should descend in magnitude as we move on the diagonal from the top left to the bottom right of the swing weight matrix. Multiple measures can be placed in the same cell with the same or different weights. If we let the letters represent the diagonals in the matrix A, B, C, D, and E, A is the highest weighted cell, B is the next highest weighted diagonal, then C, then D, and then E. For the swing weights in the cells in Table 9.3 to be consistent, value measure in a given cell must have a greater weight than a value measure in any cell to the right or below the given cell.

9.10.4.2 Assessing Unnormalized Swing Weights. Once all the value measures are placed in the cells of the matrix, we can use any swing weight technique to obtain the unnormalized weights as long as we follow the consistency rules cited above. In assigning weights, the stakeholders need to assess their trade-offs between importance and impact of the value measure scale. Again, we can use absolute or relative assessments. One absolute assessment technique would be to assign the measure in cell A (the upper left-hand corner cell) an arbitrary large unnormalized swing weight, for example, 100 ($f_A = 100$). Using the value increment approach (Kirkwood, 1997), we can assess the weight of the lowest weighted measure in cell E (the lower right-hand corner) the appropriate swing weight, for example, 1. This means the swing weight of measure A is 100 times more than that of measure E. It is important to consider what the maximum in cell A should be. Common choices are 1000 and 100. Of course, f_E can be other numbers besides 1. If we use 100 and 1, we have three orders of magnitude. If we use 1000 and 1, we have four orders of magnitude. Using a value increment approach, unnormalized swing weights can be assigned to all the other value measures relative to f_A by descending through the very important measures, then through the important measures, then through the less important measures.

A relative assessment technique for swing weights is the balance beam method (Buede, 2000). This technique uses relative judgments such as "going from the lowest to the highest score on measure 1 is equivalent to going from the lowest to the highest scores on measure 2 and measure 4." With $n - 1$ assessments (since the weights must sum to 1), we can solve the set of linear equations for the appropriate swing weights.

9.10.4.3 Calculating Normalized Swing Weights.

We can normalize the weights for the measures to sum to 1 using this equation.

$$w_i = \frac{f_i}{\sum_{i=1}^{n} f_i}, \tag{9.3}$$

where f_i is the unnormalized swing weight assessed for the ith value measure, $i = 1$ to n for the number of value measures, and w_i are the normalized swing weights from Equation 9.1.

9.10.4.4 Benefits of the Swing Weight Matrix.

We believe this method has six advantages over traditional weighting methods. First, it develops an explicit definition of importance that forces explicit consideration of impact of the measure range. Second, the consistency rules help ensure consistent swing weight assessments. Third, the matrix helps to reduce the number of measures. Suppose cell A has an unnormalized weight of 100 and cell E has an unnormalized weight of 1. It is very obvious that any measure that is placed in cell E will not impact the decision and does not need to be included in the analysis. In practice, this has resulted in significant reduction of the number of value measures. Fourth, it provides a simple yet effective framework to present and justify the weighting preference. Fifth, the approach is very flexible. When measures are added we perform one more assessment for each measure and renormalize. When measures are deleted, all we have to do is renormalize. Finally, swing weights make it easy to communicate a complicated concept to decision makers.

9.10.5 SCORING THE ALTERNATIVES

In this chapter, we assume we know the score of each value measure with certainty. In Chapter 11, we discuss uncertainty analysis. Once we have vetted the quantitative value model and developed alternatives, we must score the alternatives on the value measures. In addition to scoring our alternatives, we should include the current (or baseline) alternative and the ideal (or perfect) alternative. It is important to note that the ideal alternative may not be achievable due to conflicting objectives. Some analysts like to use an operational or realistic ideal as the benchmark. Here, we use the ideal alternative. In practice, the development of the value model and the scoring is an iterative procedure. Many times, the value model has to be revised if scores are not available for a planned value measure or if scorers identify a missing value measure. In fact, to capture this concept, we usually say that "no value model ever survives first contact with the alternatives.[10]"

> No value model ever survives first contact with the alternatives.

A major purpose of value-focused thinking is to generate better alternatives. Therefore, alternative scoring has two purposes: scoring alternatives and generating better ones. The second purpose is often more important! When we begin to score our alternatives, we identify value gaps—chances to improve the alternatives (create better scores) to achieve higher value. Chapter 11 provides more information on improving the alternatives.

There are five primary sources of scores: operational data, test data, simulation, models, and expert opinion. Typically, we will have operational data on the current products and services.

It is prudent to consider who will score the alternatives and how disagreements will be resolved. Four scoring approaches have been particularly successful: performance models, alternative champions, a scoring panel, and alternative champions reviewed by a scoring panel.

9.10.5.1 Scoring by Performance Modeling. In many decision analysis problems, the modeler scores some alternative value measures by using external performance models. For example, if modeling alternative locations for warehouses, there are so many parameters to consider that it is impossible for a panel to derive a score. Rather, value measures may be scored through simulation model outputs.

9.10.5.2 Scoring by Alternative Champions. The champion of each alternative scores his or her alternative independent of the others. This approach is useful because it sends information about values from the value function directly to alternative "champions" as they do the scoring. A disadvantage is the perception that a champion of an alternative may bias a score to unduly favor it or that scores from different champions will be inconsistent.

[10]The second author coined this phase after the famous military adage—no plan ever survives first contact with the enemy.

9.10.5.3 Scoring by a Panel. To avoid the perception of bias and potential inconsistencies, we can use scoring panels. Two types have proven useful. In the first type, we convene a panel of subject matter experts to score and improve the alternatives. Alternative champions present scoring recommendations to the panel, but the panel assigns the final score. In the second type, we have experts score and the champions review the scores. Experts for each value measure score all alternatives being considered (usually with their rationale for the score) and submit it to the analysis team to consolidate. The analysis team then vets the scores with the project champions. The champions usually disagree with something and have a chance to: (1) change the expert's mind with new data, (2) change the alternative so that it scores better (thereby improving the alternative), or (3) modify when an inconsistency is noticed. We have found virtual panels to be the best approach in large distributed organizations.

9.10.5.4 Scoring by Alternative Champions Reviewed by a Panel. Having the champion score the alternative and modify it to create more value is the essence of value-focused thinking. A review panel can then ensure the scores are unbiased and consistent.

Once we have scores, we can start evaluating the alternatives—typically through deterministic analysis and uncertainty (or risk) analysis (Chapter 11).

9.10.6 DETERMINISTIC ANALYSIS

Several types of analysis are useful to obtain insights about the alternatives, and many software packages have built-in features to do them. The following analyses will be illustrated in the data center location example.

9.10.6.1 Value Components. Stacked bar graphs are a useful way to compare alternatives. Usually, the "stacks" show the contribution for each value measure. However, we can plot the stacked bar graphs for any level in the hierarchy. If we do, then analysis usually begins top down to identify insights.

If there are trade-offs of one objective or value component versus another, here are some ways to use them to develop an improved strategy:

- Enhance a favorable item.
- Minimize or mitigate an unfavorable item.
- Optimize for a different value measure.
- Find a better trade-off between the objectives.

9.10.6.2 Value Gaps. Value gaps are one of the key insights we can extract from value component charts. They are the delta between the best alternative and the ideal alternative for each value measure. We can examine them at all levels in the value hierarchy, so they "shine a light" on areas for value-focused thinking.

9.10.6.3 Value versus Cost. When using MODA, many decision analysts and decision makers find it useful to separate cost and benefits (value), typically by plotting the value versus the cost of the alternatives. This chart helps decision makers identify the dominant alternatives and enables them to see the value added for the additional cost of the dominant alternatives.

9.10.6.4 Waterfall Chart. A waterfall chart is a visual depiction of the value contribution between two or more nondominated alternatives.

9.10.6.5 Sensitivity Analysis. Sensitivity analysis is useful for key parameters, including weights and scores. Performing weights sensitivity analysis can be done using either the assessed weights or the normalized swing weights. In either case, the weights must be normalized over the sensitivity analysis range. The usual assumption is that the weights not subject to sensitivity analysis remain in the same relative proportion. Two sensitivity analysis plots are typical—a rainbow diagram and a tornado diagram

9.11 Illustrative Examples

See Table 1.3 for further information on the illustrative examples. The RNAS example is discussed in Section 9.8. This section illustrates the application of deterministic analysis techniques to the other two examples, Geneptin personalized medicine and the data center location decision. The spreadsheet models for all three examples are available on the Wiley website.

9.11.1 GENEPTIN

The Geneptin team constructed a deterministic financial model in Excel to evaluate the strategic alternatives against the primary objective, ENPV.

The modeling was spearheaded by the decision professional, who interacted with the core team regularly to facilitate team discussions and drive consensus on the strategic alternatives, analysis/evaluation framework, and recommendations on the development strategy.

To determine the proper modeling relationships and help structure model input assumptions, the Geneptin team used an influence diagram to map out the relationships between different parameters (see Fig. 9.10).

The deterministic model was developed using a cascade of calculations, considering specific personalized medicine-related parameters, such as biomarker prevalence, patient test rate, and so on. DNA Biologics employed influence diagrams to reflect this relationship. Figure 9.11 illustrates a "drill-down" influence diagram for Drug Market Share, showing additional considerations that went into its assessment.

Based on the deterministic model, sensitivity analysis was done, enabling the team to understand the drivers of value and identify the need to reexamine

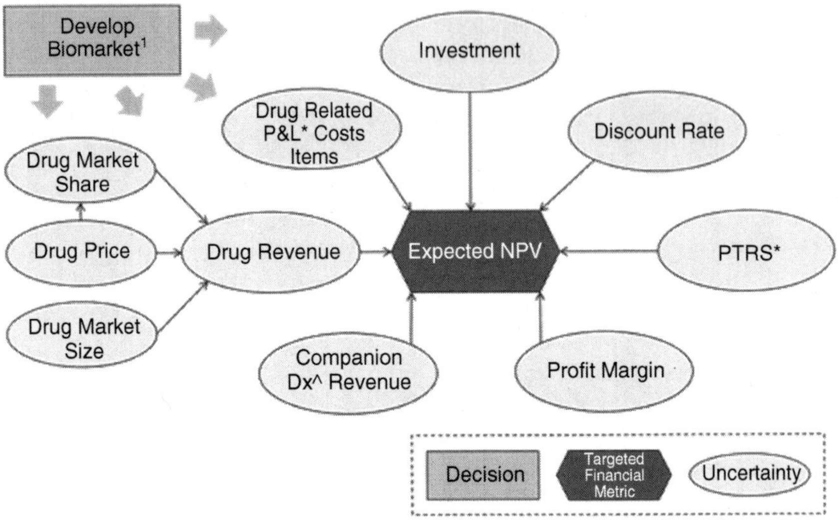

FIGURE 9.10 Geneptin influence diagram. [1]Decision to develop biomarker will impact all uncertainties shown except Discount Rate. *Probability of Technical and Regulatory Success; ^Diagnostic. [&]Profit and loss.

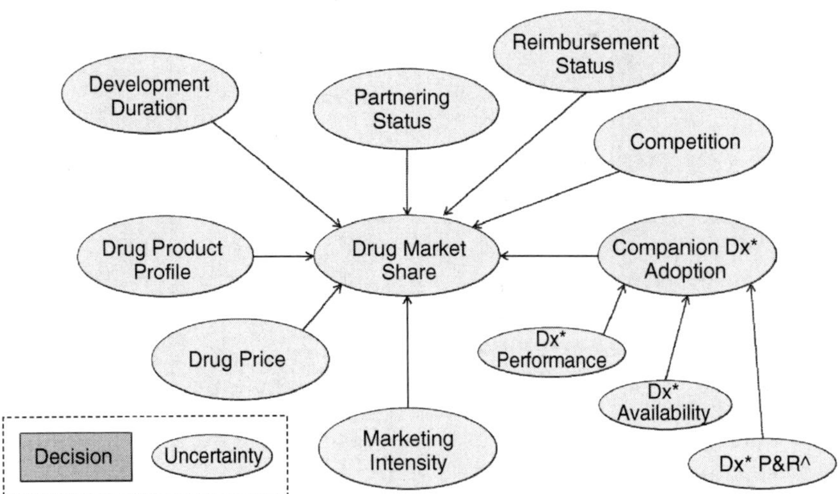

FIGURE 9.11 Geneptin drill-down ID for market share. *Diagnostic; ^Pricing and Reimbursement.

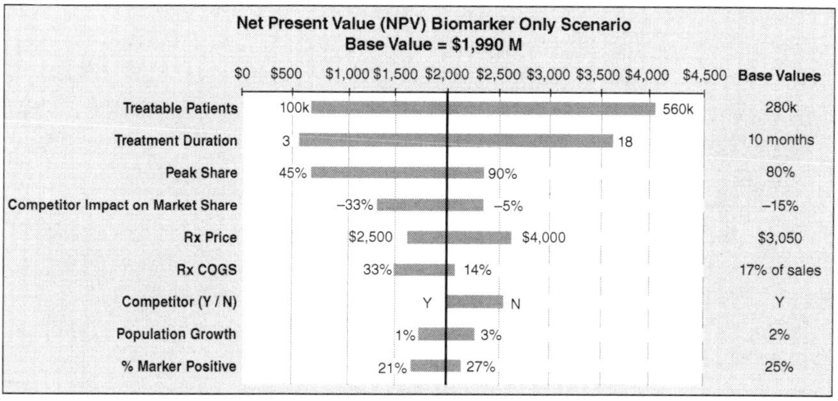

FIGURE 9.12 Geneptin tornado diagram.

assumptions that were significant value drivers. The results are shown in a tornado chart for the personalized medicine strategic alternative (see Fig. 9.12).

9.11.2 DATA CENTER LOCATION

We illustrate deterministic MODA using the data center location qualitative value model that we describe in Chapter 7. The functional value hierarchy is repeated in Figure 9.13 for ease of reference.

9.11.2.1 Additive Value Model. As this point we need to determine if we meet the preference independence assumptions of the additive value model (see Chapter 3). We constructed the data center value hierarchy to meet the preferential independence requirements.

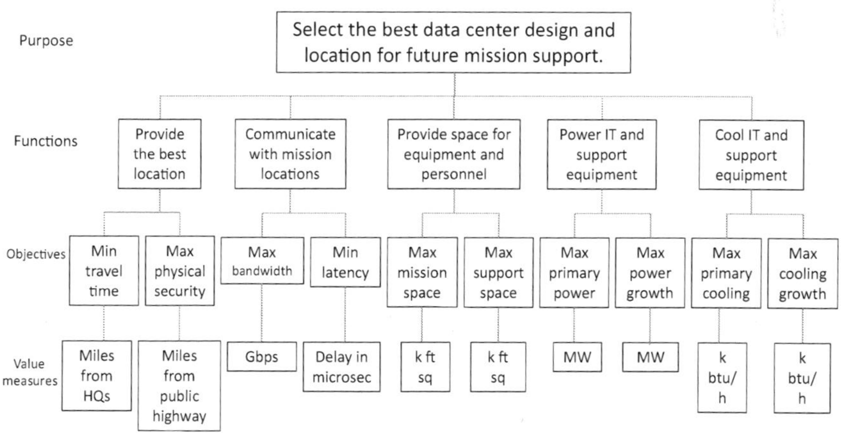

FIGURE 9.13 Data center location functional value hierarchy.

9.11.2.2 Decision Analysis Software. The mathematical calculations shown in this example were done in Excel. For the real problem with 40 measures, we used Logical Decisions (Logical Decisions, Fairfax, VA). INFORMS does a biennial survey of decision analysis software (Buckshaw, 2010) that includes several other decision analysis packages that do MODA. The selection of Excel or a software package depends on several factors including: client preferences, analyst preferences, analyst experience with the packages, funding availability, and time to perform analysis.

9.11.2.3 Value Functions. We began by working with stakeholders and SMEs to develop value functions for each of the 10 value measures. Note that we had previously developed the range of the x-axis since this was required to assess the swing weights. However, the value functions are not required for the swing weights. The value functions were now developed for each value measure. Table 9.4 summarizes notional data for the value functions that we used in this example. For example, for the function to "provide power for IT and facility," we identified two objectives: maximize primary power and maximize the potential power growth. For the maximize primary power objective, the value measure was megawatts. The single-dimensional value function for this measure was linear, ranging from 50 to 90 MW. For the maximize potential power growth objective, the value measure was again megawatts and the single-dimensional value function was linear, ranging from 25 to 45 MW. Although this example includes only natural measures, many problems require the development of constructed measures.

9.11.2.4 Swing Weight Matrix. Once the range has been determined for each value measure, we can assess the swing weights using the swing weight matrix (Section 9.10.4). The art of the swing weight matrix is defining the importance categories (top row) and the impact of the variation (first column) in terms that make sense to the decision makers and the stakeholders. The importance was defined in three categories: critical regional characteristics (e.g., the primary power sources), important site features (e.g., miles from public highway), and features that were fixable with additional funding (e.g., building more floor space or funding longer travel). The impact of site variation was defined in three categories: significant impact (e.g., primary power from 50 to 90 MWs), some impact (e.g., growth power from 25 to 45 MWs), and minor impact (miles from HQ since the facility is designed to be a lights-out facility and the primary travel will be to support the initial equipment installation which are relatively low dollars). Once the value measures are placed in the cells, the unnormalized weights (f_i) are assessed with absolute or relative preference judgments. For this study, absolute judgments were used. The normalized weights (w_i) are calculated using the equation in Section 9.10.4.

9.11.2.5 Scoring the Alternatives. In our illustrative example, we assume that eight notional locations passed the screening criteria in Chapter 7 and were scored by the decision team. The scores were obtained from SMEs who were

TABLE 9.4 **Data Center Single-Dimensional Value Functions**

Function	Objective	Value Measure	Min Acceptable Level	Ideal Level	Curve Shape	Rationale
Provide the best location	Min travel time	Miles from headquarters	2500	500	Linear	Each increment is equally valuable
	Max physical security	Miles from public highway	1	5	Linear	Each increment is equally valuable
Communicate with mission users	Max bandwidth	Bandwidth (Gbps)	7	10	Linear	Each increment is equally valuable
	Min latency	Latency (microsec)	300	100	Concave	Low latency is more valuable
Provide space for equipment and personnel	Max mission space	Mission k ft^2	150	200	Linear	Each increment is equally valuable
	Max support space	Support k ft^2	30	50	Concave	Initial increase is more likely to be required
Provide power for IT and facility	Max primary power	Megawatts (MW)	50	90	Linear	Each increment is equally valuable
	Max power growth	Megawatts (MW)	25	45	Linear	Each increment is equally valuable
Cool IT and facility	Max primary cooling	k BTU/h	205	369	Linear	Each increment is equally valuable
	Max growth cooling	k BTU/h	102.5	184.5	Linear	Each increment is equally valuable

assigned the task of scoring each alternative for each value measures. For example, a power expert determined the current primary power and potential power growth for each site. Table 9.6 lists the alternatives and provides the scores.

9.11.2.6 Single-Dimensional Value Calculations. Next, we use the scores and the single-dimensional value functions to calculate the single-dimensional value for each alternative for each value measure (Table 9.7). For example, a bandwidth of 9 Gbps has a value of 67 on the linear value function. We used a macro from Kirkwood (Kirkwood, 1997) that does a piecewise linear interpolation between the assessed points on the value function curve. It is important to include the value calculations for the ideal as a check to insure that the value calculations are always 100. The theoretical ideal has value here for verification of the mathematics and can be a guide to improvement of alternatives even if it is not achievable.

9.11.2.7 Normalized Swing Weight Calculations. Next, we calculate the normalized swing weights. These are the w_i columns from Table 9.5. We usually put them in a row in Excel so we can easily do our next calculation. Table 9.8 provides the cell referenced weights and a check to verify that they add to one per Equation 9.2.

9.11.2.8 Alternative Value Calculations. The next calculation we need to make is the value calculation for each alternative using the additive value model in Equation 9.1. First, for each alterative, we multiply the swing weight by the value for each measure score. Second, we add up the weighted values for each of the 10 value measures to obtain the total value of each alternative. The results of the first calculation are shown in the first 10 quantitative columns in Table 9.9, and the second calculation is shown in the table's last column.

9.11.2.9 Value Components. We can use Table 9.9 to plot a stacked bar graph to show the contribution for each value measure. Figure 9.14 shows the data center value components chart for total value. We typically show these charts in color, which makes it easier to differentiate the contributions of each value measure. The Ideal alternative is always shown for reference. The height of each ideal bar is equal to the swing weight × 100. For example the primary power is 27. Although the data are the same as Table 9.9, it is easier to see the contribution of each value measure to the value of each alternative in the chart. Washington State has the highest total value, followed by Tennessee and Texas. Several insights are shown is this chart. First, Washington State scores the best on several value measures, including the highly weighted power (primary power and growth power) and cooling (primary and growth) value measures. The salience of power leads us to ask whether we could create another alternative to capture this benefit, for example, located near cheap power sources (Niagara hydropower, onsite with a minemouth coal-fired plant). The Tennessee and Texas scores are very similar for all the value measures.

TABLE 9.5 Data Center Swing Weight Matrix

	Critical Regional Characteristic			Important Site Feature			Fixable with Dollars		
	Value Measure	f_i	w_i	Value Measure	f_i	w_i	Value Measure	f_i	w_i
Significant impact of site variation	Primary power MW	100	0.27	Primary k BTU/h	50	0.13	Bandwidth (Gbps)	25	0.07
							Mission k·ft^2	20	0.05
Some impact of site variation	Growth power MW	50	0.13	Growth kBTU/h	40	0.11	Support k·ft^2	15	0.04
Minor impact of site variation	Latency (microsec)	40	0.11	Miles from public highway	30	0.08	Miles from HQ	1	0.00
				Sum of f_i	371				

TABLE 9.6 **Data Center Scores on Each Value Measure**

Site Location	Miles from HQ	Miles from Public Highway	Bandwidth (Gbps)	Latency (Microsec)	Mission k ft^2	Support k ft^2	Primary Power MW	Growth Power MW	Primary k BTU/h	Growth kBTU/h
California	2450	3	7	300	150	30	50	25	205	102.5
Florida	800	5	8	300	160	30	55	27.5	225.5	112.75
Kentucky	450	7	9	200	170	30	55	27.5	225.5	112.75
Tennessee	600	10	9	100	190	45	60	30	246	123
Texas	1400	5	9	100	190	40	60	30	246	123
Washington State	2200	10	9	200	190	40	80	40	328	164
West Virginia	300	7	8	200	180	30	55	27.5	225.5	112.75
Wyoming	1530	10	8	200	170	30	60	30	246	123
Ideal	500	10	10	100	200	50	90	45	369	184.5

TABLE 9.7 Data Center Single-Dimensional Value Calculations for Each Value Measure

Site Location	Miles from HQ	Miles from Public Highway	Bandwidth (Gbps)	Latency (Microsec)	Mission k ft^2	Support k ft^2	Primary Power MW	Growth Power MW	Primary k BTU/h	Growth k BTU/h
California	2.5	50	0	0	0	0	0	0	0	0
Florida	85	100	33	0	20	0	12.5	12.5	12.5	12.5
Kentucky	102.5	100	67	80	40	0	12.5	12.5	12.5	12.5
Tennessee	95	100	67	100	80	90	25	25	25	25
Texas	55	100	67	100	80	80	25	25	25	25
Washington State	15	100	67	80	80	80	75	75	75	75
West Virginia	110	100	33	80	60	0	12.5	12.5	12.5	12.5
Wyoming	48.5	100	33	80	40	0	25	25	25	25
Ideal	100	100	100	100	100	100	100	100	100	100

TABLE 9.8 **Data Center Normalized Swing Weights**

	Miles from HQ	Miles from Public Highway	Bandwidth (Gbps)	Latency (Microsec)	Mission k ft^2	Support k ft^2	Primary Power MW	Growth Power MW	Primary k BTU/h	Growth k BTU/h	Total
Normalized Swing Weight, wi	0.00	0.08	0.07	0.11	0.05	0.04	0.27	0.13	0.13	0.11	1

TABLE 9.9 **Data Center Weighted Value and Total Value Calculations**

Site Location	Miles from HQ	Miles from Public Highway	Bandwidth (Gbps)	Latency (Microsec)	Mission k ft^2	Support k ft^2	Primary Power MWh	Growth Power MWh	Primary k BTU/h	Growth k BTU/h	Total Value
California	0.0	4.0	0.0	0.0	0.0	0.0	0.0	0.0	0.0	0.0	4
Florida	0.2	8.1	2.2	0.0	1.1	0.0	3.4	1.7	1.7	1.3	20
Kentucky	0.3	8.1	4.5	8.6	2.2	0.0	3.4	1.7	1.7	1.3	32
Tennessee	0.3	8.1	4.5	10.8	4.3	3.6	6.7	3.4	3.4	2.7	48
Texas	0.1	8.1	4.5	10.8	4.3	3.2	6.7	3.4	3.4	2.7	47
Washington State	0.0	8.1	4.5	8.6	4.3	3.2	20.2	10.1	10.1	8.1	77
West Virginia	0.3	8.1	2.2	8.6	3.2	0.0	3.4	1.7	1.7	1.3	31
Wyoming	0.1	8.1	2.2	8.6	2.2	0.0	6.7	3.4	3.4	2.7	37
Ideal	0.3	8.1	6.7	10.8	5.4	4.0	27.0	13.5	13.5	10.8	100

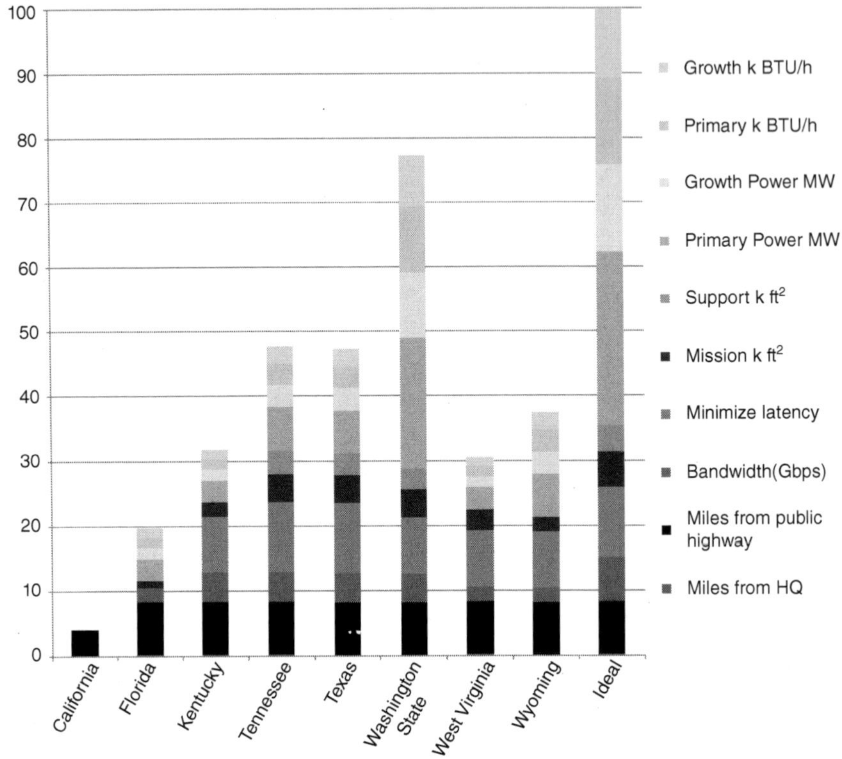

FIGURE 9.14 Data center value components chart. (See insert for color representation of the figure.)

9.11.2.10 Value Gaps. Value gaps are easy to see from the stacked bar graph. They are the delta between the best alternative and the ideal alternative. An alternative way to plot the value components chart is to show the individual bars for each measure together. The value gaps identify areas where we can attempt to improve the alternatives.

For example, the largest potential value increase for Washington State would be to increase the primary power from 75 to 100 MW. This would increase the value by 6.8 out of 100.

9.11.2.11 Value versus Cost. Since so many salient benefits were essentially financial, a life cycle cost model was used to evaluate the costs of each site over a common time period. Mathematically, the cost of the alternatives could be one of the objectives and value measures. However, many decision analysts and decision makers find it useful to plot the value versus the cost of the alternatives and look for dominated alternatives. Decision analysts use different costs for different decisions. For example, if the operating and support cost would be about the same for the alternatives the analyst might use the procurement cost.

TABLE 9.10 **Data Center Life Cycle Cost and Value for Each Alternative**

	Life Cycle Cost ($M)	Value
California	800	4.0
Florida	900	19.7
Kentucky	500	31.7
Tennessee	400	47.8
Texas	600	47.3
Washington State	750	77.3
West Virginia	700	30.6
Wyoming	750	37.4
Ideal	0	100.0

However, for the data center decision, procurement costs will be different, and the operating and support costs of the alternatives depend on several cost factors that depend on the site, for example, the power costs and the security costs. Therefore, for the data center decision, we calculated the life cycle cost (LCC) of each alternative. The life cycle cost factors were land acquisition (purchase or rental), construction/rental of buildings, power, cooling, communications, fire suppression systems, security costs, transportation, and facility and equipment maintenance. Table 9.10 provides the LCC data and the total value of each alternative.

A plot of this data (Fig. 9.15) helps us quickly identify the dominated alternatives and helps decision makers to see the value added for the additional cost of the dominant alternatives. In our data center problem, we see that Tennessee and Washington State dominate the other alternatives. Assuming that all significant factors are included in the analysis, we would not recommend a dominated alternative since we would be paying more money for less value compared with one of the dominant alternatives. As analysts, we would now focus on analysis of the two nondominated alternatives.

9.11.2.12 Waterfall Chart. Next, we focus on Washington State and Tennessee. Tennessee has a value of 47.8 and Washington State has a value of 77.3, almost a 30-point increase. A waterfall chart (Fig. 9.16) is a visual depiction of the components of the difference in value between the two nondominated alternatives. Tennessee is better than Washington State for miles from headquarters, latency, and support space. However, Washington State is significantly better than Tennessee for primary power, growth power, primary cooling, and growth cooling. For example, we see a 13.5 point advantage that Washington State has over Tennessee in Primary Power MW. This is calculated from the 50-point difference in values between them (75 for Washington State, 25 for Tennessee) multiplied by the swing weight of Primary Power (0.27)

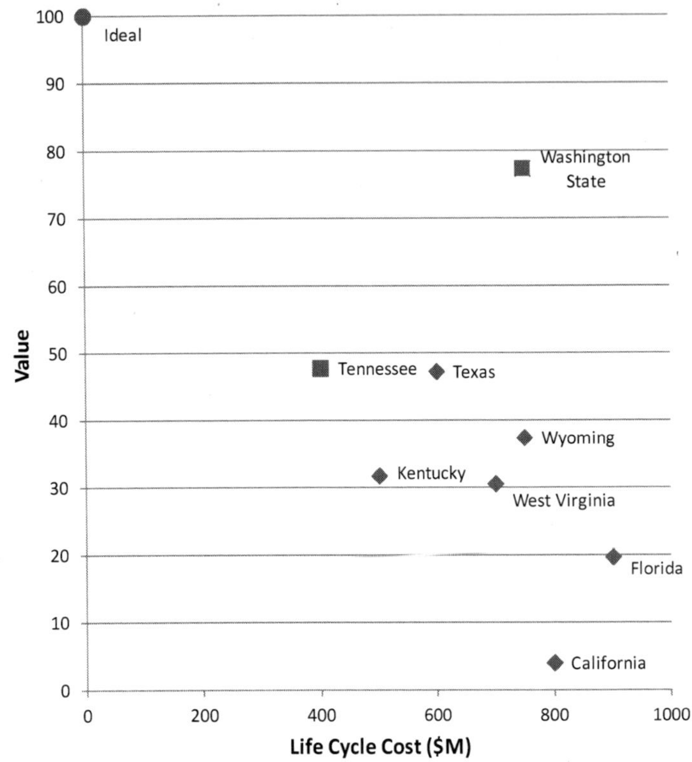

FIGURE 9.15 Data center cost versus value plot. Some analysts reverse the x-axis so best (cheapest) is on the right.

9.11.2.13 Sensitivity Analysis. Similar to the difference tornado diagram, we use sensitivity analysis to determine if our preferred alternative is sensitive to a modeling assumption. We can perform sensitivity analysis to any parameter. In MODA, sensitivity analysis can be performed to swing weights, value curve shapes, and scores. The most common sensitivity analysis is to swing weights. When we vary one swing weight, we must decide what changes to make to insure that the swing weights sum to 1. The standard assumption is that the remaining $n - 1$ swing weights remain in the same proportion. For example, Figure 9.17 shows the sensitivity analysis for the unnormalized swing weight assigned to latency.[11] The original latency unnormalized swing weight was 40. In Figure 9.17, we vary the unnormalized swing weight from 0 to 100. Since Washington State has the highest scores in many of the value measures, it is not sensitive to

[11]We performed sensitivity analysis to an unnormalized swing weight since this was the judgment that we made and varied the weight from 0 to 100. Some analysts prefer to perform sensitivity analysis to the normalized swing weight and vary the weight from 0 to 1.

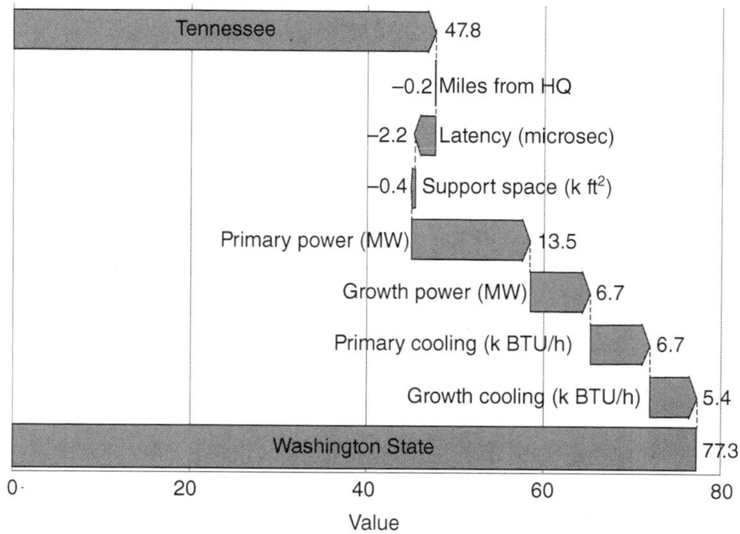

FIGURE 9.16 Data center waterfall chart.

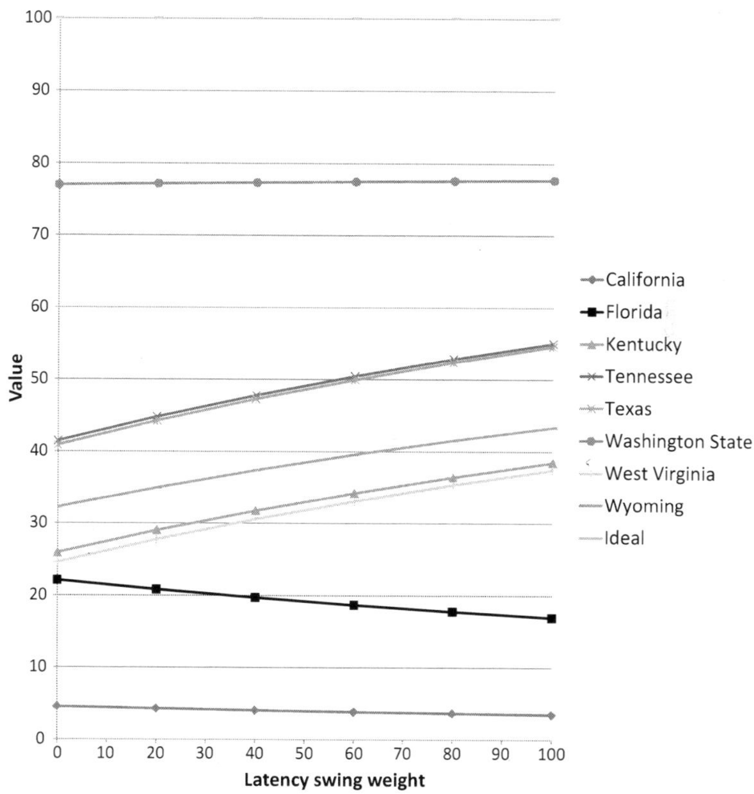

FIGURE 9.17 Data center sensitivity analysis for latency unnormalized swing weight without change in preferred alternative.

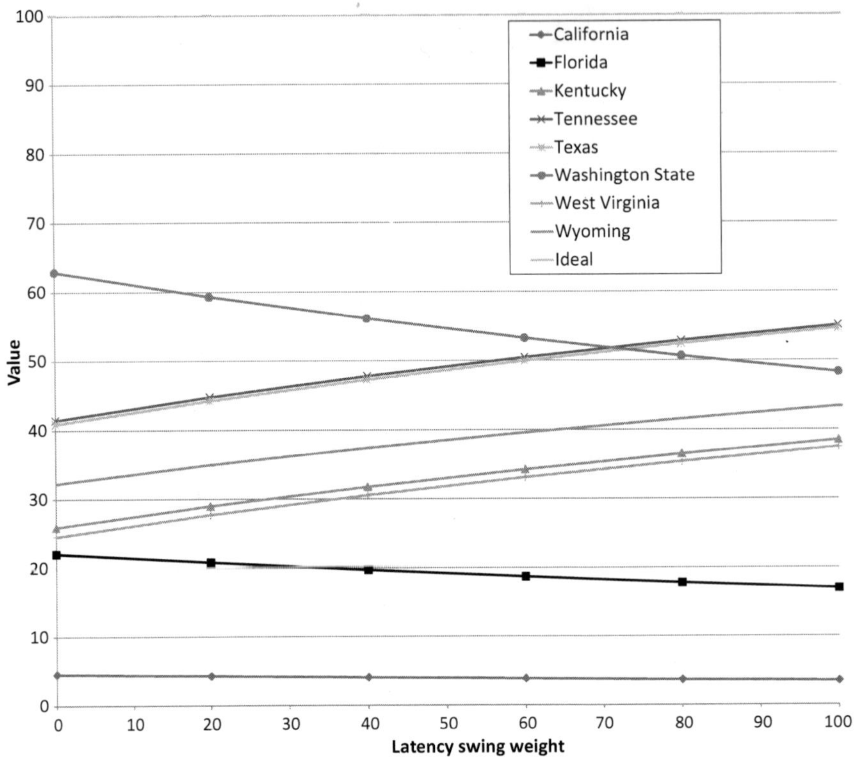

FIGURE 9.18 Data center latency swing weight sensitivity with change in the preferred alternative.

the weight assigned to latency or any other swing weight. In general, the preferred alternative will not be sensitive to the weights.

However, suppose we change the following Washington State scores: miles from public highway from 10 to 1, bandwidth from 9 to 7, and latency from 200 to 300. Figure 9.18 shows the sensitivity analysis for the unnormalized swing weight for latency after this change in scores. The preferred data center location is now sensitive to the latency weight. The original latency weight was 40. If the latency weight is less than 70, Washington State is the highest value data center location. However, if the weight is greater than 70, Tennessee is the preferred data center location.

9.11.2.14 Value-Focused Thinking. Value-focused thinking was used extensively in the data center location problem. The scores and costs presented above were *after* VFT. The following actions were taken to improve the scores and costs:

- Added higher value alternatives. Initially, there were two alternatives. Several alternatives were added. For example, the Washington alternative was included to get closer to the ideal.

- Negotiated with location owners to provide most favorable features and costs especially on the Tennessee alternative.
- Negotiated with service providers (e.g., local power companies) to obtain the most favorable power capabilities and costs especially on the Tennessee alternative.

9.12 Summary

Our decision analysis approach to evaluating complex alternatives is to develop and refine a composite perspective on how value is created, and to use this to identify even better ways to create value. After eliciting the expertise comprising the composite perspective, the next step is to embody it in a model, and analyze it to facilitate a value dialogue with experts, stakeholders, and decision makers. This chapter presents modeling and deterministic analysis. An influence diagram is an excellent graphical tool to use to identify the variables, the decisions, and their relationships to the value metric. Spreadsheets are a useful platform for decision models. We provide our guidelines for building a spreadsheet decision model and offer suggestions for organizing a spreadsheet model. For financial variables, the P&Ls addressing business units, time and value components should usually be extensively represented. Often decisions and uncertainties should be represented intensively (in strategy table and probabilistic Inputs sheets), enabling subsequent simulation and analysis. For complex models, we need to verify the calculations so we describe approaches for debugging a spreadsheet model. Next, we present important tools for deterministic analysis, including the tornado diagram and difference tornado diagram for deterministic sensitivity analysis and the waterfall chart for displaying the sources of value for an alternative or the differences in value for several alternatives. We present deterministic multiple-objective decision analysis using the most common model, the additive model. We present the mathematics and art of MODA. Finally, we illustrate MODA deterministic modeling and analysis using the data center problem.

KEY TERMS

Analysis Characterizing the emergent behavior of a simulation in terms that allow direct comparison to individuals' perspectives and generation of improved strategies. cf. Modeling, Exploration.
Base case The scenario with all continuous input parameters set at their median level, and all discrete variables set to most likely outcome.
Boolean algebra Representation of logic with numbers: 0 for false; 1 for true; addition for OR; multiplication for AND.
Business unit Segment of the entire enterprise that the decision team may want to view in isolation. This can apply not only to geographic or functional

business units that are reported distinctly in an organization's financials, but also to opportunities or projects that the team wants to consider on their own.

Clean Not misleading in any material way.

Composite perspective Perspective on value creation that is created by stitching together individual experts' perspectives. This is often represented in an influence diagram and then on the Inputs sheet of a spreadsheet model.

Conditioning We represent an expert's perspective on an uncertainty by a probability distribution. If the expert specifies different distributions depending on the outcome of some other event, this is probabilistic conditioning. Likewise, we intend for decisions to be made so as to optimize expected utility. If the choice can be made in different ways depending on the outcome of some prior event, this is decision (or informational) conditioning.

Decision analysis Creating a composite perspective on how well each alternative will meet one's objectives and create value, and using it to choose among alternatives and create even better alternatives.

Deterministic analysis Calculating results for a specific setting of all inputs (i.e., without regard to uncertainty in the inputs). cf. Probabilistic analysis.

Difference or delta tornado diagram Representation of the impact of individual variables' variation on the difference between value measures of two selected strategies. If the value measure is the primary (or only) decision criterion, the difference tornado highlights which uncertainties have the power to switch the optimal choice depending on how they turn out. cf Direct Tornado Diagram.

Dimensions of complexity The five dimensions considered here are the value components, alternatives, uncertainty, business units, and time.

Direct tornado diagram Representation of the impact of individual variables' variation on a value measure, under a selected strategy. cf. Delta tornado diagram.

Elementary distribution Distribution of an uncertainty for a specified conditioning case.

Emergent behavior Behavior of a system that is entailed by its governing principles, but not obvious, even to individuals familiar with those principles.

Expected value (EV) Sum (or integral) of outcomes multiplied by their probabilities (or probability densities).

Expertise Well-developed understanding of events pertinent to a decision.

Exploration Instantiating possible outcomes based on a model, to allow the behavior of value in response to different choices and outcomes of uncertainties to emerge. cf. Modeling, Analysis.

Extensive representation Addressing all possible cases explicitly simultaneously. cf Intensive Representation.

Functions The activities that have to be performed by a system. For complex system decisions, functions may be identified before identifying the objectives. Alternative terms are capabilities, missions, or tasks.

Influence diagram A compact graphical representation of conditioning relationships among uncertainties and decisions in a perspective on a decision situation.

Insight Conclusion about how value is created that can be reached only by systematic review of the behavior of a model.

Intensive representation Representing all cases, but only one at a time. cf. Extensive Representation.

Live Of or relating to the scenario and strategy currently instantiated.

Mathematical model A complex problem is decomposed into smaller, simpler pieces that the human brain can comprehend and then put back together in an analytic structure that is used to evaluate alternatives.

Modeling Building representations of experts' perspectives about important events, stitching them together to create a composite perspective on value creation, and representing this in a computer model. cf. Exploration, Analysis.

Net present value (NPV) The sum of a sequence of cash flows through time, with each multiplied by a discount factor that represents the value today of one monetary unit in its time period.

Objective A desirable outcome to be achieved in a decision. Example: maximize profit.

Override (verb) Directly specify a value for an uncertain input, rather than allowing it to be simulated. (noun) The value specified in this way.

Probabilistic analysis Analysis that considers results from many possible instantiations of uncertainty, and aggregates them according to their probability. cf. Deterministic analysis.

Pseudo EV The deterministic value resulting from all inputs being at their EV values; with success/failure variables treated as Boolean (so that their EV is their probability) and multiplied by their consequences in the model.

Qualitative value model A description of our qualitative values, including the purpose of the value model, functions (if used), objectives, and value measures. cf. Quantitative value model.

Quantitative value model A mathematical model that includes value functions, weights, and mathematical equation (such as the additive value model) to evaluate the alternatives. cf. Qualitative value model.

Range of a value measure The maximum variation of the scores of a value measure. For example, the value measure availability might have a range of 0.7 (minimum acceptable) to 1.0 (ideal).

Requisite decision model A model whose form and content are sufficient to solve a particular problem.

Score (level) The specific numerical rating of the value measure, such as an availability of 0.8. A score may be on a natural or a constructed scale.

Sensitivity analysis Analysis that assesses the impact of changes in a parameter on value of an alternative, or difference of value between two alternatives.

Stage-gate process Process that comprises multiple stages, with a decision between each stage whether to proceed to the next one. This typically reflects success or failure of an investigation.

Strategy A fixed set of choices, one for each decision.cf Scenario.

Swing weights The swing weight assigned a value measure depends on the measure's range. Weights are our relative preference for value measures. They must sum to one.

Tier (layer) A level in the value hierarchy.

Tornado diagram A deterministic sensitivity analysis plot that shows the impact on an alternative's value of changing parameters from their low to high settings.

Utility The fundamental value measure to be maximized, regardless of whether it was generated via MODA or ENPV calculations. If there is no explicit risk preference adjustment, this term refers to the value measure or ENPV that is used for decisions under uncertainty.

Value components The numbers that are combined via addition or subtraction to calculate overall value. In MODA, these are the contributions of the individual objectives. In financial decisions, this term can refer either to the PV of line items (revenues, salaries, operating expenses, investment, taxes, etc.), or the NPV value contributed by business units.

Value function A function that assigns value to a value measure's score. Quantitatively, value is defined as returns to scale on the value measure.

Value hierarchy (value tree) A pictorial representation of the qualitative value model.

Value measure Scale to assess how much we attain an objective. Alternative terms are attribute, evaluation measure, measure of effectiveness, measure of merit, and metric.

Value model The structure used in a decision analysis to calculate the value of any alternative in any specified scenario.

Waterfall chart A waterfall chart shows the sources of value for an alternative.

REFERENCES

Buckshaw, D. (2010, October). Decision Analysis Software Survey. http://www.informs.org/ORMS-Today/Public-Articles/October-Volume-37-Number-5/Decision-Analysis-Software-Survey, accessed October 20, 2011.

Buede, D. (2000). *The Engineering Design of Systems*. New York: John Wiley & Sons.

Clemen, R.T. & Reilly, T. (2001). *Making Hard Decisions with Decision Tools*. Belmont, CA: Duxbury Press.

Dees, R.A., Dabkowski, M.F., & Parnell, G.S. (2010). Decison-focused transformation to improve communications. *Decison Analysis*, *7*(2), 174–182.

Howard, R.A. (1983). The evolution of decision analsysis. In R.A. Howard & J.E. Matheson (eds.), *Readings on the Principles and Applications of Decision Analysis*, Vols. 1 and 2. Palo Alto, CA: Strategic Decisions Group.

Howard, R.A. & Matheson, J.E. (2005). Influence diagrams. *Decision Analysis*, *2*(3), 127–143.

Keeney, R. & Raiffa, H. (1976). *Decision with Mulitiple Objectives: Preference and Value Tradeoffs*. New York: John Wiley & Sons.

Keeney, R.L. (1992). *Value-Focused Thinking: A Path to Creative Decision Making*. Cambridge, MA: Harvard University Press.

Kirkwood, C. (1996). *Strategic Decision Making*. Belmont, CA: Duxbury Press.

Kirkwood, C. (1997). *Strategic Multiple Objective Decision Analysis with Spreadsheets.* Belmont, CA: Duxbury Press.

Logical Decisions. (n.d.). http://www.logicaldecisions.com/, accessed October 20, 2011.

Parnell, G.S. (2007). Value-focused thinking. In A. Loerch & L. Rainey (eds.), *Methods for Conducting Military Operational Analysis*, pp. 619–657. Washington, DC: Military Operations Research Society.

Parnell, G.S., Driscoll, P.J., & Henderson, D.L. (2011). *Decision Making for Systems Engineering and Management*, 2nd ed. Hoboken, NJ: John Wiley & Sons.

Phillips, L.D. (1984). A theory of requisite decision models. *Acta Psychologica, 56*, 29–48.

Powell, S.G. & Baker, K.R. (2011). *Management Science: The Art of Modeling with Spreadsheets*, 3rd ed. Hoboken, NJ: John Wiley & Sons.

Savage, S. (2009). *The Flaw of Averages*. Hoboken, NJ: John Wiley & Sons.

CHAPTER TEN

Quantify Uncertainty

ERIC R. JOHNSON and STEVEN N. TANI

Do not expect to arrive at certainty in every subject which you pursue. There are a hundred things wherein we mortals . . . must be content with probability, where our best light and reasoning will reach no farther.

—Isaac Watts

Handbook of Decision Analysis, First Edition. Gregory S. Parnell, Terry A. Bresnick, Steven N. Tani, and Eric R. Johnson.

10.1 Introduction

Our aim is to help decision makers identify choices that can add substantial value when facing complex, uncertain, novel, high-stakes decision situations. Our decision analysis approach is to develop a composite perspective on how value is created based on the best available expertise, and to employ it in a value dialogue with experts, stakeholders, and decision makers aimed at providing insight to decision makers to aid their decision making.

We develop the composite perspective probabilistically, but not in the spirit of resignation indicated by the Isaac Watts quote above. Like Watts, we acknowledge that perfect knowledge is almost never available; but we rejoice in the availability of expertise, and set out to marshal it and bring it to bear on the decision problem at hand.

Previous chapters discuss framing and structuring of our composite perspective. This chapter discusses how to further structure and populate it with the knowledge of experts about the important uncertainties. An expert has a well-developed understanding of events pertinent to the decision, an understanding that usually includes uncertainty about future events. Our approach is to encode the range of outcomes that the expert believes could occur probabilistically. Chapter 3 and Appendix A discuss why probabilistic encoding is so compelling. Chapters 9 and 11 discuss how to model the consequences of the expertise elicited here, and use this model to facilitate a dialogue aimed at finding value.

Consideration of historical data can play an important role in the formation of expertise. However, in a novel decision, the identification of the proper use of historical data requires expertise. The best practice is for the decision analyst to ask carefully crafted assessment questions directly to the experts. If the experts wish to consult historical data, that is within their professional prerogative. It is not a good idea for the decision analysts to usurp the expert's role and directly analyze historical data, unless the decision maker acknowledges that the analysts also have pertinent domain expertise and wants them to employ it. When dealing with multiple experts, it may be important to ensure that they all have access to pertinent historical data.

The perspective on value creation usually must be developed by eliciting and combining the expertise of multiple experts, because it is rare that a single expert has all the requisite expertise in the novel decision situations we address. We call this a composite perspective. In this chapter, we discuss how to elicit the expertise that we stitch together into a composite perspective, and how to use an influence diagram to think clearly about and display its structure. We address both the technical skills of structuring information so that it can be integrated into the decision model, and the soft skills required to elicit expertise from experts. Indeed, the construction of a composite perspective can be viewed as an approach to building consensus on value creation by decomposing the situation according to expertise and allowing each expert to have control within their own area, and trusting that they will defer to others in areas about which they are not expert.

Section 10.2 describes how to use an influence diagram to map out which uncertainties to model, and what conditioning structure to use. Section 10.3 discusses how to address challenges in eliciting expertise authentically. The discussion throughout refers to the Roughneck North American Strategy (RNAS) illustrative example introduced in Chapter 1. Other uncertainty assessments for the illustrative examples are discussed in Section 10.4.

10.2 Structure the Problem in an Influence Diagram

When eliciting expertise about uncertainties, it is important that we represent it in a way that is relevantly similar to the expert's point of view; that is, it represents the underlying expertise in ways that seem appropriate to the expert in all relevant cases. Doing so requires correct structuring of assessments, and good technique in assessment. We discuss the former here, and the latter in Section 10.3.

Specification of probabilities can be difficult when there are many uncertainties, and when the probability of one depends on the outcome of another. As discussed in Appendix A, we can reduce the burden of assessments and improve their authenticity by identifying independence and assessing probabilities conditionally when appropriate. As discussed in Appendix B, these considerations led to the development of influence diagrams (Howard & Matheson, 1983). An influence diagram (ID) represents the independence and conditioning relationships among uncertainties that should be included in a probabilistic model of a common state of information. Arrows into an uncertainty node indicate that its probabilities will be specified conditional on the predecessor uncertainties at the other end of the arrows. An ID is also useful in identifying the uncertainties to be considered explicitly in the composite perspective. It is helpful to develop the ID at a meeting of key experts and stakeholders during the kickoff of a decision project.

It is important to remember that there is a common state of information for an ID.

An ID creates a roadmap for expert assessments. To underscore the roadmap metaphor, we typically put the decision(s) on the left, and objective(s) on the right, and try to arrange nodes so that arrows point from left to right.

There must be at least one decision node, corresponding to the strategies among which we will choose. If we wish to explore other combinations of decisions besides those encoded in the strategies, we can prepare for this by creating one node for each decision that we will want to consider independently. Using multiple distinct decisions requires experts to consider more cases for subsequent uncertainty nodes, but in some cases, we will find that the additional work is modest compared with the potential value added, or that assessment of the consequences of distinct decisions leads to additional clarity of thought in the assessment process.

As discussed in Chapter 8, for RNAS, there were decisions to be made for each of the five major business units: exploration and production (E&P), coalbed methane (CBM), tar sands, enhanced oil recovery (EOR), and power. Figure 10.1 shows the decision nodes in the evolving RNAS ID.

In a multiple objective problem, a node must be present for each objective, along with a value node to calculate the total value (e.g., the additive value model presented in Chapter 9). In a single objective problem, a node representing utility must be present, along with nodes for any other objectives that key stakeholders will need to consider.

Figure 10.2 shows the addition of nodes for objectives to the evolving Roughneck ID, as discussed in Chapter 7.

Frequently, stakeholders have opinions about which uncertainties play an important role in the evaluation of alternatives. It is a good idea to explicitly incorporate these in the model. A good way to do so is to ask for the "win scenario" of each strategy. This is the state of affairs under which the strategy is expected to be more valuable than the others. Sometimes it can also be helpful to probe for "lose scenarios." We then take note of the uncertainties that are

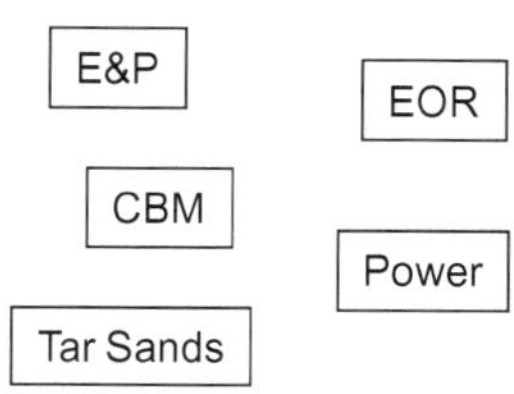

FIGURE 10.1 Roughneck ID fragment: decisions.

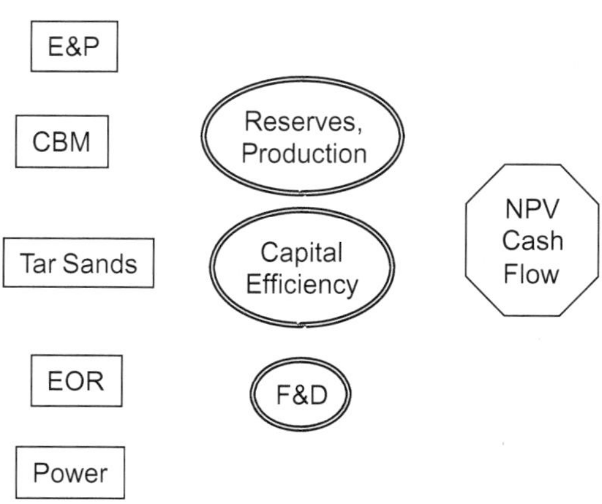

FIGURE 10.2 Roughneck ID fragment with decisions and objectives.

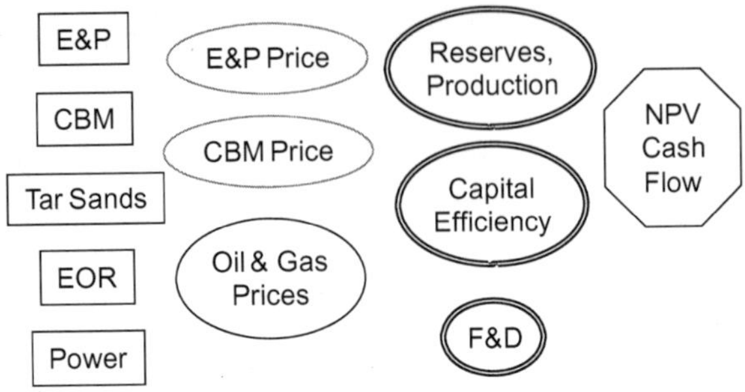

FIGURE 10.3 RNAS ID fragment with decisions, win-scenario variables, and value node.

salient in the description of these scenarios, and ensure that these are included in the ID explicitly.

For RNAS, the success or failure of the maintain, growth, and leader strategies turned on the oil price. This was not surprising; it would certainly have been included in any model. However, the success scenario for the Divest strategy involved prices for oil & gas properties being high. It was initially surprising to the RNAS team that this should be in the model, as they had wanted to assume that real estate transactions would be made "at a fair price" and leave it at that.

Figure 10.3 shows the addition of these nodes to the RNAS ID, when we are focused on NPV cash flow as the key objective.

When building an ID for a decision, we typically start from the value node, and work from right to left, adding predecessors and arrows to the ID that enable us to assess each node in turn (R. Howard, 1983). Here is a protocol to ensure this for a selected node, which we call the target node:

- "Which of the nodes already in the ID (decisions, as well as uncertainties) influence the target node? Do we need to add any nodes to represent uncertainties that could influence the target node?"
- For each of these potential predecessors, ask "Would you want to condition the target node on that predecessor directly?"
    - If yes, add an arrow from the predecessor to the target
    - If no, "What uncertainty between here and there would be more helpful in assessing the target?" Then add a node for the uncertainty, and an arrow from it to the target node

While the semantics of an ID is probabilistic, not causal, expertise can usually be represented more authentically in a causal direction (Shachter & Heckerman, 1987). This is the reason we use the causal term "influence" in the assessment protocol.

Figure 10.4 shows the RNAS ID after the predecessors of NPV Cash Flow have been identified.

Figure 10.5 shows the next pass, identifying the predecessors of Investment/ Divestment proceeds.

Specifying multiple conditioning arrows into a node has the potential to improve the specification of that uncertainty, by bringing more information to bear on its assessment. But this must be balanced against the increased workload associated with creating a distinct joint assessment for each possible combination

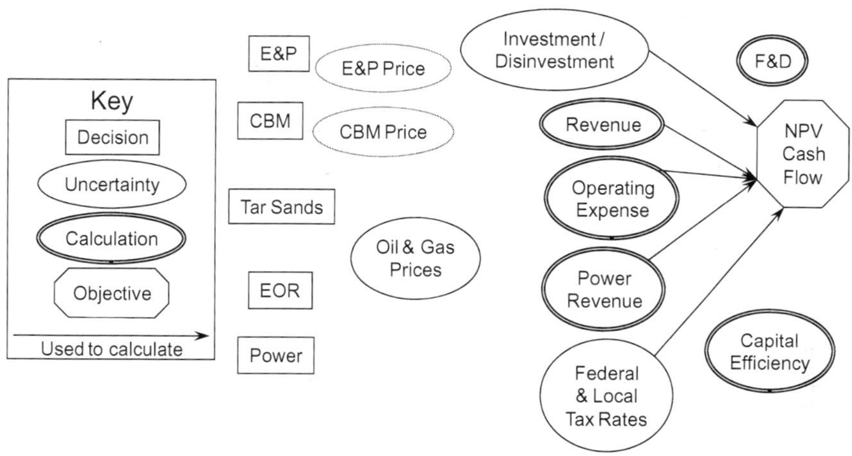

FIGURE 10.4 RNAS ID fragment with predecessors of NPV Cash Flow.

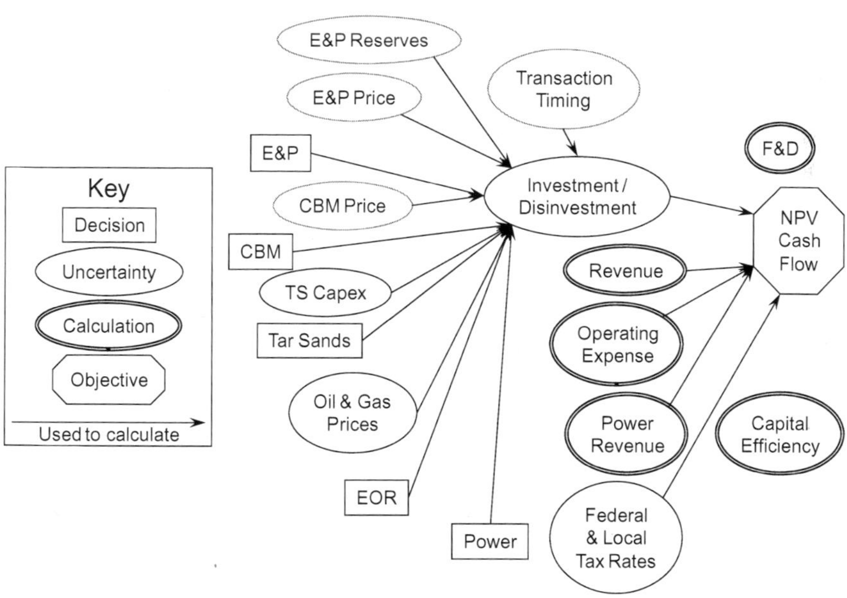

FIGURE 10.5 RNAS ID fragment with predecessors of Investment/Divestment proceeds.

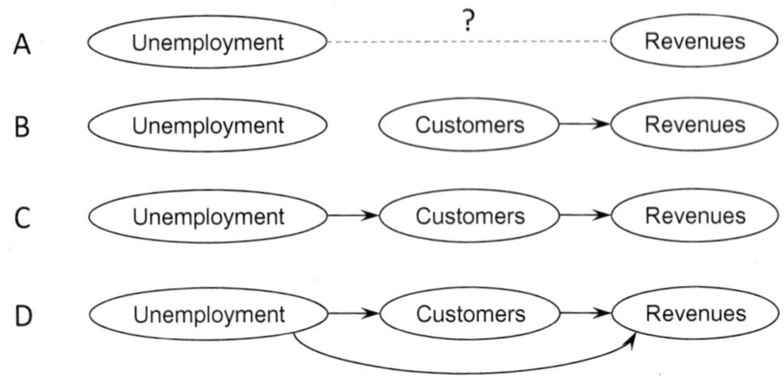

FIGURE 10.6 Restaurant ID fragments.

of outcomes of the predecessor uncertainties. In the cases of RNAS NPV and Investment/Divestment proceeds, the nodes were the sum of components, so the multiple predecessors brought better information without increased assessment complexity.

If one uncertainty is thought to influence another, it may be an indirect predecessor instead of a direct one. For instance, if we are trying to assess Revenues at a *prix fixe* restaurant, we may think that the local Unemployment rate influences our Revenues. This corresponds to an ID fragment with Revenues and Unemployment rate (see Fig. 10.6A). As we try to assess the impact on the Revenue target node, we would naturally find ourselves thinking about how Unemployment rate influences the number of Customers entering the restaurant, and how this influences Revenue. We represent this by adding a node for Customers and begin by adding an arrow from Customers to Revenue (Fig. 10.6B), which represents multiplying the number of Customers by the price they pay to calculate Revenues. Then when Customers is the target node, we add an arrow from Unemployment to Customers (Fig. 10.6C) and assess it accordingly. Note that we no longer need Unemployment rate to assess Revenues; all we need is Customers. So we would not need to add an arrow from Unemployment to Revenues.

However, if the items on our menu had a wide range of prices, we would still want to start our assessment of Revenues by considering the number of Customers entering the restaurant. But we might also believe that the ones who enter might buy lower-priced items in times of high Unemployment. In this case, we would need arrows into Revenues from Unemployment rate, as well as Customers (Fig. 10.6D).

In summary, in the case of the *prix fixe* restaurant, Revenue is not independent of Unemployment, but it is conditionally independent, conditioned on Customers, and we can calculate Revenue directly from the number of Customers. In another restaurant, this independence may not hold, and the calculation or assessment of Revenues would need to consider (and be conditioned on) the Unemployment rate as well as the number of Customers.

If our experts wish to specify their expertise in a way that creates a directed cycle in the ID, we must find the weakest link (arrow) in the cycle and break it. We call an arrow the "weakest link" if we have considered all the arrows comprising a directed cycle, and identify one whose impact we judge to be smallest and easiest to work around. We say we have "broken a link" if we ignore a potentially relevant conditioning. Sometimes, experts may find it acceptable to assess from an ID where the problematic arrow is reversed; this acknowledges the relationship between the uncertainties in question, and allows it to be addressed in a tractable way.

When an ID is nearly complete, we test it by asking whether each line of thought about the decision can be represented there. For example, if we find that the only path from a decision node to the value node is via cost, this means that our analysis will amount to cost minimization. In such cases, we must ask ourselves whether a cost-minimization analysis would be compelling, or whether there are also other benefits associated with the alternatives.

Figure 10.7 shows the complete ID for RNAS. There are many possible decisions. All decisions are made simultaneously at the beginning, except for Tar Sands; arrows among decisions are suppressed for readability. NPV cash flow and capital efficiency were the objectives most often used in analysis, and for decision making; this ID shows NPV Cash Flow as the value node. Note that there are examples of indirect predecessors shown here. For instance, EOR Peak Production influences Reserves/Production for RNAS, but it does so only by

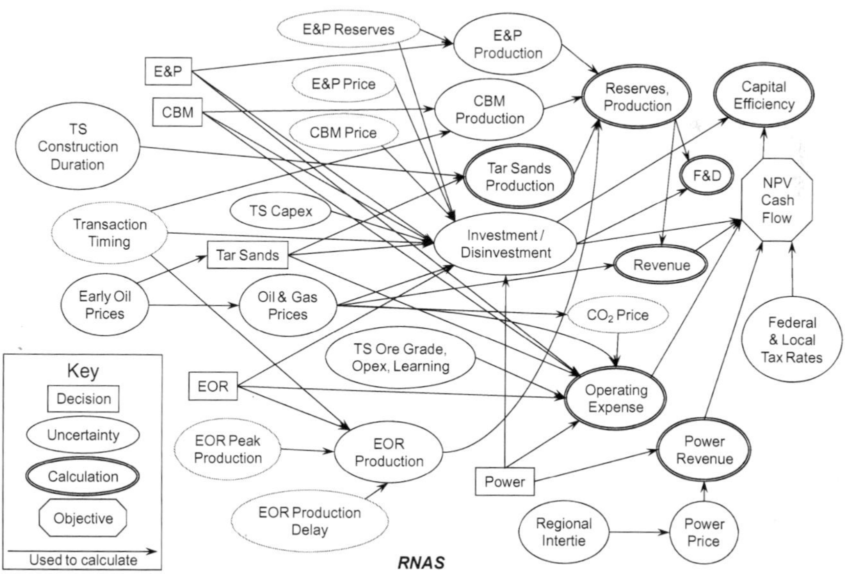

FIGURE 10.7 Complete RNAS influence diagram.

virtue of its impact on EOR Production, so there is only an arrow to EOR Production, but not all the way to [RNAS] Reserves/Production.

To summarize, an excellent ID should have the following attributes:

- Decision nodes
  - There is at least one decision node, reflecting the choice among strategies.
  - If we have the opportunity to make decisions at different time points, there is a decision node for each decision time point.
  - If exploration of independent decisions is desired, these decisions must be in distinct decision nodes.
- Uncertainty/value nodes
  - Expertise is available for each uncertainty.
  - Uncertainties that stakeholders expect to drive the decision are represented explicitly.
  - Each important objective is represented by a node.
  - The value node represents a utility function.
- Conditioning arrows
  - The arrows depict all conditioning necessary to represent the expertise at each uncertainty node.
  - Arrows into decision nodes indicate uncertainties whose outcomes will be known when the decisions are made, but not any other uncertainties.
  - Every line of thought about the impact of decisions or key uncertainties on our objectives is represented.
  - There is no directed cycle.

10.3 Elicit and Document Assessments

This section discusses the challenges inherent in capturing expertise probabilistically, and shows how to address them.

10.3.1 HEURISTICS AND BIASES

Decision analysis was developed in the 1960s. A key aspect of the discipline was the use of probability distributions to represent expertise. In the 1970s (Tversky & Kahneman, 1974), it was found that experts use certain heuristics to simplify thought, and that these lead to predictable biases, some of which are discussed in Chapter 2, including:

- Motivational—too much weight on what the expert or their management wants
  - When the expert is scoring alternatives in which they have a personal or professional interest for a MODA, this is called "scoring bias."
- Anchoring—probabilities of outcomes near a central anchor are too high.
- Availability, representativeness—too much probability on outcomes that are similar to recent, salient, or easily visualized ones (e.g., graphic images of a recent catastrophe that stick in one's mind).
- Unstated assumptions, leading to assessments that overlook cases where the assumption fails.
- Confirmation—de-emphasis of data that is inconsistent with one's preconceptions.

To avoid these biases when we obtain probabilities from experts, we should:

- Use explicit conditioning, to remove[1] unstated assumptions.
- Use prospective hindsight, to remove unstated assumptions.
- Discuss extreme outcomes first, to avoid anchoring.
- Surface availability bias, and explicitly look for other older or less graphic evidence.
- Surface motivational bias, and consult experts "from both sides" to counteract it.
- Use explicit Bayesian updating to combine baseline and new specific data, to combat confirmation and representativeness bias.

Decision analysts at Stanford Research Institute (SRI) responded to the biases by suggesting a five-step probability elicitation procedure sometimes referred to as the SRI protocol: motivate, structure, condition, encode, and verify (Spetzler & Staël von Holstein, 1975). More recent treatments (Kirkwood, 1996), (Hora, 2007) still follow this outline. This section gives recommendations that employ many of the bias-fighting techniques listed above and an elicitation protocol based on the SRI protocol.

10.3.2 REFERENCE EVENTS

The simplest probabilistic assessment is to state the probability that a discrete event (one that can occur or not, but nothing in between) occurs. An important way to elicit such judgments is for the expert to compare the probability of the

[1]It is not enough to merely capture or document assumptions. To create a robust decision, we must identify them and cease to assume them. We must treat them as part of the uncertainty to be characterized. In this way, we put ourselves in a better position to work around them or exploit them.

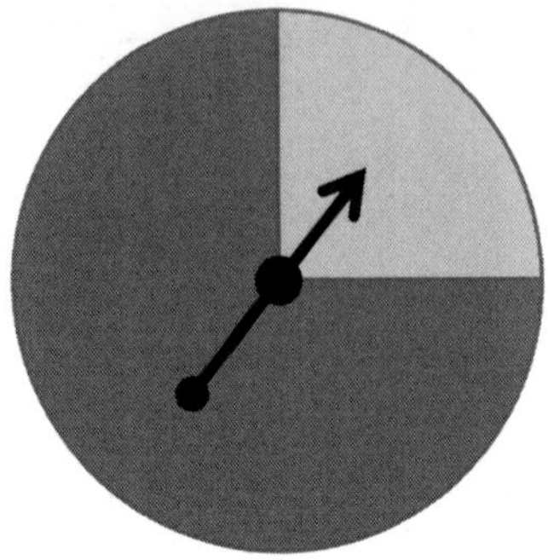

FIGURE 10.8 Probability wheel.

event to that of a reference event whose probability is understood. In the early literature, authors referred to pulling a ball of a specified color from an urn that holds that colored ball with the desired frequency. Raiffa (Raiffa, 1968) introduced the basic reference lottery ticket for this purpose, to give finer-grained resolution. Today, the most common reference event is based on a probability wheel (see Fig. 10.8). The wheel is made of two concentric discs of contrasting colors (e.g., blue and red), each slit along a radius and interleaved. By rotating one of them, we can create a blue slice in a red pie representing any fraction from 0 to 1. The reference event is that an arrow centered at the wheel's center is spun, and points to the blue area when it stops. Inscriptions on the back indicate the probability of this event.

Experts find it particularly difficult to assess the probability of rare events, and a wheel may not be helpful. Discussion in terms of odds (1 in 200) is generally easier to comprehend than a probability (0.005).

Reference processes other than the wheel have been used successfully for probability assessment as well, such as events where the probabilities are well known. For example:

- the probability of heads on a coin flip is 1/2;
- the probability of the roll of a 6 on a die is 1/6
- the probability of "snake eyes" (1 and 1) on the roll of two dice is 1/36
- the probability of 10 heads in a row on a coin flip is roughly 1/1000, and
- The probability of a royal flush in five-card stud is approximately 1/65,000.

By taking an uncertain event and comparing it to known probabilities such as those above, we can "converge" on the probability for the event (e.g., "I think the event is less likely than a six on a die, but more likely than snake eyes") that brackets the probability between 1/6 and 1/36. By introducing more known-probability events, we can bracket even more, but we must be careful of the "pedigree" of the probabilities. We may occasionally read in the paper an assessment of the odds of being struck by lightning or of being killed by a shark while swimming. While in theory, we can use such events as reference processes, we must have confidence that the probabilities are stated correctly, that we understand what they were conditioned upon, and that the data were collected in an acceptable manner.

10.3.3 ASSESSMENT PROTOCOL

Assessment of uncertainties can be viewed as part of the value dialogue. In that process, a preliminary composite perspective is generated, and we critically compare it with individuals' perspectives, aiming to improve one or the other as necessary. This means that it is entirely reasonable for preliminary assessments to be done quickly. The value dialogue discussed in Chapter 11 gives us the opportunity to identify which assessments need to be revised, and which are adequate as they stand. As discussed in Chapter 9, a preliminary assessment can take as little as a few minutes per uncertainty.

Some nodes in an ID are deterministic. This indicates that we can calculate the outcome of the underlying uncertainty directly. In this case, the expert interview is brief; the expert merely needs to define the uncertainty clearly, and specify how its outcome is calculated.

When we do a careful assessment of an uncertainty, an elicitation template like the one in Figure 10.9 can serves as a reminder of the steps of a full elicitation, as well as a template to capture its results. Thoughtfully addressing these questions takes roughly an hour.

A key point to notice is that the experts are asked to give substantial verbal input before assessing numbers. This discussion helps to debias the assessments. For experts who are impatient with these preliminaries, it may be helpful to note that professional athletes always warm up before doing any heavy lifting. Here are the steps of an assessment:

Uncertainty Name 1	
Experts: 2	**Date:** 2
Expert Involvement: 3	
Definition: 4	
Predecessors: 5	
Discussion: 8	
Factors making it low: 6a •	**Factors making it high:** 6b •

	P10 7a	**P50** 7c	**P90** 7b
variable			

FIGURE 10.9 Expert assessment template.

First, discuss the context of the assessment: the importance of the decision, the decision maker's request for help from the expert(s), and the time frame within which the expertise must be gathered, understood, and used to formulate a sound decision.

1. Record on the template the name given to the uncertainty in the ID.

2. Record on the template the names of the experts and the date.

3. Ask the experts about their relationship to the uncertainty. This serves to establish their credibility, and can be a more comfortable way to begin an interview than diving into difficult content questions. It can also surface any motivational bias, if the experts want the uncertainty to have a particular outcome.

4. Have the experts define the uncertainty and its possible outcomes clearly. A definition is clear if independent ideal observers[2] would report the outcome the same way in all cases. Probing and clarifying gray areas or unusual cases ensures that the relationship among uncertain outcomes in the composite perspective is consistent. We must ensure that the definition of an uncertainty is usable by the experts for successor uncertainties conditioned on it. Sometimes a joint meeting with experts on successor uncertainties is required to define a mutually workable definition.

5. Note conditioning predecessors that have been identified in the ID. Ask whether these are appropriate, and whether any others should be used.

6a. Document other factors that could lead to a very good (or bad) outcome of the uncertainty. A good way is to employ prospective hindsight. First, ask for a very high outcome. Then postulate that we are looking back from the future, and an even more extreme outcome had occurred. Ask the experts to explain how this came to pass. Take notes of the factors identified, in bulletpoint form.

6b. Then reverse the question, Ask for a very low outcome. Postulate that a low outcome has occurred and ask for it to be explained. Document the factors that explain it.

If the experts have given only a few factors, press for more. If the experts have given many factors, ask them to survey the list, and identify a few that are the most important or influential. Move these to the top of the list and highlight them.

Or, if the event is discrete, elicit and document factors making it more or less likely.

7a. For a continuous-valued uncertainty, elicit the outcome such that the experts would assign a 10% probability that the uncertainty's actual

[2]We define an ideal observer as a person who is asked to apply a criterion or definition to a situation, understands the language in which it is stated, and is able to perceive all pertinent aspects of the situation, but who has no specific expertise in the matter at hand.

outcome is below it. We call this outcome the 10th percentile, often denoted "P10." Various techniques can be employed here, including:

- Use the term "surprising" to refer to an event having only 10% probability. Contrast this with a "shocking" event that has much lower probability. Then ask for a "surprisingly low" estimate of the quantity in question.
- Set up a comparison of gambles between the uncertainty and a reference event of 10% probability: "Suppose I will give you a bar of gold if the reference event occurs, or if the outcome is below *X*; which would you take?" Then iteratively adjust *X* until they are indifferent.
- Ask the experts to consider 100 independent cases like this one, and report the 10th-lowest outcome from among them.[3]
- If the experts have identified three or four sensible-looking low factors, ask them for an outcome typifying those conditions, and then test whether it is at the 10th percentile.

7b. Elicit the P90 in analogous fashion to P10.

7c. Finally, elicit the P50, such that the experts believe it equally likely that the actual outcome will be above or below that quantity. The experts will be tempted to jump to this number much earlier in the interview, but it is best to dissuade them from doing so, because if they state a central tendency, the anchoring bias will invariably lead them to report a P10–P90 range that is narrower than it would be than if they had thought first about the more extreme possibilities.

Or, if the event is discrete, assess it using a probability wheel. Repeatedly ask whether the event is more or less likely than the reference event, and adjust the reference event probability until the experts judge the reference event equally probable to the event in question.

If there are other conditioning cases, repeat steps 7a–7c to address them.

8. Capture any important discussion that does not fit elsewhere in the "Discussion" field.

Figure 10.10 shows the documentation of an assessment from the Roughneck North American Strategy (RNAS) project. Per step 6, the most important factors have been highlighted in **boldface**.

10.3.4 ASSESSING A CONTINUOUS DISTRIBUTION

As discussed in Chapter 11, for most cases, modeling using only this P10-P50-P90 assessment will be sufficient to identify the best decision, and to guide

[3]This approach calls on a frequentist interpretation of probability. While this interpretation does not contribute to a compelling argument regarding sound decision making, it does generate probability assessments that are consistent with other approaches, and it can be helpful for certain experts.

Experts: [experts' names]	**Date**: June 8, June 22, 2005
Definition: West Texas Intermediate oil price, US $, in 2020; and escalation rate	
Discussion: 6/8 Team assessed 25 - 75 - 150. 6/22 [with additional experts] Inflation could be 3%, but it's more likely to be 1.5-2%. Price is mean-reverting, flat in real terms. Use 20-50-80 instead.	
Factors making it low: • Recession • Big discoveries • Alternate sources (biodiesel) • GDP issues • Hydrogen economy • Efficiency improvements decrease demand	**Factors making it high**: • Mideast problems • Growing third-world demand (China, India) • Major war • Disaster in an alternate fuel • Global warming (more effect on escalation than on 2020 price)

	P10	P50	P90
oil price	20	50	80
escalation	−3%	0	3%

FIGURE 10.10 RNAS documentation of oil price assessment.

thinking about improving it. Even in cases where we need the distribution of a variable to have nonzero density throughout a plausible interval (e.g., when searching for the optimal level of a price threshold to be used in a decision rule), we do not necessarily need to assess a continuous distribution from the expert. Often, fitting a continuous distribution to a P10-50-90 assessment will support this analysis just as well.

If a variable is found to be important, and there is reason to believe the specific shape of the distribution matters, we can assess a continuous distribution for it. When assessing a continuous distribution, we hypothesize a number of possible outcomes and ascertain the percentile of each. To do this, we ask the experts to compare the probability of the variable being less than *X* with the probability of the reference event (probability wheel). We then adjust the probability shown on the wheel until the experts are indifferent. Having done so for one hypothesized outcome, *X*, we test other outcomes. It can be worthwhile to invert the direction of the comparison question from one question to the next: if we have the experts compare the probability that the variable is less than *X* to the wheel, for the next point, we should have them compare the probability that the variable is greater than *Y* to the wheel. While they may find this confusing at first, the inversions neutralize certain judgment biases, and the intellectual "reset" required to address the question ensures that each set of responses calls upon new independent thought, ensuring the broadest possible consideration. Having elicited multiple points, we plot them, show them to the experts, and invite revision and rationalization. Finally, we draw a curve through the elicited points. See the reference by Henrion and Morgan (Henrion & Morgan, 1990) for examples.

Assessing the Probability Distribution of a Chemical Spill

This story of the uncertainty in a chemical spill quantity is an example of when it is useful to assess more than three points in the distribution. Our decision analysis team was performing a cost–benefit analysis to identify the best alternatives to clean up a chemical spill at a Department of Energy nuclear production facility. The spill had occurred over many years, but was discovered only recently when an area of ground caved in above a section of drainage pipe. A hole at the bottom of the pipe had allowed the chemical to drain directly into the ground. The initial project estimate of the spill was 7500 gallons which seemed to be very large.

Our project had monthly meetings. At each meeting, the facility team members provided an update on the assessment of the spill. It soon became apparent that the size of the spill was greater than the initial estimate. When questioned, the engineers who had provided the initial estimate stated that they had "wagged" the 7500 gallons based on sparse information in their records. Since we had learned that the remediation alternatives varied greatly with the size of spill they were designed to clean up, we needed better information on the spill uncertainty. We asked if the individual who was familiar with the equipment causing the spill could be interviewed to assess the potential magnitude of the spill.

Later that day, we spent 1 hour with the equipment expert using the techniques described in this chapter to assess a probability distribution using seven points. We began by understanding how large amounts of the chemical could have leaked from the drainage pipe. We started at the extremes and worked our way to the middle. We had the expert describe the major considerations that would cause a very large spill and a very small spill. The entire project team observed the probability assessment. At the end of the hour, all participants agreed that the maximum spill could be 10,000–500,000 gallons and that the expected value (from the distribution) was approximately 100,000 gallons.

For a continuation of this story, see the Robust Alternative sidebar in Chapter 13.

Parnell was one of the decision analysts on this project.

10.3.5 CONDITIONING CASES

If a target ID node has conditioning uncertainties (predecessors), we must specify its value in each possible combination of the predecessors' outcomes. If all the predecessors have discrete outcomes, we do a distinct assessment conditioned on each possible combination of predecessors' outcomes.

A common example of this is decision predecessors that essentially "turn off" a node's impact on the target node. This is seen many times in RNAS. For instance, E&P Reserves influences E&P Production, but its impact is turned off if the E&P Decision is to "Divest," representing the fact that the assets bearing those reserves would be divested, and could not be used for production. If the E&P Decision is not "Divest," E&P Production is calculated based on E&P Reserves and standard decline curve logic.

If a predecessor has a continuum of possible outcomes, there are three possible approaches:

1. Specify how the target can be calculated from the values of the predecessors.
2. Identify a few possible ranges of predecessor outcomes, and assess distributions of the target conditioned on which range contains the actual outcome
3. Agree on a parametric distribution form (e.g., Gaussian or lognormal) for the target, and have the experts assess its parameters in terms of the outcome of the predecessor. Then pick some test cases and review them to ensure that they make sense.

Approach 1 was frequently used in RNAS, where nodes with multiple predecessors could often be calculated from the predecessors. For instance, Investment/ Divestment proceeds is calculated as the sum of terms stemming from some of its uncertainty predecessors, but only if they were enabled by appropriate decisions in the decision predecessors.

When assessing multiple conditioning cases for an uncertainty per Approach 2, it can be helpful to document them all on one assessment template if they fit, because the variable name, experts, date, and definition are common to them all. Two modifications should be made to the template. First, if there are not too many cases, simply add additional rows to the assessment table for each case, and assess P10 and P90, then P50 for each case in turn. Second, when documenting factors that could make the assessed value high or low, take note of which factors explicate the nature of the conditioning, and which are intrinsic to the variable being assessed. The former give guidance in differentiating the conditioning rows of the assessments, while the latter help distinguish the assessments in the P10-P50-P90 columns.

Decision analysis gives us three excellent ways to model the relationship between a pair of uncertainties:

- Completely independent
- Completely dependent (i.e., we might assess European and North American price of oil to be distinct, but perfectly correlated: whenever one is P90, so is the other)
- Conditioned assessment, which allows intermediate levels of dependence.

Some experts may wish to use a fourth approach: specifying the degree of correlation among variables. Modeling based on correlation is problematic: the intended behavior when one variable is fixed or influenced by a third variable is unclear. This undermines sensitivity analysis and the calculation of value of information and value of control (see Chapter 11). Furthermore, correlation is apt to derail the value dialogue, because its impact on value can be difficult to explain, insofar as explanation is essentially causal, while correlation is not. Although Monte Carlo simulation packages allow the generation of correlated distributions, we recommend that correlation not be used in a decision analysis.

10.3.6 THE RELUCTANT EXPERT

Experts are sometimes reluctant to give a timely answer to questions they are asked, preferring to do additional analysis first. A multipronged approach can be helpful for these situations:

- Note the importance of this effort to the decision maker, who is presumably an important person in the organization.
- Note the faith that is being placed in this expert.
- Note the timeline within which the analysis must be completed.
- Propose to asses a placeholder range of values, which can be refined later if needed.
- Note that the use of ranges reduces the pressure to be exactly right. Using a wide range to ensure that it captures the actual outcome is perfectly acceptable.

If the above suggestions do not work, the problem may be that the reluctance is due to the "expert" not being an expert. A solution may be to find another expert.

Verbal Probabilities

In a decision analysis about litigation tactics, the key uncertainties were assessments by lawyers about whether the client might be found liable for various offenses in various venues. The lawyers did not want to specify an actual probability number in a document that the adversary might be able to get via a discovery motion, for fear that the transgression would then be judged willful and thus be subject to triple damages. They were, however, willing to use a fixed set of nine verbal terms, including Very high probability, Probably, Possibly, and Unlikely.

We gave the lawyers sticky notes with the nine terms and an axis on the wall from "it happens" to "it doesn't happen." They were able to rank-order them and place them along that axis. We then did the analysis with probabilities drawn from simply measuring the position of the stickies along the axis. We tested other mappings of terms to probabilities that were variants of the ones found by physical measurement (warped up or down, but preserving order), and found that the recommended tactics and rationale did not change. The client agreed that the recommended approach (a combination of counter-threat and negotiation) was a good one.

Johnson was a decision analyst on this project.

10.4 Illustrative Examples

We have described the RNAS case throughout this chapter. The data center case did not require uncertainty assessment. The uncertainty assessments for the Geneptin example are described below. See Table 1.3 for further information on the illustrative examples.

10.4.1 GENEPTIN

The Geneptin decision professional was responsible for generating modeling input assumptions with other core team members and other stakeholders across multiple functional areas, such as commercial (global/strategic marketing), clinical development, pricing and market access, biomarker development, regulatory, and medical affairs and business development (for potential diagnostics partnership aspects). The Geneptin team generated the input assumptions by leveraging published information, drug analogs, and collective team experience and by assessment of experts.[4] The decision professional conducted P10, P50, and P90 assessment sessions for most uncertainties.

10.5 Summary

If uncertainty is represented explicitly, it should be represented probabilistically. An ID helps us map out which uncertainties to model explicitly, and what conditioning structure to use for each assessment. When doing a probabilistic

[4]Input assumptions for this case study were based on a published Herceptin case study (Trusheim et al., 2011).

assessment of uncertainties, we must bear in mind biases that experts are apt to exhibit. An assessment protocol that requires clear definition of the variable and verbal identification of influencing factors before assessing P10, P90, and P50 aids in authentic assessment of expertise. For conditioned assessments, we assess a simple distribution for each possible combination of outcomes of conditioning variables.

KEY TERMS

Authentic Relevantly similar to the expert's intention. We say a formal assessment of a probability distribution is authentic if it represents the underlying expertise in ways that seem appropriate to the expert in all relevant cases.
Clear Defined so that independent ideal observers would report its outcome the same way in all cases.
Composite perspective Perspective on value creation that is created by stitching together individual experts' perspectives. This is often represented in an ID.
Conditioning Specifying distinct distributions for possible combination of outcomes of predecessor events.
Deterministic Calculating results for a specific setting of all inputs (i.e., without any uncertainty).
Discrete event Event that either occurs or does not.
Expertise Having better-developed or more reliable understanding of events pertinent to a decision.
Ideal observer Person who is asked to apply a criterion or definition to a situation, understands the language in which it is stated, and is able to perceive all pertinent aspects of the situation, but who has no specific expertise in the matter at hand.
Independent Two events A and B are said to be independent if the probability of A Pr(A|&) is the same as the probability of A assuming B is true Pr(A|&,B).
Influence diagram (ID) A graphical representation of conditioning relationships among uncertainties and decisions in a perspective on a decision situation.
Insight Conclusion that can be reached only by review of the behavior of a model.
Modeling Building representations of experts' perspectives about important events, stitching them together to create a composite perspective on value creation, and representing this in a computer model.
Objective (adjective) True without reference to point of view. cf Subjective.
Objective (noun) Something that stakeholders would like to get or achieve, for example, more profit.
Prospective hindsight Envisioning oneself in the future looking back on a specified outcome of an uncertainty, and explaining how it came to be.
Subjective Based on a point of view. cf Objective.
Target node Influence diagram node currently being assessed.

Uncertainty Event whose outcome is explicitly considered in the composite perspective, and not known with certainty.
Value dialogue Critical comparison of the emergent behavior of value in a model of the composite perspective to individuals' perspectives. It aims to improve one perspective or the other where they differ, and then to use the resulting perspective to identify even better ways to achieve value.

REFERENCES

Henrion, M. & Morgan, G. (1990). *Uncertainty: A Guide to Dealing with Uncertainty in Quantitative Risk and Policy Analysis.* Cambridge: Cambridge University Press.

Hora, S. (2007). Eliciting probabilities from experts. In R. Edwards, R. Miles, & D. von Winterfeldt (eds.), *Advances in Decision Analysis.* Cambridge University Press.

Howard, R. (1983). The evolution of decision analysis. In R. Howard & J. Matheson (eds.), *The Principles and Applications of Decision Analysis.* pp. 5–16. Menlo Park, CA: Strategic Decisions Group.

Howard, R. & Matheson, J. (1983). Influence diagrams. In R. Howard & J. Matheson (eds.), *The Principles and Applications of Decision Analysis*, pp. 719–762. Menlo Park, CA: Strategic Decisions Group.

Kirkwood, C. (1996). *Strategic Decision Making.* Belmont, CA: Duxbury Press.

Raiffa, H. (1968). *Decision Analysis: Introductory Lectures on Choices Under Uncertainty.* Reading, MA: Addison-Welsey Publishing Co.

Shachter, R. & Heckerman, D. (1987). Thinking backward for knowledge acquisition. *AI Magazine, 8*(3), 55–61.

Spetzler, C. & Staël Von Holstein, C.-A.S. (1975). Probability encoding in decision analysis. *Management Science, 22*, 340–358.

Trusheim, M.R., Burgess, B., Hu, S.X., et al. (2011). Quantifying factors for the success of stratified medicine. *Nature Reviews Drug Discovery, 10*, 817–833.

Tversky, A. & Kahneman, D. (1974). Judgment under uncertainty: Heuristics and biases. *Science, 185*, 1124.

CHAPTER ELEVEN

Perform Probabilistic Analysis and Identify Insights

ERIC R. JOHNSON and STEVEN N. TANI

An organization's ability to learn, and translate that learning into action rapidly, is the ultimate competitive advantage.

—Jack Welch

Handbook of Decision Analysis, First Edition. Gregory S. Parnell, Terry A. Bresnick, Steven N. Tani, and Eric R. Johnson.

11.1 Introduction

Previous chapters discuss initial development of a composite perspective on what drives value for the decision. To improve and exploit the composite perspective, we must find ways to understand the behavior of value that emerges from this model and test these against individuals' perspectives. For decisions involving significant uncertainty, explicitly modeling uncertainty with probability gives us an improved perspective on the value and risk of the alternatives. This is a key part of the learning that Jack Welch emphasizes in our headline quote. This chapter shows how to achieve that learning and use it to improve the value and/or mitigate the risk of the recommended alternative(s).

The chapter is organized as follows. Section 11.2 presents the two main ways to incorporate uncertainty in our decision model: decision trees and simulation. Both approaches can create complex emergent behavior that manifests itself across all the other dimensions. Section 11.3 shows how to generate comprehensible analyses that facilitate critical comparison of perspectives and development of improved strategies. We refer to this process as a value dialogue. Section 11.4 is a discussion of how to incorporate risk attitude in a decision analysis when appropriate. Our Roughneck North American Strategy (RNAS) illustrative example (introduced in Chapter 1) is referenced throughout the chapter. Section 11.5 reviews how probabilistic analysis was done in the other two illustrative examples, and the chapter is summarized in Section 11.6.

It is important to note that all the techniques described in this chapter can be applied to single- and multiple-objective decision analysis. The examples we use are single-objective decision analyses. See Parnell et al. (2011) for examples of multiobjective decision analysis (MODA) with Monte Carlo simulation and decision trees.

Some people may feel that the purpose of decision analysis is to make the right choice among a given set of options, and view the development of even better courses of action as an optional activity. While the relative amount of value created by these two activities varies widely, depending on the situation, simply making the right choice from the initial set of options can leave substantial value on the table, particularly if none of the original options satisfies the fundamental objectives very well. Hence, this chapter aims to support strategy improvement as well as identification of the optimal choice among the initial set of strategies.

This chapter and Chapter 13 both discuss communication with stakeholders, but the spirit of the communication in the two is different. Here, value implications of the composite perspective are communicated in the spirit of critical inquiry; Chapter 13 discusses communication of a chosen strategy in the spirit of building understanding of insights that support decision making and commitment to action.

Much of the discussion here is framed as finding or adding value. The other side of this coin is identifying and mitigating risk. While these have different connotations, approaches discussed in this chapter are equally applicable to both: create a composite perspective, understand what drives value and risk, and hypothesize and test strategies that may achieve higher value and/or reduce risk.

11.2 Exploration of Uncertainty: Decision Trees and Simulation

As discussed previously, our recommended tool to identify the uncertain variables and their probabilistic interactions is the influence diagram (see Chapter 10 and Appendix B). As discussed in Chapter 9, the two ways to represent uncertainty are extensive (representing all possible combinations simultaneously) and intensive (representing one combination at a time). For an extensive decision tree representation of uncertainty, the method of analysis is to roll back the tree (see Section 11.2.1). For an intensive representation in an Excel model, the choice of analytic method is between decision trees and simulation.

11.2.1 DECISION TREES

A decision tree (Luce & Raiffa, 1957) is an extensive representation of all possible outcomes of uncertainty. In a decision tree, each possible outcome is depicted by a path through a tree. Conventionally, trees flow rightward from a point of origin ("root") at the left, with a branching point ("node") for each uncertainty and decision. Uncertainty nodes are conventionally shown as circles, and branches emerging from them are possible outcomes. The conditional probability of each outcome is noted on the branch. A decision node is shown as a square, and branches to its right represent possible choices for that decision. Decisions are arranged chronologically with earliest at the left and latest at the right. Uncertainties to the left of a decision must be resolved by the time the decision is made, so each decision node represents a decision in a distinct state of information. The value of an outcome corresponding to a particular path through the tree (a "scenario") is shown at the far right end of the path (the "leaves" of the tree).

The optimal choice and its value at each decision node can be calculated using a rollback procedure (Raiffa, 1968). In it, we start at the right and calculate the value at each node in turn. If it is an uncertainty, we substitute a leaf whose value is the EV (expected value) of the branches for that subtree. If the node is a decision, we note which alternative gives the best value in that circumstance, and substitute the value of the optimal choice for the subtree. This process continues until the entire tree has been simplified to a leaf, indicating the value of the optimal set of choices, and the optimal choices in all states of information have been noted.

Software packages are available to process an intensive representation of uncertainty and alternatives in decision tree fashion. These packages make it easy to specify the structure of the tree, including conditional probabilities. Typically, these packages cache and aggregate only a single value metric; some can handle multiple value measures in the rollback. The size of a tree, and the computational time to evaluate it, rises geometrically with the number of uncertainties included, so the tree should contain only the most important uncertainties. We use tornado diagrams, as explained in Section 9.8.2, to identify which uncertainties to put into the tree.

Another interesting use of trees is for game theory situations. In game theory, we consider multiple rational decision makers ("players") in a situation where the actions of each player influence the outcomes of the others. Formulating such models requires attention to the states of information of the participants when they make their decisions, and the solution approach or software must ascertain the value realized by each player in each scenario. The notion of what behavior is optimal is more complex in a game theory situation. The most common insight from game theoretic analysis is that delivery of a credible threat can sometimes influence the behavior of another player in ways favorable to oneself. Discussion of game theory is beyond the scope of this handbook. See Papayoanou (2010) for a recent practitioner's treatment, and Bueno de Mesquita (2002) for a discussion of recent game theoretic view of American foreign policy.

11.2.1.1 Example Decision Tree Application. To see how a decision tree can be used to resolve an important decision situation, consider the example of a company planning to build a manufacturing plant for a new product due

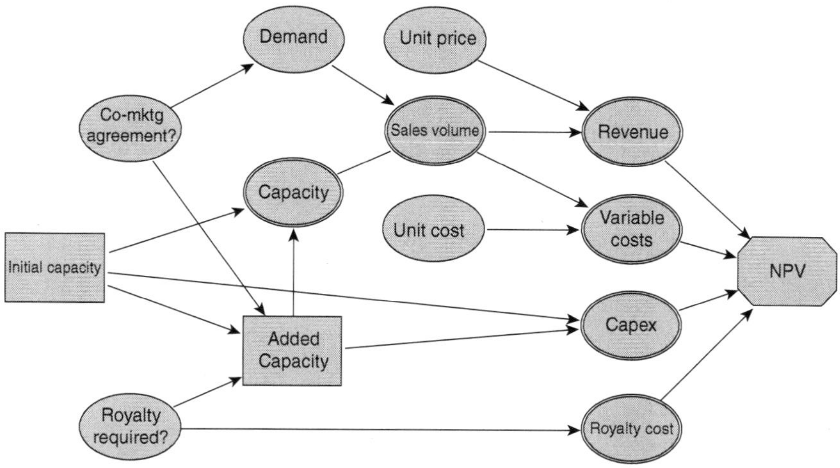

FIGURE 11.1 Influence diagram for capacity planning example.

to be launched in2 years. We find it helpful to develop an influence diagram such as the one shown in Figure 11.1 as the first step in developing the decision tree.

The key decision under consideration is how much capacity to build into the new plant. This decision is made difficult by several sources of uncertainty. The future level of demand for the new product is quite uncertain, ranging from a low (10th percentile) of 200K units per year to a high (90th percentile) of 1000K units per year, and is made even more uncertain by the possibility that a comarketing agreement will be signed with a TV cable channel that would boost demand substantially (by about 40%). And the degree of profitability of the product is uncertain because of uncertainty in both unit cost and selling price, as well as the chance that a pending patent infringement lawsuit will be lost, requiring that a royalty be paid on all sales of the product.

The decision team plans to build out capacity in two stages. The initial level of capacity must be chosen now, but capacity can be expanded 4 years from now. Capacity installed initially has a lower per-unit cost than capacity added later. But waiting to add capacity has the advantage that two key uncertainties will be resolved by the time that decision is made—whether the comarketing agreement is signed and whether the patent infringement lawsuit is won or lost. For initial capacity, the team has identified three alternatives:

1. High capacity—1000K units per year (equal to the high demand estimate)
2. Low capacity—500K units per year (equal to the base case estimate of demand)
3. Flexible capacity—500K units per year initially, but with a low-cost option to expand later

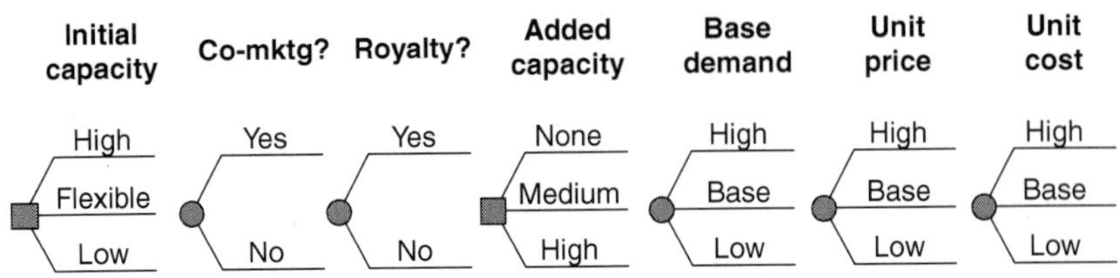

FIGURE 11.2 Schematic decision tree for capacity planning example.

For the capacity expansion 4 years from now, the team has chosen three levels to analyze:

1. None
2. Medium—additional 500K units per year
3. High—additional 700K units per year.

Clearly, the optimal amount of capacity to add will be driven by how much capacity was built initially, the demand forecast (given whether or not the co-marketing agreement has been signed), the cost of adding capacity (given whether the Flexible alternative was chosen initially), and the projected profitability of the product (given whether or not a royalty will be required). Because of this "downstream" decision, a decision tree approach is the best way to analyze the situation.

The decision tree for the capacity planning decision is illustrated schematically in Figure 11.2.

This decision tree has 972 unique paths and is small relative to many trees developed in professional practice, which can have many thousands of paths. But it is sufficient to provide good decision insight in this situation.

To evaluate the tree, a spreadsheet model is built to calculate the net present value (NPV) for the product for each path through the tree (scenario). This model is then connected to decision tree software, which performs the rollback calculations. The results of the rollback are shown in the partial tree display in Figure 11.3.

Each of the 36 NPVs at the right-hand end of the partial tree is the expected value of 27 NPVs for different combinations of demand, unit cost, and selling price. We can see from the tree that the optimal decision policy is to choose the Flexible alternative for initial capacity and then to choose high added capacity if the comarketing agreement is signed and medium added capacity if it is not. The optimal policy has an expected NPV of $64 versus $62 million for starting with low initial capacity and $57 million for starting with High initial capacity.

Note that the optimal choices for the downstream decision may not be easy to predict intuitively (see Table 11.1).

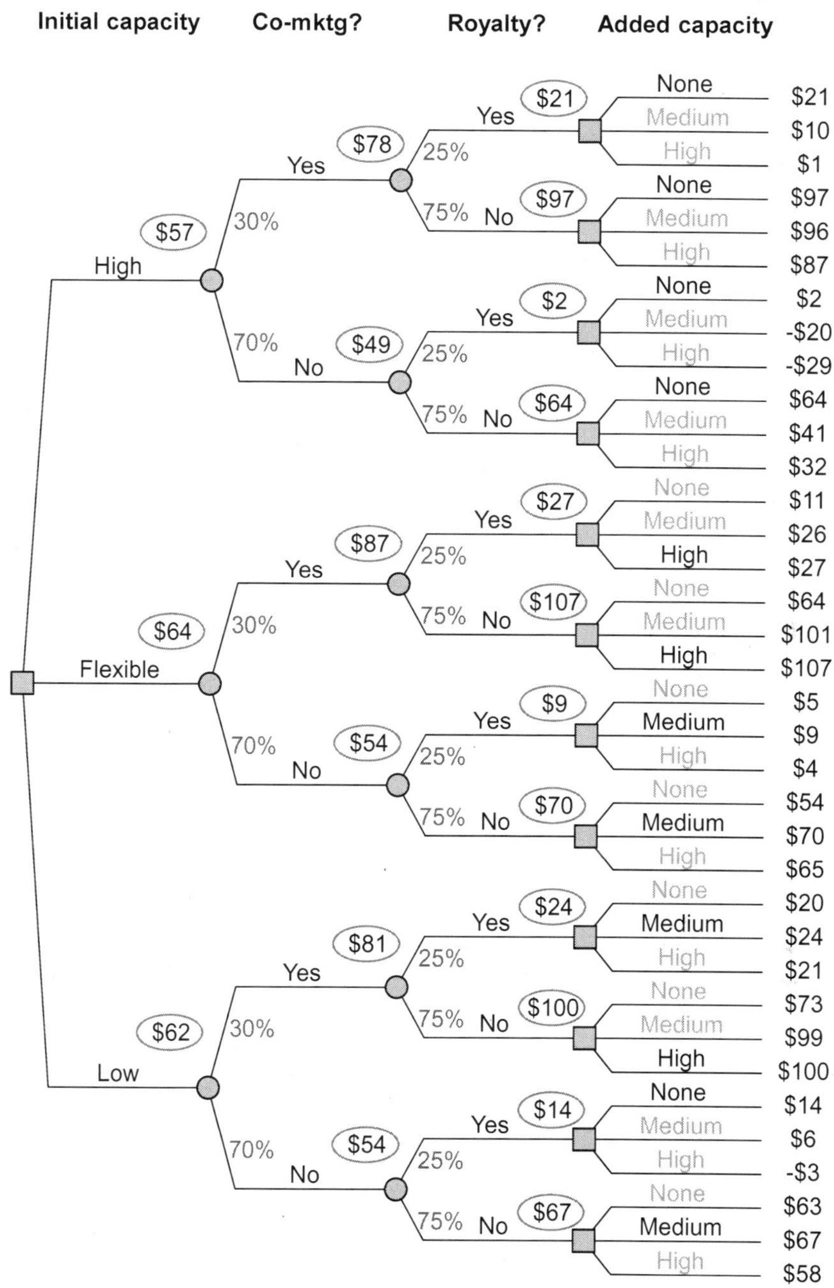

FIGURE 11.3 Partial display of evaluated decision tree for capacity planning example (values in $ million).

TABLE 11.1 **Optimal Downstream Decisions in Capacity Planning Example**

		Initial Capacity		
Comktg?	Royalty?	High	Flexible	Low
Yes	Yes	None	High	Medium
Yes	No	None	High	High
No	Yes	None	Medium	None
No	No	None	Medium	Medium

11.2.2 SIMULATION

We refer to a particular combination of outcomes of all uncertainties as a scenario. In a simulation approach, we sample repeatedly from possible scenarios with probability distributions driven by the composite perspective and record the aggregated results. A description of marginal and conditional probabilities in an influence diagram (see Fig. 10.2 in Chapter 10) and encoded in an Input sheet of a spreadsheet (see Fig. 9.2 in Chapter 9) is an intensive representation of the composite perspective. Chapter 9 discusses construction of a spreadsheet model that embodies this representation in a form that can be manipulated.

Simulation is viewed as a sub-discipline of operations research, and it can play an important role in decision analysis. Different kinds of simulation have been developed, indicating different areas of emphasis:

- **Monte Carlo simulation.** Realization of a sequence of outcomes by repeated random sampling of input scenarios according to probability distributions. In decision analysis, we obtain these distributions from experts.
- **System dynamics.** Simulation of flow through a network of transformations over time, typically focusing on the implications of positive or negative feedback loops in the network (Forrester, 1961).
- **Agent-based simulation.** Simulation of the behavior of a community of interacting "agents" or actors, each of which has objectives and a behavioral repertoire represented by deterministic or probabilistic rules.
- **Discrete-event simulation.** Simulation of a system that is represented as a chronological sequence of events. Each event occurs at an instant in time and marks a change of state in the system (Robinson, 2004).

When the term "simulation" is used in this handbook, we have in mind Monte Carlo simulation, but other variants of simulation should be employed if appropriate for the situation at hand.

The normal way to aggregate value across simulated scenarios is to calculate the expected value (EV), because this is the required way to handle utility, and it is frequently informative for other value measures. Calculation of EVs requires caching and aggregation (or on-the-fly aggregation) of the instantiated values.

Commercially available simulation packages support caching and aggregation of a large number of value measures.

One characteristic of Monte Carlo simulation is that it introduces a certain amount of noise into results. This can be managed by choosing an adequate number of iterations, or "trials." In practice, the number of trials required is not strongly influenced by the number of uncertain variables. If there are rare events with important consequences, a standard Monte Carlo will require a larger number of trials to give reliable results. Variants, such as importance sampling (Asmussen & Glynn, 2007), can be employed in such cases.

11.2.2.1 Monte Carlo Sampling. In Monte Carlo simulation, it is important to select a sampling distribution for each uncertain parameter that fairly represents the impact of plausible extremes (10th percentile, denoted "P10," and 90th percentile, denoted "P90"), as well as the central tendency of value. But it may not be important for the cumulative distribution curve of a sampling distribution to look plausible. For instance, for many intrinsically continuous uncertainties, a three-point discretization (which has a stairstep cumulative probability curve) is often adequate. We discuss a few reasonable sampling distributions here, but others will suffice.

A good and simple approach is to use a discrete sampling distribution that gives 30/40/30 probabilities to the expert's P10-50-90 inputs. This "extended Swanson–Megill" distribution preserves the mean and variance of a variety of utility functions (Keefer, 1994). Some people prefer 25/50/25 probabilities. While this is apt to understate the variance of the resulting distribution of value, the understatement may not be large, so this is also a good choice.

For a variable that we may want to use as a precondition for a future decision, it can be important that the sampling distribution gives some nonzero probability density over a range of values (i.e., a continuous support). For these cases, the most authentic representation is to elicit a full continuous distribution from the expert, but this can be time consuming, especially if there are multiple experts. Often we can develop the insights we need by fitting a continuous curve to the P10-50-90 assessment we have and using it as a sampling distribution. Various distributions, including the skew logistic (Lindley, 1987) or a distribution we call the Brown–Johnson, can be used in this way. Brown–Johnson is piecewise uniform in the four ranges, from its P0 to P10 to P50 to P90 to P100.[1]

For example, in the Roughneck North American Strategy (RNAS) illustrative example (see Section 1.5.1 for the case introduction), the model makes extensive use of 30/40/30 probabilities of P10-50-90s. However, a continuous

[1]If the expert has assessed only P10, P50 and P90, we choose P0 ≡ 2.5*P10 − 1.5*P50 and P100 ≡ 2.5*P90 − 1.5*P50. These formulas for P0 and P100 were chosen so that the resulting fractile curve resembles Gaussian, lognormal and other common distributions. The derivation of the P0 and P100, and their use to generate a distribution that is similar to other well-known distributions is due to Rob Brown. Brown generated a smooth fractile curve by fitting an increasing spline to the five fractiles. Brown–Johnson substitutes piecewise linear.

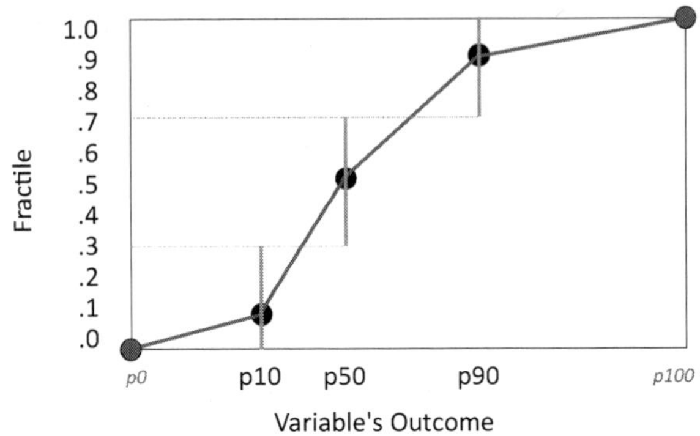

FIGURE 11.4 Extended Swanson–Megill and Brown–Johnson distributions.

distribution on future oil price is needed to test various levels of the go-forward threshold for Tar Sands full-scale development. The analysis uses a Brown-Johnson distribution for oil price based on carefully assessed P10-50-90 points.

11.2.2.2 Cumulative Probabilities and Uncertainties. Generation of an EV tornado diagram using the approach described later in this chapter (see Section 11.3.4) requires the cumulative probability at which each uncertainty is sampled for each trial. One way to provide this is to generate the cumulative probability first by sampling from a Uniform[0,1], and then to calculate the corresponding fractile within its sampling distribution. This is straightforward for both extended Swanson–Megill and Brown–Johnson, as shown in Figure 11.4.

Typically in Monte Carlo simulation, a distinct sample is generated for each uncertainty in each iteration/trial. In cases where experts believe that two or more variables are strongly linked (e.g., by specifying the same "Factors Making It High" for those variables in the assessment interview), these variables should use the same sampled cumulative probability in each trial of the simulation. In such a case, it is helpful and usually straightforward to find a name for the common sample that reflects this use. For instance, in the RNAS model, the sample named CBM Cost is used for six variables, relating to CBM opex and capex under various strategies. The CBM Cost bar in the tornado diagram represents the combined impact of all six variables moving in concert.

If we are simply comparing EV results from different strategies, it does not matter whether the same samples are used across all strategies. However, for delta EV tornado diagrams (see Section 11.3.5), if the experts believe that a variable's outcome is unaffected by strategy choice, the same sample can be used for simulating the variable under all strategies. This constitutes perfect correlation across strategies, and is reasonable if the same Factors Making It High apply across all strategies. If the Factors Making It High are different from one strategy to the next, independent samples should be used, indicating that these variables are independent. If a target variable has been specified using probabilistic conditioning, samples should be drawn from the distribution that is appropriate for the predecessor's value. Some of the Monte Carlo software packages support the specification of probabilistic conditioning on input variables.

11.2.2.3 Running a Simulation Model. Having built the model and set up the sampling logic, the next step is to run a simulation. If strategies are also specified intensively, this requires defining strategies, specifying overrides (if any), and specifying the number of strategies and trials per strategy.

Override values are values for one or more uncertainties to be used in all trials, regardless of the sampling distribution.

Strategies are specified in a strategy table data structure. Initially, the strategies generated in the structuring workshop are tested. As hybrid strategies are developed, these can also be specified and simulated. If we are testing for the optimal parameter for a decision rule, we can specify a number of strategies all identical except for the decision rule parameter. For instance, for RNAS Tar Sands, we specified various oil price thresholds below which the full-scale plant would not be built and compared their results. In addition, when we want to map out an efficient frontier for portfolio analysis, it can be useful to formulate and simulate stylized strategies that test each possible level of investment at each business unit. By doing this, we can then use the results to calculate the incremental cash flow for each opportunity.

Selecting the number of trials to run in a Monte Carlo simulation gives us control over the amount of simulation noise introduced into the results. If we run a simulation with sufficiently many trials, the size of this simulation error can be made small enough that it does not interfere with the critical comparison and exploration required to develop insights and make decisions. But runtime increases if we run more trials, so this runtime–precision trade-off must be managed.

One must understand how much simulation noise is present to avoid over-interpreting noise. While the mathematics of simulation variances is beyond our scope, one key fact to bear in mind is that noise is reduced by the square root of the number of trials. Running four times as many trials cuts noise in half. Running nine times as many trials cuts noise to one-third. It can be helpful to employ one or more dummy variables in the model to understand the impact of simulation noise on EV tornado diagrams. A dummy variable is a variable with a sampling distribution, but with no impact on the calculation. The true width of its tornado bar, if a sufficiently large number of trials were employed, would be zero. This means that any observed variation of value attributable solely to a dummy variable in the tornado is simulation noise. If a variable's tornado bar is narrower than that of a dummy variable, there is no reason to think that uncertainty in such a variable has a material impact on uncertainty in value. If this creates difficulty, run more trials.

11.2.3 CHOOSING BETWEEN MONTE CARLO SIMULATION AND DECISION TREES

Monte Carlo simulation and decision trees are two different methods to solve the same problem—given a number (*N*) of important uncertain inputs, find the probability distribution on the value metric for each decision alternative. Both methods accomplish this by examining points in the *N*-dimensional space defined by the *N* uncertainties. The methods differ in the way those points are selected. In Monte Carlo simulation, the points are selected "randomly,"[2] based on the probability distributions of the inputs. In the decision tree method, the points to be examined are preselected as every possible combination of the values of the discrete inputs.

When done well, both methods produce comparable results. However, there are some characteristics of the decision situation that may influence the choice of method.

11.2.3.1 Downstream Decisions. The decision tree method is better suited than the Monte Carlo method to handle downstream decisions (i.e., decisions that will be made in the future after some uncertainty is resolved). An important class of decision situations with this characteristic is called "real options." As illustrated in the capacity planning example (Section 11.2.1.1), the decision tree method handles downstream decisions "automatically" by determining in the rollback procedure the optimal choice for each such decision in the tree. To handle downstream decisions in the Monte Carlo method, a side analysis must be conducted to determine the optimal choices for the downstream decision given each possible conditioning scenario. These optimal downstream choices must then be "hardwired" into the simulation analysis of the original decision alternatives. (This approach is illustrated for the RNAS Tar Sands decision in Section 11.3.8.)

11.2.3.2 Number of Uncertainties. The number of uncertain inputs to be included in the probabilistic analysis can push the choice toward either method, depending on whether the number is small or large.

If the number of uncertain inputs is small (say, fewer than eight), the decision tree method is computationally better than Monte Carlo. Consider a case in which there are just five important uncertain inputs. Assuming that each input is discretized with three possible settings, the total number of unique scenarios to be examined is 243 (= 3^5). Because the decision tree method examines each scenario only once, it would require only 243 evaluations of the value function. But if the analysis were done with Monte Carlo simulation using only 243 trials, there is no guarantee (and very little likelihood) that each of the 243 scenarios would be examined. In fact, a great many more trials need to be run in the Monte Carlo simulation for the probability of each scenario to be accurately portrayed. So, with a small number of uncertainties, the decision tree method is more efficient in computational time than Monte Carlo.

However, if the number of uncertain inputs is large (say, more than 11), then the Monte Carlo method is computationally better than decision trees. The

[2]On a computer, the Monte Carlo samples are actually generated via well-designed pseudo-random number algorithms whose initial "seed" is usually taken from a hard-to-predict source, such as the system clock.

decision tree method *requires* that every possible scenario be examined. So, for example, in an analysis having 12 uncertain inputs (each discretized with three possible settings), the decision tree method would need to evaluate the value function for over a half million scenarios, requiring substantial computational time. By contrast, the Monte Carlo method allows the analyst to make an explicit trade-off between running time and accuracy of results by choosing the number of simulation trials. Of course, the analyst should avoid the delusion that running a Monte Carlo simulation with just a few thousand trials always produces valid results. Instead, the analyst should monitor the mean square error (MSE) statistic produced by most Monte Carlo software packages to gauge whether the number of trials is sufficient.

What if there are many uncertain inputs defined for a decision but not all of them are really important contributors to overall uncertainty in value? In this case, the Monte Carlo method often gets the nod because of ease of use. With the decision tree method, one would need to produce a tornado diagram to identify the most important uncertainties and then specify those to the decision tree software. With Monte Carlo, one could simply include all uncertainties in the simulation. The important uncertainties are sure to be included, and the fact that many unimportant uncertainties are included does not matter.

11.2.3.3 Anomalies in the Value Function. If the value metric is smooth and "well-behaved" over the range of possible input settings, then discrete approximations can be made for all of the inputs in the probabilistic analysis and the results will be quite close to those that would be produced using continuous distributions on the inputs (assuming that there are more than a few inputs with a material impact). However, if there are anomalies in the value function, then discrete approximations are not appropriate and the Monte Carlo method using continuous input distributions is a much better choice than the decision tree method. An anomaly exists if, for some region in the N-space of inputs, the value metric changes rapidly for small changes in the inputs (i.e., a "sink hole" or "pinnacle"). An example of a decision situation with a value function anomaly is a company making choices that could lead to illiquidity. For a setting of inputs that define a scenario in which the company is close to having insufficient cash balances, a small change in inputs could result in a big and disastrous impact—insolvency and shutdown.

It is a wise decision analyst who has both Monte Carlo and decision trees in his or her tool bag. As described above, depending on the situation, one or the other method may be much preferred. And for the situations when either method will serve well, the choice can be made based on the analyst's preferences and familiarity with the software tools.

11.2.4 SOFTWARE FOR SIMULATION AND DECISION TREES

Software packages are available to process an Excel model in either decision tree or simulation fashion. The professional society INFORMS periodically publishes a thorough review of DA software (http://www.informs.org/ORMS-Today/Public-Articles/October-Volume-37-Number-5/Decision-Analysis-Software-Survey). Software used for the illustrative examples reported in this book includes Excel with VBA, Analytica® (Lumina Decision Systems, Los Gatos, CA; http://www.lumina.com), and @Risk (Palisade Corporation, Ithaca, NY; http://www.palisade.com/decisiontools_suite) for RNAS; Excel and Crystal Ball (Oracle Corporation, Redwood Shores, CA; http://www.oracle.com/us/products/applications/crystalball/index.html) for data center; and Excel for Geneptin Personalized Medicine.

Some decision analysis software offers the capability to model with influence diagrams and/or decision trees and then, if the computational time is too long, to solve with Monte Carlo simulation.

An additional feature of some influence diagram, Bayesian net, and decision tree software is automatic Bayesian calculations. In this software, the user inputs marginal and conditional probabilities for one order of conditioning, and the software calculates the corresponding probabilities for any other order of conditioning.

11.3 The Value Dialogue

The value dialogue (based on analysis using either decision trees or Monte Carlo simulation) has two phases. First, we use analyses of results to compare the composite perspective to individual perspectives, aiming to improve one or the other where they differ. Then we use the improved perspectives to search for ideas that could further improve the strategies by increasing the value and/or mitigating the risk. We will illustrate the value dialogue with the RNAS illustrative example. See Table 1.3 for further information on the illustrative examples.

If comparison shows an individual's perspective to be inferior to the composite perspective, it is the analyst's job to explain the latter in terms that are readily interpretable. If the model is found to be inferior, but in a way that will not affect decisions or insights, we chalk it up to the inaccuracy inherent in modeling and move on. This is sometimes difficult for clients, who lose sight of the fact that the sole purpose of the model is to generate insights. They want it to be perfect. If the model is found to be inferior to an individual's perspective in a material way, we must improve it. Sometimes, this is as simple as revising the P10-50-90 assessment of an uncertainty (especially if the initial assessment was generated rapidly). Other times, the deficiency in the model requires that additional variables and/or additional logic be added to the model.

This handbook presents many ways to analyze results; some of them can be uninteresting in a specific decision situation, and others may not be applicable at all. As we discuss in Chapter 9, the prudent analyst runs as many of these analyses as feasible, but shows only those that are interesting to the decision team.

If a promising new strategy is developed, we evaluate it alongside the others, eliciting additional expertise and updating the spreadsheet models as necessary.

Additional spreadsheet logic is often required for the analyses described here. While it may seem a large task to build analysis tools like these for every decision problem, the tools are oriented to dimensions of decision complexity that persist across many decision contexts. This ensures that a set of tools built for one decision context (such as the RNAS model) can be modified without too much difficulty to be applicable to other situations.

11.3.1 P&L BROWSERS

We use the term P&L (profit and loss statement) to refer to the formatted calculation of NPV value, based on monetary costs and benefits across time. At this level of generality, the P&L is analogous to a tabulation of the components of a MODA value model.

It is often useful to be able to understand the behavior of the P&L as it is affected by the dimensions of complexity discussed in Chapter 9. This is best handled by maintaining and analyzing a multidimensional data structure, showing the P&L in context of some or all of the dimensions of complexity. In a well-designed simulation platform, a multidimensional structure representing objectives, strategies, uncertain scenarios, business units and time can be browsed in a variety of display modes. Analytica meets these requirements, with an influence diagram user interface, a Monte Carlo simulation engine, and a facility for browsing multidimensional arrays in many display modes. In Excel, one of these dimensions frequently ends up spread across multiple worksheets, and there is no native support for uncertainty display modes. Accordingly, decision analysts using Excel must build "P&L Browser" capabilities explicitly.

The RNAS model employs intensive representation of both strategies and uncertainties, so only one scenario for only one strategy is calculated at a time. Business-unit (BU) P&Ls for the five major business units are each on separate sheets, and each sheet has a selector for strategy. Figure 11.5 shows the initial years of the P&L for the E&P (exploration and production) business unit under the Growth strategy with uncertainty averaged out. A summary P&L appears at the top, with drill-down information below. Operating expenditure and capital expenditure are referred to as opex and capex.

To create "browsing" capability for P&Ls, we need to display summary values always, or on demand. In a conventional P&L, value components and other line items are in rows and time periods in columns. In this case, we collapse the time series in a line-item row by showing its present value (PV) at the left, and collapse the line items in a given time period by showing their net cash flow at the top.

11.3.2 TOTAL VALUE AND VALUE COMPONENTS

For financial decisions, detailed data displays like the P&L can be boiled down to what we call value components by considering only the PVs of the time series (i.e., rows).

E&P	2006	2007	2008	2009	2010	2011	2012	2013	2014	2015
Revenue	1090	1024	963	892	873	872	863	867	868	832
Opex	43	41	39	36	35	34	34	35	36	37
Taxes	431	403	377	347	331	324	314	309	303	283
Capex	32	34	52	189	193	214	237	248	248	152
Cash Flow	584	547	495	320	313	300	278	275	281	360
Revenue										
Gas Revenues	1014	942	881	810	798	804	818	848	869	848
Royalties - Basin_2	199	193	182	168	160	157	138	121	106	91
Royalties from Sales	-	-	-	-	-	-	-	-	-	-
Legacy Oil Revenues	41	43	45	47	45	44	40	37	33	30
Royalty Paid	(164)	(154)	(144)	(134)	(131)	(132)	(134)	(138)	(141)	(137)
Opex										
Conventional Opex	41	38	34	31	29	26	24	22	20	18
Opex at ENP_4	1	3	4	4	4	3	3	3	2	2
Exploration Opex	0	0	0	1	2	4	7	10	14	16
Opex at Basin_3 Acquisition	0	0	0	0	0	0	0	0	0	0
Taxes										
Depreciation	0	4	9	15	39	61	82	104	124	141
Net Capital	32	61	105	279	433	586	741	886	1,010	1,021
Federal Taxes	332	310	290	266	252	244	233	226	218	200
State Taxes	100	93	87	81	80	80	81	83	85	83
Capex										
Sale of Land and WI	0	0	0	0	0	0	0	0	0	0
Conventional Capex	32	32	32	32	32	32	32	32	32	32
Capex at ENP_4	0	0	0	118	0	0	0	0	0	0
Exploration Capex	0	2	21	39	59	80	103	115	114	121
Capex for Downspacing	0	0	0	0	102	102	102	102	102	0
Basin_3 Acquisition Cost	0	0	0	0	0	0	0	0	0	0
Basin_3 Acquisition Capex	0	0	0	0	0	0	0	0	0	0

FIGURE 11.5 RNAS E&P profit and loss statement—growth strategy (amounts in $ million).

Table 11.2 shows the value components for RNAS. The business units are Exploration and Production (E&P), Enhanced Oil Recovery (EOR), Coalbed Methane (CBM), Tar Sands (TS), and Electric Power. Revenues are positive, and the Opex and Taxes are negative. Capex is usually negative; however, the positive numbers for Capex for the Divest strategy reflect the divestiture proceeds.

The value components chart presented in Section 9.8.1 allows the value components of multiple strategies to be shown in one chart by displaying all components for a strategy in just two columns, one for negative and one for positive components. Figure 11.6 shows the components of expected value for each strategy in the RNAS example.

TABLE 11.2 **RNAS Value Components ($ Billions)**

	Divest	Harvest	Growth	Leader
E&P Revenue	3.9	5.7	7.8	8.7
CBM Revenue	1.0	5.9	6.3	11.7
EOR Revenue	1.3	6.3	9.1	11.2
TS Revenue	–	–	1.8	2.7
Power Revenue	–	–	.3	.8
E&P Opex	(0.1)	(0.2)	(0.3)	(0.4)
CBM Opex	(0.1)	(0.5)	(0.5)	(1.0)
EOR Opex	(0.3)	(1.1)	(1.7)	(2.1)
TS Opex	–	–	(0.5)	(0.7)
Power Opex	–	–	(0.1)	(0.2)
Taxes	(3.8)	(6.5)	(8.6)	(11.5)
E&P Capex	1.9	(0.3)	(1.1)	(2.0)
CBM Capex	1.4	(0.8)	(0.8)	(2.6)
EOR Capex	0.7	(1.0)	(1.9)	(2.9)
TS Capex	–	–	(.8)	(1.2)
Power Capex	–	–	(0.1)	(0.3)
Cash Flow	5.9	7.6	8.8	10.3

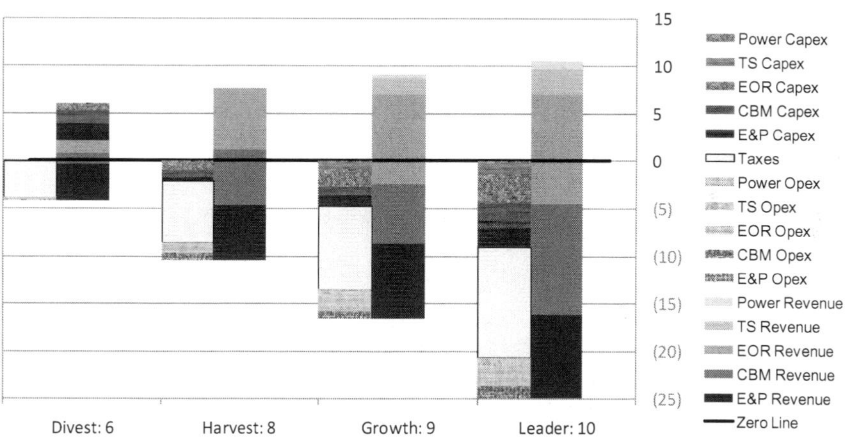

FIGURE 11.6 **RNAS value components chart.** (See insert for color representation of the figure.)

The first thing to do with the value components chart is to test the composite perspective it portrays:

- Do we believe the value components for each strategy?
- Do we believe the relative value components across strategies?
- Do we believe the value ranking that emerges from the parameters in the value model?

In the value components chart for RNAS, we can see that:

- Strategies with larger revenues require larger investments.
- Taxes are the most significant cost factor.
- The Leader alternative looks best based on total value.
- The Divest alternative looks worst based on total value.

Additional decision insights can be generated by creating a chart showing the *difference* in value components between strategies. In this way, value components that are common to all strategies cancel out, allowing the ones that differentiate the strategies to stand out. To do this, select a reference strategy, and then for each value component of each strategy, subtract out the corresponding value component of the reference strategy, and then plot in the two-column fashion as before. Typically, it is informative to choose a reference strategy that is fairly good, but not the best. A "status quo" or "momentum" strategy can also be a helpful reference point.

Figure 11.7 displays the RNAS delta value components, where the Growth strategy is used as the reference strategy. For Divest and Harvest, the revenue components are less than the reference (Growth), so in the delta chart, they are negative and show up in the left column.

The point of this graphic is that a relatively small number of blocks have large size, and these are immediately apparent. For RNAS, the most salient blocks are the high CBM revenues for Leader, and the low (absent) CBM revenues and EOR revenues for Divest. This tells us that the decision is mostly about CBM. For Divest, the substantial size difference between the CBM or EOR revenues

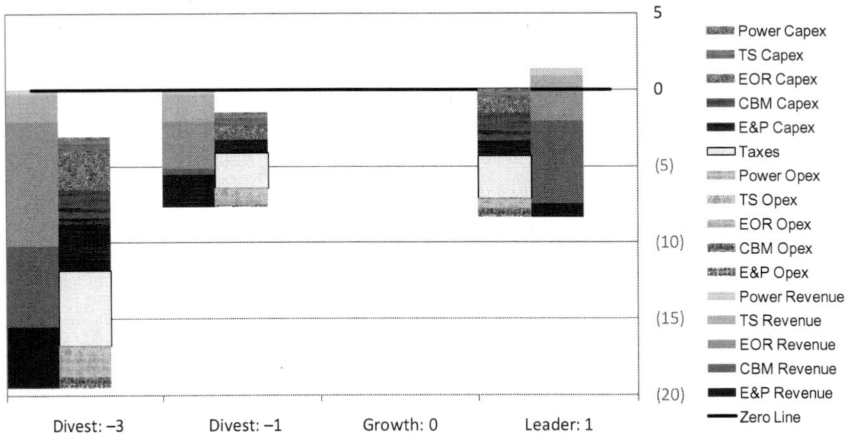

FIGURE 11.7 RNAS value components, as compared with the Growth strategy. (See insert for color representation of the figure.)

forgone and the corresponding Capex benefits (= sales proceeds) suggest that sale of those assets is not advantageous.

When possible, it is helpful to use colored graphics for delta value components charts, because the components can shift sides between positive and negative.

As the RNAS experts considered an early version of this chart, they felt that the model yielded too optimistic a portrayal of EOR and CBM. This led them to identify key uncertainties that could reduce value: the probability of outright failure at EOR fields, and the possibility of substantial dewatering delay at CBM. After these issues were addressed, these BUs still looked attractive, and points of weakness in the analysis that had material impact were improved upon.

If there are tradeoffs between one objective or value component and another, here are some ways to use them to develop an improved strategy:

- Enhance a favorable item.
- Minimize or mitigate an unfavorable item.
- Optimize for a different value measure.
- Find a better compromise among the objectives implicated in the tradeoff.

Note that a richer and more varied set of strategies is apt to unearth a broader variety of ways to find value. These will show up as advantages in one Value Component or another, and give additional opportunities to create an improved hybrid strategy (Keeney, 1992).

11.3.3 CASH FLOW OVER TIME

While senior executives normally accept PV discounting, the results of time discounting and the rank-ordering this gives to the strategies may not be immediately obvious or compelling to them. In addition, it abstracts away specific timing of events, which may be of interest. If strategies have distinctly different time profiles of cash flow, it can be helpful to show the cash flow profiles of the strategies together on one chart. This way we can ask the decision team:

- Do you believe the time profile for each strategy?
- Do you believe how the profiles change from one strategy to the next?
- Do you affirm the value ordering implied by your stated time discount rate?

For most of the analyses discussed here, EV of uncertainties is the cleanest approach. For cash flow over time, the EV approach appropriately represents the aggregate impact of benefits when there is a probability of failure, or downstream optionality. However, this must be balanced against the tendency of an EV cash flow to give a misleading view of the duration of investment when the timing of the investment is uncertain—if the investment takes place in a short time period but the time when it starts is uncertain, the EV chart will spread the investment amount over a range of years. A pseudo EV (see Section 9.5.3) chart

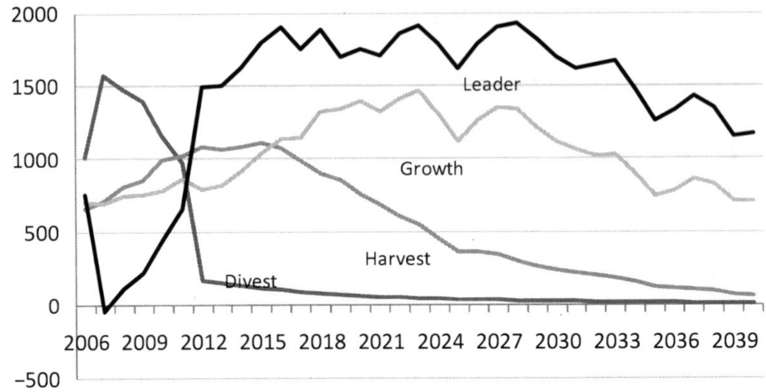

FIGURE 11.8 RNAS EV cash flows.

can be advantageous in such cases, because it distributes investment over a typical time duration starting at a typical time point, rather than spreading it across all possible years.

Figure 11.8 shows the expected cash flow profiles of the four initial RNAS strategies through time. This chart shows that the Leader strategy, which looks very favorable, delivers little positive cash flow during early years. To RNAS management, this was a jarring realization, insofar as North America had been viewed as a source of cash to fund overseas investment.

Having affirmed this view of the strategies, we can look for ways to improve: Can we accelerate benefits? Can we delay costs? How can we obtain early information on key uncertainties before we commit to major capital and operational expenses?

11.3.4 DIRECT EV TORNADO DIAGRAM

As discussed in Section 9.8.2, an informative way to understand the impact of uncertainty is via sensitivity analysis, which identifies the impact on value of varying each uncertainty in turn from low input to high input. To manage the other dimensions of decision complexity, we choose a strategy, take NPV value across time (or some other metric of interest), and aggregate Business Units.

We display the results of sensitivity analysis in the form of a tornado diagram. In an EV tornado diagram, the value axis is horizontal, and each uncertainty is represented by a horizontal bar from the conditional EV generated by its "low" input to the conditional EV generated by its "high" input. These bars are then sorted with the largest bars at the top, to call attention to variables with the greatest impact on expected value. We call a tornado diagram addressing a chosen strategy a "direct" tornado diagram.

With Monte Carlo simulation, it is not necessary to run a separate simulation to calculate the conditional EV for each end of each tornado bar. Instead, the complete EV tornado diagram can be calculated from the tabulated inputs

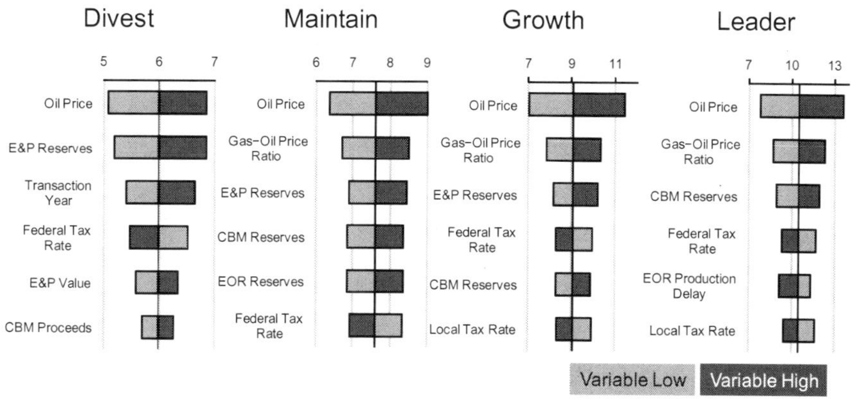

FIGURE 11.9 RNAS direct tornado diagrams, $B.

and outputs of one set of simulation trials. For each uncertain input, the subset of trials in which that input is at its "low" setting is identified, and the expected value of the output value metric is calculated for that subset (by summing the values and dividing by the number of trials in the subset). This expected value is one end of the EV tornado bar for the input parameter. The same procedure is used for the "high" setting of the input to get the other end of the tornado bar. For purposes of this procedure, if an input is sampled from a continuous probability distribution, we define as a "low" input setting anything below the P30 (if the 30/40/30 discretization is used) or P25 (if the 25/50/25 discretization is used). Similarly, a "high" input setting is anything above the P70 or P75.

The direct tornado diagram is a standard feature of most decision analysis software. The user specifies low and high settings for the input parameters and the tornado diagram is calculated and displayed. However, while deterministic tornados are widely available, fewer packages offer EV tornado diagrams.

Figure 11.9 shows the most sensitive variables from the EV tornado diagrams for the four RNAS strategies.

It was not surprising to find the prices of oil and gas at the top of the direct tornado diagram for all the RNAS strategies. Fairly wide ranges of uncertainty about effective taxation rate represented the belief that substantial tax relief might be available.

After developing an EV tornado diagram, review it with the decision team. Compared with the deterministic tornado, whose center line is the base case value or pseudo EV (see Section 9.8.2), the EV tornado helps the decision makers become more familiar with the true probability-weighted average value of the strategy, which is the ultimate decision criterion.

Compare the large bars of the EV tornado to the factors the decision team thinks should be important, and identify anything additional or missing in the tornado. Check the direction of each bar against intuitions about which direction it should point. Each point where intuitions differ from the result of the analysis constitutes a "tornado puzzle." Tornado puzzles are a great way to develop insights or to identify model errors. If a result does not make sense initially, investigate how it came about and determine whether this makes sense. If it seems wrong, figure out how to improve the model so that it reflects the team's expertise more closely. If the result holds up, this is an insight for the team. Understand it clearly and explain it to them. As an example, it is interesting to note that in the RNAS Divest Strategy tornado (see Fig. 11.9), the transaction year is a sensitive parameter. This suggests that a later divestment is better than an early divestment. This is consistent with the overall picture that divestment of all assets is less attractive than retaining, developing, and producing the assets; hence, putting off the divestment is attractive.

11.3.5 DELTA EV TORNADO DIAGRAM

An important variant of the tornado diagram is created when we specify two strategies and show the impact of uncertainties on the *difference in value* between them. The resulting tornado diagram, called the "delta" tornado, highlights the variables that drive the choice between the two strategies. Typically, it is most informative to investigate the delta between two leading strategies. The delta tornado diagram is introduced in Section 9.8.2.

In the RNAS case, having seen the favorable financials for CBM and EOR, the decision team formulated a hybrid strategy with leader-level investment in CBM and EOR, and growth-level investment in E&P. While the ENPV value of this hybrid was very favorable, its investment requirement was high. Accordingly, a similar hybrid was created, identical in all regards except that the legacy E&P assets would be divested instead of being developed. These strategies, which were named Team Hybrid and Divest Hybrid, had very similar ENPVs, indicating that the proceeds from divestment would be as attractive as continued development, as measured by ENPV. Furthermore, as shown in the left and center of Figure 11.10, the direct tornado diagrams for the two strategies were very similar, leaving the team to wonder whether the analysis could shed any light on the choice between them. However, a delta EV tornado (Fig. 11.10, right) showed that the strategies were different and identified what drove the difference. Oil price, tax rates, and variables related to CBM and EOR affected both strategies equally and "canceled out" in the delta tornado, leaving at the top only the variables that drove the difference in value for this decision. The variables affecting the relative value of the Team Hybrid and Divest Hybrid were the gas price that would be realized if the E&P assets were retained and the market value of E&P assets if they were divested.

A large bar in a delta tornado diagram that crosses zero indicates that additional information about this variable might cause us to change our decision. Accordingly, it can be worthwhile to convene experts to consider whether there is some way to gather information about any of the leading variables in the delta tornado diagram before the decision must be made. If there is an information-gathering approach, we reformulate the decision model to reflect the option to gather the information, and test whether a strategy embodying this option is worthwhile.

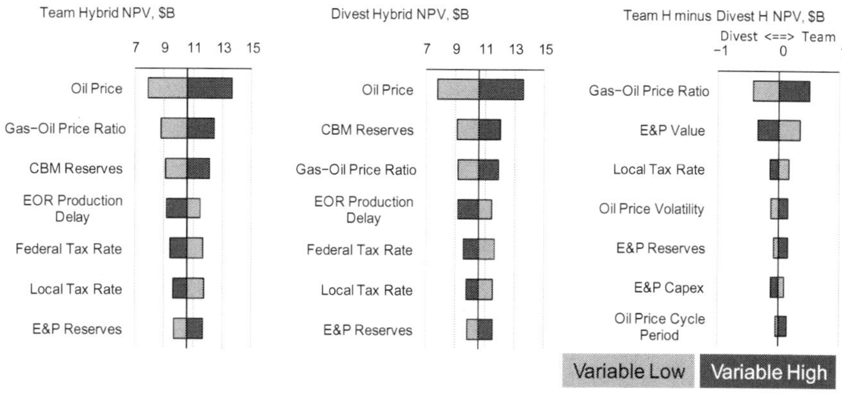

FIGURE 11.10 Direct and delta tornado diagrams for team hybrid and divest hybrid.

The value of being able to make a decision in light of information to be gained in the future is at the heart of the "real options" approach to valuation (Smith & Nau, 1995). While this was not possible for E&P divestment at Roughneck, Section 11.3.8 shows where this was worthwhile.

11.3.6 ONE-WAY AND TWO-WAY SENSITIVITY ANALYSIS

The delta tornado diagram provides a comparison across many input parameters of how changes in those parameters affect value. It is sometimes insightful to focus on just one input parameter and see how the values of strategies are impacted by changes in that parameter. This is called one-way sensitivity analysis.

Figure 11.11 displays an example of one-way sensitivity analysis. It shows how the values of the two hybrid strategies in the RNAS case vary with changes in the input parameter Gas Price. In this example, the range of uncertainty in the parameter is divided into five equally likely intervals ("quintiles"), and the conditional ENPV is shown for each of the two strategies given that the parameter is in each quintile. We can see that both strategies increase in value as the Gas Price increases but that the value of the Team Hybrid strategy rises faster than does that of the Divest Hybrid strategy, because Team Hybrid has more gas production to sell.

The idea behind one-way sensitivity analysis can be extended to show how the values of strategies are affected by changes in two input parameters taken together. This is called two-way sensitivity analysis. A graphical representation of two-way sensitivity analysis would be a three-dimensional chart showing a value surface for each strategy above a two-dimensional space of combinations of settings of the two parameters. A two-dimensional rendition of such a chart would likely be too complicated to produce useful insights. Instead, the results of two-way sensitivity analysis are best shown in tabular form, subtracting one strategy's value from the other's.

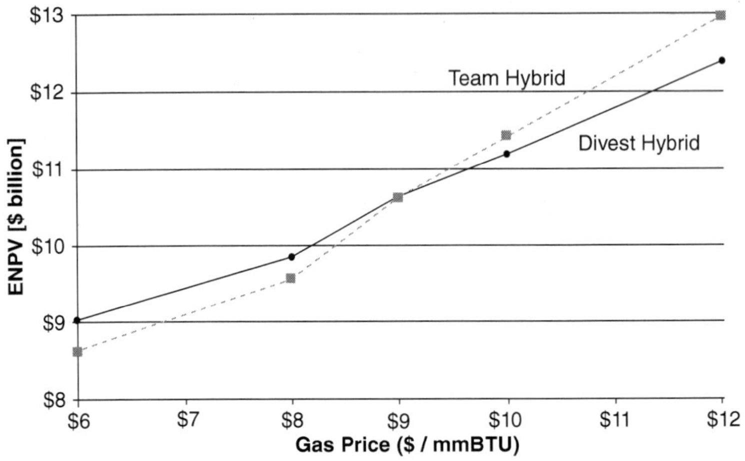

FIGURE 11.11 One-way sensitivity analysis of RNAS hybrid strategies to Gas Price.

TABLE 11.3 Two-Way Sensitivity Analysis of RNAS Hybrid Strategies to Gas Price and E&P Value

	Delta ENPV (Team Hybrid Minus Divest Hybrid)				
	Gas Price ($/mmBTU)				
E&P Value ($/boe)	$6	$8	$9	$10	$12
$15.0	$0.1	$0.2	$0.2	$0.7	$1.1
$17.5	−$0.3	−$0.3	$0.2	$0.4	$0.7
$20.0	−$0.4	−$0.3	$0.2	$0.1	$0.8
$22.5	−$0.5	−$0.3	−$0.3	$0.0	$0.3
$25.0	−$1.0	−$0.8	−$0.2	−$0.1	$0.2

Table 11.3 shows an example of two-way sensitivity analysis. This table shows the difference in conditional ENPV for the two RNAS hybrid strategies for each of 25 combinations of settings of two input parameters: Gas Price and E&P Value. These variables were chosen to represent the top variables from the delta tornado diagram. Gas Price itself was used, rather than the Gas–Oil Price ratio, to make the chart more easily comprehensible. As in the one-way sensitivity analysis example, the range of uncertainty for each parameter is divided into five quintiles, so the probability that the actual settings of the two parameters will be in any given cell of the table is 4%.

We can see from the table that the value delta between Team Hybrid and Divest Hybrid is greatest when Gas Price is high and E&P Value is low (upper right-hand corner). This indicates that Team Hybrid does better when gas price is high (because it retains more gas production), while Divest Hybrid does better when the price paid for divested assets is high.

The shaded cells in the table indicate the combinations of settings of the two parameters for which the Divest Hybrid has the higher ENPV.

11.3.7 VALUE OF INFORMATION AND VALUE OF CONTROL

Two powerful concepts in decision analysis that should be a part of every value dialogue are the value of information and the value of control.

11.3.7.1 Value of Information. An important question to ask is the following: How much more valuable would this situation be if we could resolve one or more of the key uncertainties before making this decision? The answer to this question is the value of information. We make a distinction between perfect and imperfect information. We often call perfect information *clairvoyance*, because it is what we would learn from a *clairvoyant*—a person who perfectly and truthfully reports on any observable event that is not affected by our actions (R. Howard, 1966). Whereas clairvoyance about an uncertain parameter eliminates uncertainty on that parameter completely, imperfect information reduces but does not eliminate the uncertainty. Despite the fact that almost all real-world information about the future is imperfect, we tend to calculate the value of clairvoyance first because it is easier to calculate than the value of imperfect information and it sets an upper limit on the value of imperfect information in the current decision context. If the value of clairvoyance is low or zero, we know that we need not consider getting any kind of information on that parameter for the decision under consideration. Of course, information that has zero value for one decision might have positive value for another. For instance, information on oil price periodicity may have no value for making long-term strategic decisions for a petroleum company, but it could have high value for a decision about whether to delay the construction of a petrochemical plant.

Calculating the value of clairvoyance is fairly straightforward. It involves restructuring the probabilistic analysis so that the choice of strategy is made optimally after the uncertainty in question is resolved. If the analysis is structured as a decision tree, this is equivalent to moving the uncertain node in front of the decision node. This may require using Bayes' rule to redefine conditional probabilities if this uncertainty is dependent on another. In most cases,[3] the value of clairvoyance is the difference between the optimal value in this restructured tree and the optimal value in the original tree.

Figure 11.12 shows the calculation of value of clairvoyance on whether or not royalty payments are required in the capacity planning example presented in Section 11.2.1.1. The figure shows only the start of the restructured decision

[3]The simple calculation of the value of clairvoyance is valid if the decision maker is risk neutral or is risk averse and accepts the delta property (see Section 11.4.1).

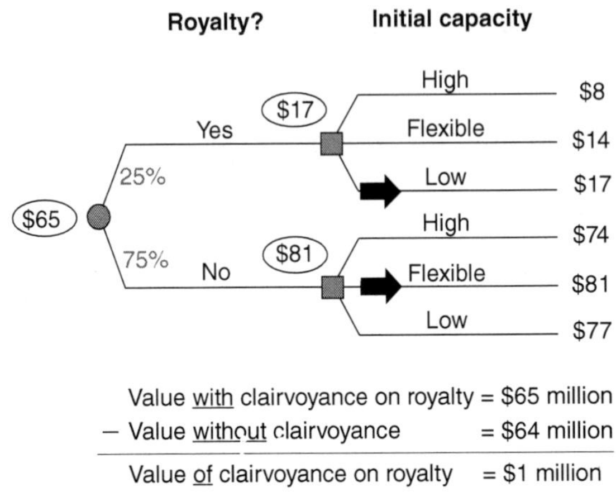

FIGURE 11.12 Calculating value of clairvoyance on royalty in capacity planning example.

tree, with the royalty uncertainty now coming before the decision on initial capacity. We see that if we know in advance that royalties will be required, we would choose Low initial capacity; but if we know that royalties will not be required, we would stick with the original optimal choice of Flexible initial capacity. The expected NPV of this new situation is $65 million, compared with $64 million for the situation without clairvoyance. So the value of this clairvoyance is the difference, $1 million.

It can be helpful to systematically calculate the value of clairvoyance on each uncertain parameter. Typically, the value is zero for many parameters. It may also be worthwhile to calculate the value of clairvoyance on combinations of two or three parameters together, particularly if such combinations correspond to actual information-gathering possibilities. The value of information cannot be negative. And the value of information on a combination of uncertain parameters together is not always equal to the sum of the values of information on those parameters taken individually.

Some influence diagram/decision tree software will automatically calculate the value of information on each uncertain variable in the decision tree.

Virtually all information-gathering opportunities considered in professional practice involve information that is imperfect rather than perfect. Calculating the value of imperfect information uses the same logic as calculating the value of clairvoyance—determine how much the overall value of the situation increases because of the information. The actual calculation is somewhat more involved, generally requiring the use of Bayes' rule to calculate conditional probabilities in a restructured tree. Section 11.3.8 illustrates how the price of oil current at the time of a decision can be used as valuable imperfect information on long-term oil prices.

Delta EV tornado diagrams give us some indication of which parameters have positive value of clairvoyance. On a delta EV tornado diagram comparing the optimal strategy with another strategy, any parameter whose bar crosses the zero delta value line has positive value of clairvoyance.

In the value dialogue, we can use the calculated value of clairvoyance to improve the decision in several different ways. The first is obvious—create a new strategy that includes initial information gathering on a parameter with high value of clairvoyance before the main decision is made. This might be a diagnostic test, a market survey, an experimental study, or a pilot plant. When evaluating such a strategy, it is important to characterize the degree to which the actual information gathered will be imperfect—how will the probabilities of the possible outcomes of the parameter in question change with the gathered information? Also, it is important to include the cost, if any, of delaying the decision in such a strategy.

Another possibility is to design a new strategy that builds-in flexibility, effectively deferring the decision until after the uncertainty is resolved. For example, in a production planning decision, if sales demand has high value of clairvoyance, it might make sense to use a temporary means of production (e.g., outsourcing) for a few years and then commit to investing in internal capacity with better knowledge of demand. (See also the discussion of real options in Section 11.3.8.) Also, it might be possible to incur a cost to speed up the resolution of an uncertainty with high value of clairvoyance.

11.3.7.2 Value of Control. The value of control is a concept analogous to that of the value of information. Here, we ask the question: How much more valuable would this situation be if we could make a key uncertainty resolve *in our favor*? The mythical character in this case is the Wizard—the one who can make anything happen at our request.

It is a simple matter to calculate the value of perfect control (assuming that the decision maker either is risk neutral or adheres to the delta property—see Section 11.4.1). Fix the parameter in question to its most favorable setting and run the probabilistic analysis for all strategies, observing the highest EV among them. The difference between that EV and the EV of the best strategy without control is the value of control. The value of perfect control can be found quite easily from direct EV tornado diagrams—simply subtract the original EV from the highest value among the right-hand ends of the parameter's tornado bars for all strategies. Similarly, if the value of clairvoyance on a parameter has been calculated by restructuring a decision tree, the value of perfect control can be found by observing the EV of the most preferred branch of the parameter in the restructured tree. For example, in the restructured tree used to calculate the value of clairvoyance on whether or not royalty payments are required in the capacity planning example (see Fig. 11.12), we see that if royalties are not required, the EV is $81 million, so the value of perfect control on the royalty issue is $17 million ($81–$64 million).

We can use the calculated value of control to guide our thinking about how to create a better strategy. For each parameter with high value of control, we ask: What could we do to make the favorable outcome more likely to occur? Sometimes, the range of uncertainty that we have assessed for a parameter is based on an assumed constraint that, in reality, can be eased at some cost. For example, the range of uncertainty on how long it will take to get a new pharmaceutical product to market might be based on an assumed fixed budget for clinical trials. If the value of control on time to market is high (as it often is in that industry), then it might be that the range on time to market can be shifted in our favor by increasing the clinical trials budget. A better strategy can be created in this case if the increase in expected value due to earlier entry exceeds the cost of additional budget.

In the capacity planning example, the realization that the value of the situation increases by $16 million if we can ensure that royalties are not required may stimulate the creation of a new alternative—offering to settle the patent infringement lawsuit out-of-court for a payment of less than $16 million.

11.3.8 REAL OPTIONS

For Roughneck, Tar Sands[4] was a new technology. There was a large risk that the economics might not pan out, so Roughneck wanted to take a cautious incremental approach, beginning with construction and operation of a pilot plant, which takes many years. Subsequent large investment in full-scale development would be made only if a favorable trigger event were seen. The decision team proposed many triggering factors: new technology, regulatory support, improved oil pipeline infrastructure, or successes of competitors in analogous efforts. However, preliminary analysis did not find much value in any of these triggering strategies.

If Roughneck were required to precommit to full-scale Tar Sands development, the program's value increment would not be favorable enough to be funded, so there would be no reason for preliminary technology development or construction of a pilot plant. Even so, we analyzed this simple precommitment strategy to identify its value drivers and used these insights to develop a more valuable plan. We found that consideration of the business-unit-specific tornado for this strategy (Fig. 11.13) allowed us to create a contingent approach that increased the value of the program substantially. High oil price can make the value increment substantially positive. By the time the pilot plant is finished, Roughneck will have additional information about the long-term level of oil price. This suggested a contingent (or real option) strategy that would build the pilot plant, but only move to full-scale development if the then-current oil price exceeds a threshold.

At the time of the analysis of precommitment to full-scale development, the model of oil price was simple: a range of possible oil prices at a fixed time point, and a range for the oil price escalator in subsequent years. We realized that testing contingent strategies against this orderly and predictable model of prices would overstate the value of the information seen at the decision point

[4]Tar sands (also called oil sands) is a combination of clay, sand, water, and bitumen, a heavy black viscous oil. Tar sands are mined and then the oil-rich bitumen is extracted.

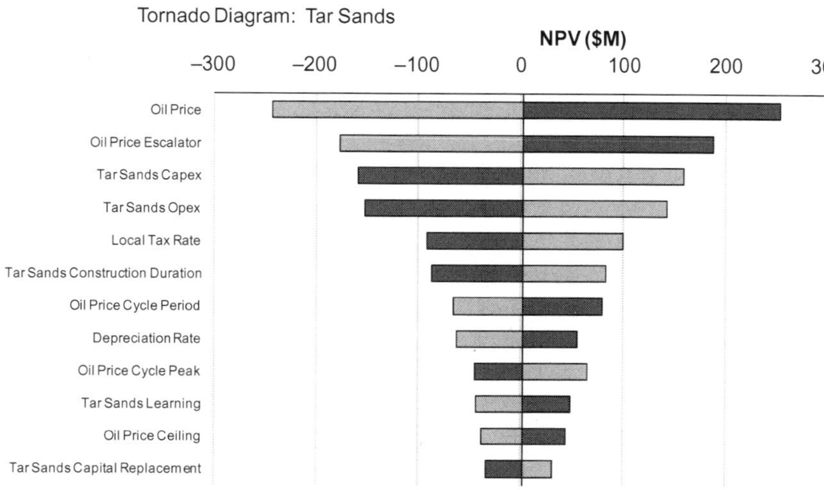

FIGURE 11.13 Tar sands tornado diagram.

for full development, and thus overstate the value of the contingent strategy. So we worked with the oil price experts to more appropriately reflect the factors that inform the company's view of future long-term oil prices: cyclicality (with partially understood period, amplitude, and phase), and volatility. Roughneck experts judged that a simulation that handled these issues properly would have enough noise that the information conferred by observing the value at the time the decision has to be made (10 years in the future) would be informative, but without overstating the value of that information. Hence, we added parameters for these uncertainties, assessed their plausible ranges, and added them to the simulation.

We then simulated strategies that would build the full-scale Tar Sands facility only if the then-current oil price exceeded a threshold. We added a decision column to the strategy table to represent the threshold oil price at which Tar Sands would go forward, and added logic to the model to implement this decision rule. We then created strategies that differed only in the Tar Sands threshold, in order to test thresholds of \$40, \$50, \$60, \$70, and \$80 dollars per barrel. We simulated these strategies and noted the Tar Sands ENPV value of each option. A threshold of \$60/bbl seemed best, improving the program from break-even to ENPV of roughly \$50M. See Figure 11.14. As we discuss in the next section, this level of threshold reduces the likelihood of unfavorable investment while still allowing favorable investment to be made.

11.3.9 S-CURVES (CUMULATIVE PROBABILITY DISTRIBUTIONS)

When a favorable strategy has a big downside, we can use the S-curve of the value measure to understand how much risk we are facing, and how well a proposed response may reduce that risk.

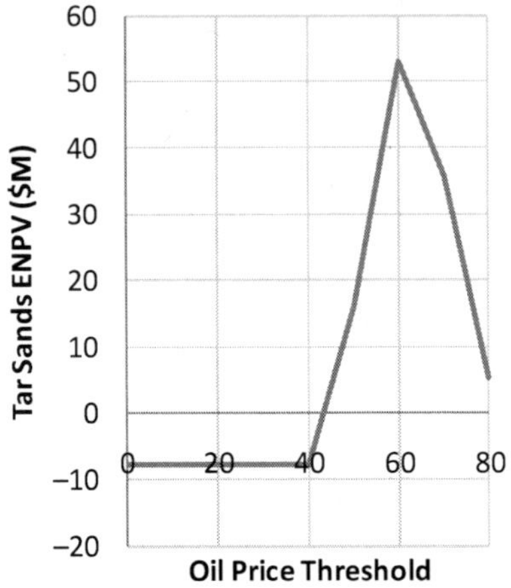

FIGURE 11.14 Tar sands construction threshold exploits optionality.

An "S-curve" (or cumulative probability distribution) depicts the probability that value achieved (or any other objective) is at or below any given level, that is, its fractile. In the conventional layout for S-curves, value is on the *x*-axis, and the cumulative probability is on the *y*-axis. Hence, in an S-curve, the "upside" is to the right, not up. To chart an S-curve, sort in order from lowest to highest the instantiated values from Monte Carlo simulation and juxtapose these with the position number, going from 1 to *N*, the total number of values. Then calculate the cumulative probability for each value in the list by dividing its position number by *N*. An S-curve is a line plot of these value-cumulative probability pairs. S-curves are automatically generated by decision tree and Monte Carlo software.

S-curves are a very useful analytical tool. However, it is important to understand that decision makers and stakeholders may have difficulty understanding the S-curve. In organizations without a culture of using probabilistic decision analysis, it will be important to clearly explain the S-curve. Some decision makers are more comfortable looking at the flying bar charts (next section) or probability density functions, which can show similar information.

Once we see the downside risk of our best alternatives, we look for ways to modify the strategies to mitigate the risk.

If the S-curve for the leading strategy is wholly to the right of the others, it is said to be "stochastically dominant."[5] This means that it is more likely to deliver any specified level of value than any other strategy. It also means that the strategy is preferred over the others regardless of risk attitude.

[5]Early works also refer to the notion of "deterministic dominance," which is seen when the 0th percentile of one strategy exceeds the 100th percentile of another. Situation exhibiting such extreme superiority rarely arise, so the term is not often useful in practice.

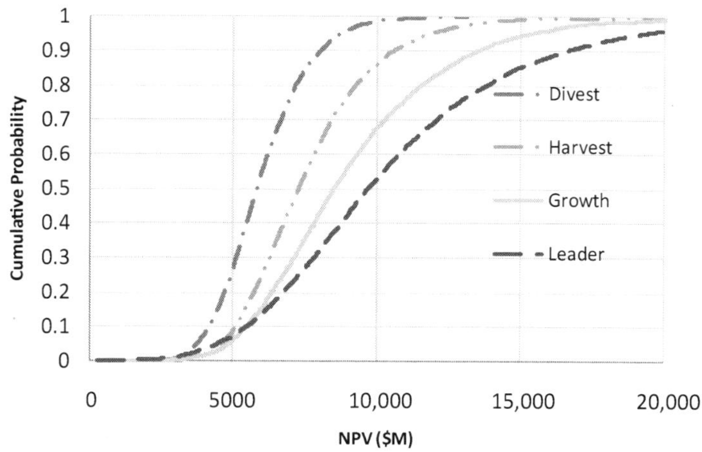

FIGURE 11.15 RNAS S-curves.

If two S-curves cross over and there is a large downside on the NPV distribution of one, and if the decision maker feels this downside constitutes a substantial risk, this would call into question using ENPV as the decision criterion. In this case, we would need to take risk attitude into account explicitly by assessing a true utility function (see Section 11.4), rather than using ENPV as a surrogate utility function.

Figure 11.15 shows the S-curves for NPV value for the four RNAS strategies. We found that the strategies' S-curves did not cross significantly, which meant that there was no risk–value tradeoff to be made. More ambitious strategies generated more upside, but their downsides were not worse. Some decision team members found this counterintuitive, because larger strategies require higher investment. However, for RNAS, the investment was incremental through time, with plenty of feedback, so there was never much investment truly at risk. None of the downsides was viewed as serious enough to call for a utility function to be formulated; so maximizing ENPV was used as the decision criterion.

As always, we start by asking the decision team the following questions:

- Do you believe the value profile for each strategy?
- Do you believe how the value profiles change from one strategy to the next?
- Do you affirm the ordering implied by your stated risk attitude (if pertinent)?

Having done so, we ask whether we can enhance the upside or mitigate the downside (reduce the risk) of a leading strategy. On its own, an S-curve gives little guidance where to look for such an enhancement. We should look instead at the tornado diagram (or the "Contribution to Variance" output provided by some Monte Carlo software packages) to identify the biggest drivers of uncertainty in value, which typically are the ones that drive the downside risk.

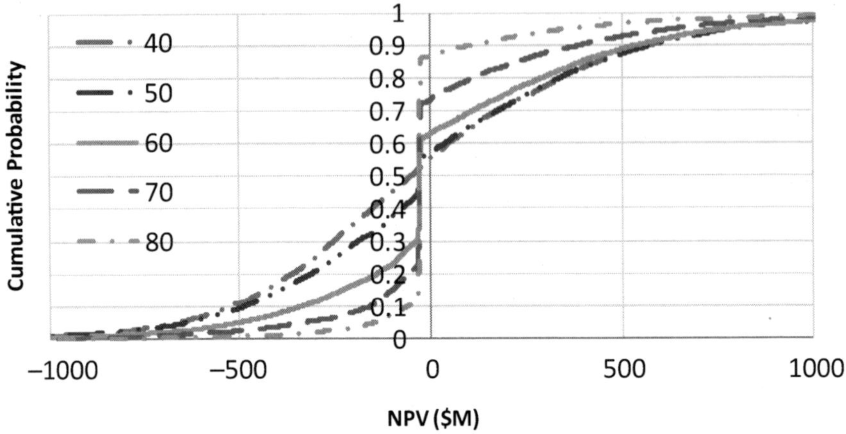

FIGURE 11.16 Tar sands value-risk profiles.

An S-curve is more useful as a description of a strategy's potential outcomes than as a spur to creativity. For example, we used business-unit S-curves to understand how an oil price threshold added value to Roughneck Tar Sands strategies.

The S-curves in Figure 11.16 show the value profiles of Tar Sands development, using various thresholds. The wedge between the vertical axis and an S-curve on the right measured the upside, while the wedge between the axis and the curves on the left measured the downside. From $40 to $50 to $60, reductions of downside (from eliminating unfavorable investment) were large, with little reduction to upsides (investment that would have been favorable), and so expected value improved. But past $60 to $70 and $80, there was not so much downside to mitigate, and the reductions of upside predominated, and so expected value decreased.

For a complex decision situation like RNAS Tar Sands, there were too many uncertainties and decision opportunities to be displayed intelligibly as a full decision tree, and representing them all at requisite granularity in the analysis software would require a huge multidimensional data structure. However, once we had made the decision and wanted to show its logic to senior stakeholders, a simplified decision tree was useful. This is discussed in the Chapter 13.

11.3.10 FLYING BAR CHARTS

A chart showing S-curves for several different strategies might not be the best way to convey insights to decision makers, particularly if the S-curves cross each other, because the insights are obscured by the detailed information in the chart. Instead, a simpler chart, called a "flying bars" chart,[6] might be a more effective

[6]These charts are commonly called box and whisker charts or stock charts in Excel.

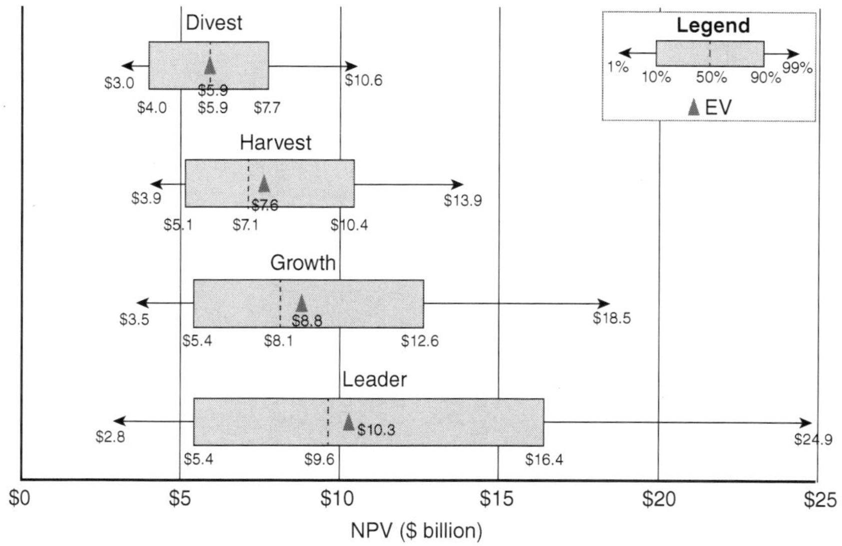

FIGURE 11.17 Flying bars chart for RNAS strategies.

way to display meaningfully a comparison of the uncertainty and risk in the strategies. The flying bars chart shows only a small subset of the information in an S-curve: a few percentiles plus the EV. Figure 11.17 shows the flying bars chart for the four original RNAS strategies that are displayed as S-curves in Figure 11.15. The ends of each bar in the chart are the P10 and P90 NPVs for the strategy, while the ends of the arrows are the P1 and P99 NPVs. The P50 NPV is shown as a dashed line, and the EV is indicated by a triangular shape. For an even simpler chart, the P1 and P99 arrows and the P50 line may be omitted.

11.4 Risk Attitude

Many decisions can be made using the criterion of maximizing EV. However, if the S-curves for the strategies reveal significant downside risk and if none of the strategies stochastically dominates the others (i.e., has an S-curve that is completely to the right all other S-curves), then the risk attitude of the decision maker must be taken into account explicitly to find the optimal strategy. To do this, we go back to the Five Rules stated in Chapter 3.

The Five Rules together imply the existence of a utility metric. To be consistent with the Five Rules, decisions should be made to maximize the expected value of that utility metric. But the Five Rules do not specify the form of the utility function that maps value outcomes to the utility metric. The characteristics of the utility function depend on the preferences of the decision maker regarding risk-taking, which we call *risk attitude*. A person who always values an uncertain deal at its expected value is said to be *risk neutral*. A person who values an uncertain deal at *more* than its expected value is said to be *risk seeking*. This person would willingly pay money to increase the level of risk. In professional practice, we almost never encounter a decision maker who is truly risk seeking. A person who values an uncertain deal at *less* than its expected value is said to be *risk averse*. This person is willing to give up some value to avoid risk. The degree of a person's risk aversion can be measured quantitatively.

11.4.1 DELTA PROPERTY

An appealing addition to the Five Rules is called the delta property, which states that if a constant amount of money is added to every outcome of an uncertain deal, the value of the deal increases by that amount. As an example, let us suppose that someone values at \$4000 an uncertain deal that offers equal chances of winning \$10,000 and winning nothing. How would that person value the uncertain deal if we add \$100 to each outcome (i.e., the modified deal offers equal chances of winning \$10,100 and \$100)? If the delta property applies, the person would value the modified deal at \$4100. This seems to make a lot of sense, since the modified deal is equivalent to the original deal (worth \$4000) plus a sure \$100. Clearly, it is hard to argue against the delta property as long as the delta amount does not significantly change the person's wealth.

For someone who wants to make decisions consistent with the Five Rules plus the delta property, the utility function must have one of two forms—either linear or exponential (Howard, 1983).

A linear utility function is appropriate for a decision maker who is *risk neutral*. This person values every uncertain deal at its expected value, regardless of the level of risk. In other words, a risk neutral decision maker is not willing to give up any expected value to avoid risk.

11.4.2 EXPONENTIAL UTILITY

An exponential utility function is appropriate for a person who is risk averse and who wants to be consistent with the Five Rules plus the delta property. The functional form of the exponential utility function is:

$$u(x) = A - Be^{(-x/R)}, \tag{11.1}$$

for $B > 0$ and $R > 0$.

The essential properties of any utility function are preserved in a linear transformation (i.e., when constants are multiplied and/or added to the function) (R. Howard, 1983). So the parameters A and B in Equation 11.1 are arbitrary. There is only one parameter (R) that matters for an exponential utility function. This parameter is called *risk tolerance*, and it is expressed in the same units as the value metric, which is usually in monetary units. (The reciprocal of risk tolerance is given the name *risk aversion coefficient* in the literature.) The larger the risk tolerance, the smaller the degree of risk aversion. An infinitely large risk tolerance indicates risk neutrality.

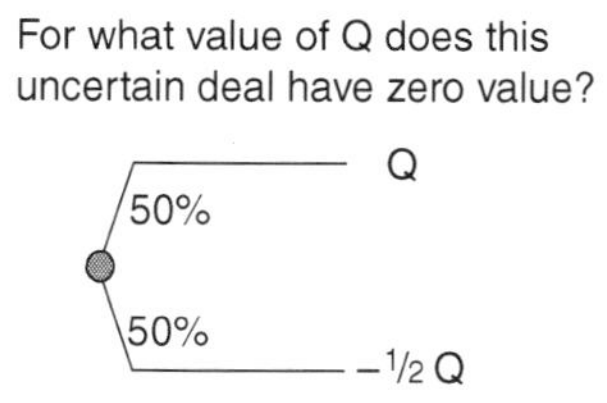

FIGURE 11.18 Assessing risk tolerance.

11.4.3 ASSESSING RISK TOLERANCE

The approximate value of risk tolerance for any decision maker can be assessed via the answer to just one question (see Fig. 11.18): For what amount of value (e.g., money) Q is the decision maker indifferent between having and not having an uncertain deal offering equal chances of winning Q and losing one-half of Q (McNamee & Celona, 2001)? For small values of Q, the downside risk is small so the positive expected value of the uncertain deal ($Q/4$) makes it look attractive. But for large values of Q, the 50% chance of losing $Q/2$ makes the deal look too risky to be attractive.

The value of Q that puts this deal on the boundary between attractive and unattractive is a good estimate of the decision maker's risk tolerance. (The actual risk tolerance is about 4% bigger than Q.)

Using an Appropriate Risk Tolerance Within a Large Corporation

Note: The role of decision analysis is to help decision makers select the best choices consistent with their information and preferences. The discussion that follows goes beyond this role in that it suggests what those preferences should be, at least regarding risk-taking.

The risk attitude expressed by decision makers for business organizations typically varies with the size of the organization. Large businesses tend to be more willing to take big risks than small ones. In the parlance of Section 11.4.3, risk tolerance tends to increase with the size of the organization. This is perfectly appropriate if the risk tolerance is related to the size of the entire company. But we believe that it is not appropriate if a decision maker for a division within a large corporation uses a risk tolerance that is based on the size of the division rather than on the size of the corporation.

Imagine a division head choosing between two alternative investments, a risky one called "Big Bet" and a much safer one called "Small Step." Big Bet has a much higher EV than Small Step, but because of its substantial risk, it has a lower certain equivalent when calculated using the divisional risk tolerance. So the division head chooses Small Step. When viewed from corporate headquarters, however, this appears to be a bad decision. When calculated using the much higher corporate risk tolerance, the certain equivalent of Big Bet is greater than that of Small Step. In the eyes of corporate management, the division head has destroyed value by making the safer choice.

This is a problem of corporate governance. By allowing its divisions to act as if they were small companies, the corporation gives up one of the main advantages of large size—the ability to pool risks across divisions and thereby take on value-creating ventures that smaller players reject as too risky. This is an important source of competitive advantage. A corporation that gives up that advantage by playing it too safe will very likely be outperformed by one that appropriately uses its greater appetite for risk.

The remedy is for corporate headquarters to establish a clear policy (with enforcement) that all "big bet" decisions must be made consistent with the corporation's risk tolerance (just as it should insist that all PV calculations use the same corporate discount rate). Prudence might dictate that decisions that seem too risky from the divisional perspective be reviewed and authorized by corporate headquarters. And divisional decision makers should be supported by a corporate culture that rewards appropriate risk-taking and does NOT punish bad outcomes that are beyond their control.

It is important to remember that discounting for risk means giving up value in exchange for safety. The more one opts for safety, the more value one loses. The best practice is to find the right balance between the two by always using the appropriate risk tolerance.

11.4.4 CALCULATING CERTAIN EQUIVALENTS

Once the value of a decision maker's risk tolerance has been assessed, it can be used to calculate his or her *certain equivalent* (CE) for any uncertain deal. The utility metric is calculated for each possible outcome using the utility function. For computational simplicity, the best choice of the two arbitrary parameters for the exponential utility function in Equation 11.1 are $A = 0$ and $B = 1$, yielding the utility function:

$$u(x) = -e^{(-x/R)}. \tag{11.2}$$

The probability-weighted average of the utility metric (i.e., the expected utility) is calculated for the uncertain deal and the certain equivalent is found via the inverse of the utility function:

$$\text{CE} = -R\ln(U), \tag{11.3}$$

where U is the expected utility.

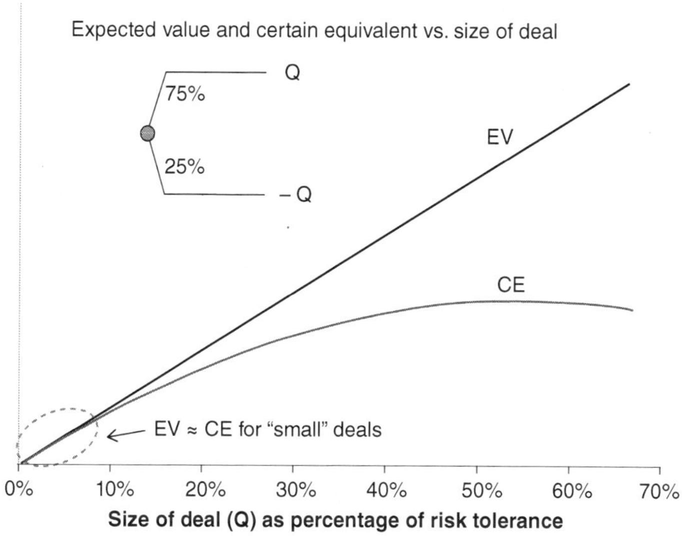

FIGURE 11.19 EV and CE versus size of deal.

11.4.5 EVALUATING "SMALL" RISKS

Knowing the appropriate value of risk tolerance, we can easily calculate the certain equivalent for any uncertain deal for which the delta property applies.

Consider the chart in Figure 11.19, which shows the calculated certain equivalent and the expected value for uncertain deals of different "sizes." The deal is a 75% chance of winning an amount Q and a 25% chance of losing that amount Q. The chart shows the EV and CE for different amounts Q expressed as a percentage of the risk tolerance. Note that for deals that are quite "small" relative to the risk tolerance (<10%), the expected value and the certain equivalent are very close. This suggests that for "small" risks, we do not need to calculate the certain equivalent but instead can use the expected value as a very good approximation of it.

But how do we know if an uncertain deal is "small" enough to use EV rather than CE as the decision criterion? If we have assessed the risk tolerance, we can apply a simple rule of thumb—If the range of outcomes (from best to worst) in a deal does not exceed 5% of the risk tolerance, use the EV; otherwise, calculate the CE. If we have not yet assessed the risk tolerance, we can sometimes make a very rough guess as to what it would be if we were to assess it. One study (McNamee & Celona, 2001) suggests that a company's risk tolerance is roughly equal to 20% of its market capitalization. If we use this rough estimate along with the rule of thumb stated above, we should feel comfortable using EV as the decision criterion for uncertain deals whose range of outcomes is less than 1% of the company's market capitalization. However, if in doubt, always check with the decision maker.

11.4.6 GOING BEYOND THE DELTA PROPERTY

Very occasionally, we may encounter a decision situation with a range of possible outcomes so big that the delta property no longer applies. This would be a situation in which at least one of the possible outcomes would significantly change the wealth of the decision maker and thus change his or her attitude toward risk-taking. In such a situation, the utility function would have a different form than the exponential and assessing it would require more than one question.

We believe that the risk preferences of most decision makers would be characterized by decreasing risk aversion as wealth increases. As an illustration, consider an uncertain deal offering equal chances of winning \$200,000 and winning nothing. A person with relatively little wealth might prefer to receive a certain \$50,000 instead of the uncertain deal because of its substantial risk of paying out nothing. But if that same person suddenly became a multimillionaire, he or she might then prefer the uncertain deal to the certain \$50,000 because the expected value of the deal is twice as great.

A particularly interesting form of decreasing risk aversion is called the one-switch rule (Bell, 1988). This rule states that for every pair of alternatives whose ranking depends on wealth, there exists a wealth level such that one of the alternatives is preferred below that level and the other is preferred above it. The rule seems to be quite easy to accept. Consider two uncertain deals: Deal A offers equal chances of winning \$100,000 and winning \$50,000 (EV = \$75,000). Deal B offers equal chances of winning \$200,000 and winning nothing (EV = \$100,000). We would not be surprised if a decision maker prefers Deal A at low levels of wealth but then switches to preferring Deal B at a higher level of wealth. But we would find it quite surprising if that same decision maker switches back to preferring Deal A at an even higher level of wealth.

The form of utility function that is consistent with the one switch rule is the linear plus exponential:

$$u(w) = Aw - Be^{(-Cw)}, \tag{11.4}$$

for $A \geq 0$, $B > 0$ and $C > 0$.

11.5 Illustrative Examples

See Table 1.3 for further information on the illustrative examples. The probabilistic analysis of our RNAS is described in Section 11.3. The next sections describe probabilistic analysis for the Geneptin and data center examples.

11.5.1 GENEPTIN EXAMPLE

The Geneptin team conducted probabilistic analysis using an Excel model with Crystal Ball©. The simulation gave probabilities of 25/50/25 to the P10, P50, and P90 inputs assessed by the experts. The resulting P10-P50-P90 range of

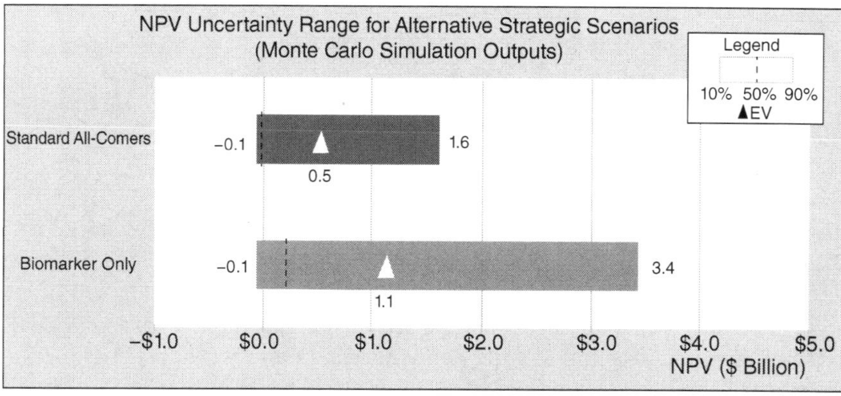

FIGURE 11.20 Geneptin flying bar chart.

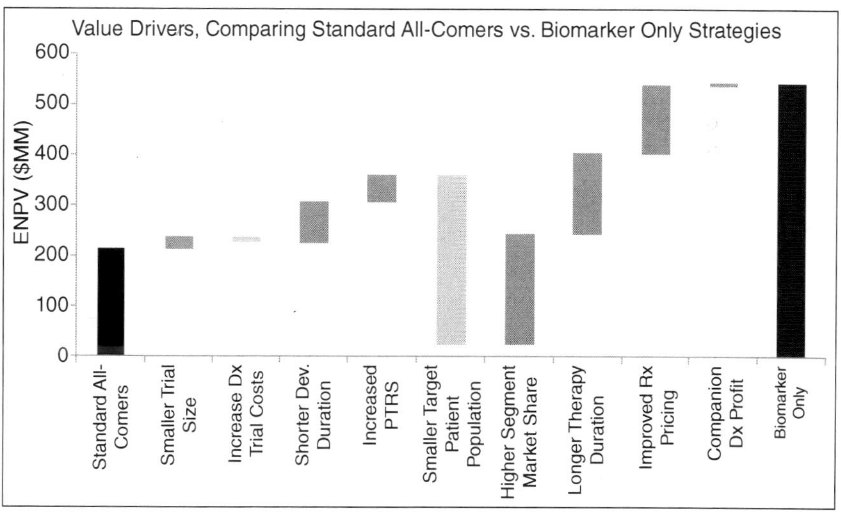

FIGURE 11.21 Geneptin waterfall chart. (See insert for color representation of the figure.)

NPV and mean (EV) for the two strategic alternatives are shown in the flying bar chart in Figure 11.20.

To communicate with stakeholders from multiple functional areas, the DA team members used tools like the waterfall chart in Figure 11.21 to illustrate how much each value driver impacts the value Geneptin generates.

11.5.2 DATA CENTER

In the real data center problem, probabilistic analysis was not performed because of tightness in the decision schedule and because some of the key uncertainties were reduced by actions taken by the implementation team.

In a different setting, a decision tree or a Monte Carlo analysis could have been performed on the data center problem. When doing probabilistic analysis for a MODA model, we need to consider if the uncertainties are dependent or independent. If the major uncertainties are the scores and these are independent, we typically use Monte Carlo simulation. If there are major interdependent uncertainties that impact several scores, a decision tree may be a more effective tool (Parnell et al., 2011).

11.6 Summary

There are two important ways to add value: choose correctly among identified strategies and improve upon the selected strategy by increasing the value and/or reducing the risk of the downside consequences. Once we have developed a preliminary version of a composite perspective on what drives value, we undertake a value dialogue with the decision team aimed at verifying the authenticity of the composite perspective, and use it to choose correctly and find even more valuable and less risky strategies.

Of the five dimensions of complexity in a decision situation (value components, alternatives, uncertainty, business units, and time), the most difficult to manage is uncertainty, since it requires more information to analyze and is more difficult to explain to the decision team. There are two main methods to analyze uncertainty: decision trees and simulation. A decision tree sets all but a small number of parameters to fixed values, discretizes the rest, and considers all possible combinations, developing an approximate EV and set of contingent decisions for the problem as structured. The simulation method samples randomly from inputs for all uncertain parameters, giving EV results that can be made arbitrarily close to the true EV answer if a sufficient number of samples is chosen. It gives guidance on construction of contingent decision rules; however, these are more difficult to obtain than with decision trees.

Either way, a large number of scenarios are considered, alongside complexity from the other four dimensions. To manage this complexity and facilitate the value dialogue, we take various two- or three-dimensional slices of value outcomes, while collapsing the other dimensions. While there are various ways to collapse dimensions, the typical approaches are:

- Look at value and value components of each strategy, or look at the delta between two strategies.
- Take EV across uncertainty outcomes.
- Aggregate across business units or opportunities.
- Calculate the PV cash flow across time.
- Aggregate value measures/objectives to utility.

There is a set of analyses that are frequently useful. For each analysis, the value dialogue has two phases:

- Compare the composite perspective with individual perspectives of the decision team, aiming to improve one or the other where they differ.
- Use the improved perspectives to search for ideas that could further improve the actions we can take.

Here are some important analyses and their purposes:

- Value components chart, to show how much of each value component is created by each strategy, and spur thought on how to combine different strategies' successful approaches
- Cash flow profiles of strategies, to highlight any time tradeoffs and spur thought on timing of costs or benefits.
- Direct tornado for a leading strategy, to identify opportunities for improvement by controlling a key uncertainty.
- Delta tornado of the best strategy versus another leading strategy, to ascertain whether the outcome of any uncertainty could change the decision, to guide information gathering or construction of a contingent strategy.
- S-curves (cumulative probability curves) for strategies' NPVs, to show whether there is any risk-value tradeoff.

If a promising new strategy is developed, we sometimes need to elicit additional expertise to ascertain its consequences properly, but typically this effort is small compared to the overall analysis effort.

KEY TERMS

Agent-based simulation Simulation of the emergent behavior of a community of interacting "agents" or actors, each of which has objectives and a behavioral repertoire.

Analysis Characterizing the emergent behavior of value in a model in terms that allow direct comparison to individuals' perspectives and generation of improved strategies.

Average Expected value.

Business unit Segment of the entire enterprise that the decision team may want to view in isolation. This can apply not only to business units that are reported distinctly in an organization's financials, but also to opportunities or projects that the team wants to consider on their own.

Capital investment (capex) Money invested to initiate and enable a project.

Cumulative probability curve Curve showing the fractile for any given point in a distribution. Also called S-curve.

Decision tree An analysis tool that extensively represents all scenarios' results and probabilities, with attention to the sequencing of decisions and revelation of information.

Delta tornado diagram Representation of the impact of individual variables' variation on the difference between value measures of two selected strategies.

Delta The difference of value or utility between two specified strategies. cf. Direct.

Deterministic analysis Calculation of the value under a single ostensibly representative scenario.

Dimensions of complexity The five dimensions considered here are the value measures, alternatives, uncertainty, business units, and time.
Direct tornado diagram Representation of the impact of individual variables' variation on a value measure, under a selected strategy.
Direct The value or utility of a specified strategy. cf. Delta.
Discrete-event simulation Simulation of a system that is represented as a chronological sequence of events. Each event occurs at an instant in time and marks a change of state in the system.
Expected net present value (ENPV) EV of the NPV of a cash flow stream.
Expected value (EV) Sum (or integral) of outcomes multiplied by their probabilities (or probability densities).
Extensive representation Addressing all possible cases explicitly simultaneously.
Intensive representation Articulation of a point of view, applied as needed.
Iteration Trial.
Mean Expected value.
Modeling Building representations of experts' perspectives about important events, stitching them together to create a composite perspective on value creation, and embodying this in a way that allows specified cases to be evaluated.
Monte Carlo simulation Realization of a sequence of outcomes by repeated random selection of input scenarios according to experts' probabilities.
Net present value (NPV) Present value of a cash flow stream having both positive and negative terms.
Operating expense (opex) Ongoing expense associated with a project.
Override (verb) Specify the value of an input variable for a specific inquiry; (noun) the value so specified.
Present value (PV) The amount of money today that is equally valuable as a stream of cash flow in the future. Typically calculated by weighting each period's cash flow by a discount factor that decreases exponentially through time at a specified discount rate.
Pseudo EV The value resulting from all inputs being at their EV values; with binary variables treated as Boolean, so that their EV is their probability.
Sampling distribution Distribution used for sampling in Monte Carlo simulation.
Scenario A fixed combination of outcomes of uncertainties. This chapter does not employ other connotations of "scenario," such as a chronological "story" lending coherence to the combination.
S-curve Cumulative probability curve for a value measure.
Second-order simulation Simulation in which the probabilities of events are randomly sampled in a first phase, and then the real-world consequences are sampled using these probabilities.
Simulation Realization of possible outcomes, based on a perspective embodied in a model.
Strategy A fixed set of choices from all pertinent decisions.
Support The set of outcomes with positive sampling density.

System dynamics Simulation of flow through a network of transformations, typically focusing on the implications of positive or negative feedback loops in the network.
Tornado diagram Representation of the impact of individual variables' variation on a value measure.
Trial Instantiation of a scenario in a simulation.
Utility The fundamental value measure to be maximized, regardless of whether it was generated via MODA or ENPV calculations. If there is no explicit risk preference adjustment, this term refers to the value measure or ENPV that is used for decisions under uncertainty.

REFERENCES

Asmussen, S. & Glynn, P. (2007). *Stochastic Simulation: Algorithms and Analysis.* New York: Springer.

Bell, D. (1988). One-switch utility functions and a measure of risk. *Management Science, 34*(12), 1416–1424.

Bueno De Mesquita, B. (2002). *Predicting Politics.* Columbus: Ohio State University Press.

Forrester, J. (1961). *Industrial Dynamics.* Waltham, MA: Pegasus Communications.

Howard, R. (1966). Information value theory. *IEE Transactions on Systems Science and Cybernetics, 2*(1), 22–26.

Howard, R. (1983). The evolution of decision analysis. In R. Howard & J. Matheson (eds.), *The Principles and Applications of Decision Analysis,* pp. 5–16. Menlo Park, CA: Strategic Decisions Group.

Keefer, D. (1994). Certainty equivalents for three-point discrete-distribution approximations. *Management Science, 40*(6), 760–773.

Keeney, R. (1992). *Value-Focused Thinking.* Cambridge, MA: Harvard University Press.

Lindley, D. (1987). Using expert advice on a skew judgmental distribution. *Operations Research, 35*(5), 716–721.

Luce, D. & Raiffa, H. (1957). *Games and Decisions.* New York: John Wiley & Sons.

McNamee, P. & Celona, J. (2001). *Decision Analysis for the Professional.* Menlo Park, CA: SmartOrg, Inc.

Papayoanou, P. (2010). *Game Theory for Business.* Sugar Land, TX: Probabilistic Publishing.

Parnell, G.S., Driscoll, P.J., & Henderson, S. (eds.). (2011). *Decision-Making,* 2nd ed. New York: John Wiley & Sons.

Raiffa, H. (1968). *Decision Analysis: Introductory Lectures on Choices Under Uncertainty.* Reading, MA: Addison-Welsey Publishing Co.

Robinson, S. (2004). *Simulation: The Practice of Model Development and Use.* Chichester: John Wiley & Sons.

Smith, J. & Nau, B. (1995). Valuing risky projects: Option pricing theory and decision analysis. *Management Science, 41*(5), 795–816.

CHAPTER TWELVE

Portfolio Resource Allocation

TERRY A. BRESNICK, ERIC R. JOHNSON, and GREGORY S. PARNELL

Good portfolio decisions begin with a conducive culture, effective behaviors, a decision quality framework, and a good process

—Michael Menke

(*Continued*)

Handbook of Decision Analysis, First Edition. Gregory S. Parnell, Terry A. Bresnick, Steven N. Tani, and Eric R. Johnson.

12.1 Introduction to Portfolio Decision Analysis

In earlier chapters, we show how to use decision analysis to select the best single choice from a set of well-defined alternatives. But in many cases, the goal is not to select the single best alternative, but rather, to select the combination, or portfolio of alternatives that best meets the articulated goals of the leadership. Many names are used for the alternatives being considered: programs, projects, assets, and opportunities. This is a more complicated problem in which issues such as dependencies (e.g., Project A's performance depends on Project B funding), ordering (e.g., Project C must be done before Project D), and resource constraints (e.g., budget and/or manpower) must be considered. If there are dozens, or even hundreds of projects, many of which are interdependent, Chapter 6–11 give good guidance on evaluating each individual project, but the complexity of modeling their interactions may be overwhelming. In this chapter, we summarize several approaches that have been used successfully in government, oil and gas, and pharmaceutical industries among others to make the portfolio problem manageable. See Salo et al. (2011) for discussion of Portfolio Decision Analysis and examples.

This chapter is organized as follows. Section 12.2 describes the socio-technical challenges with portfolio decision analysis. Section 12.3 presents portfolio decision analysis with a single financial objective and capital constraints and illustrates the greedy algorithm approach using the Roughneck North American Strategy (RNAS). Section 12.3 presents a multiple objective decision analysis with resource constraint approach using an incremental benefit/cost portfolio analysis and illustrates the approach with the data center portfolio example.

12.2 Socio-Technical Challenges with Portfolio Decision Analysis

One of our major themes of this book is that decision analysis is a social-technical process. If decision analysts approach portfolio decision analysis as a technical problem, they seldom get to implement their approach. As we describe in this chapter, portfolio decision analysis significantly increases both the technical and the social challenges. In this chapter, we focus on the technical challenges, but the success of portfolio decision analysis depends much more on overcoming the social challenges of decision making in large organizations (see Chapter 4). In addition, the increased social complexity typically requires the soft skills of an experienced decision analyst.

In Chapter 9 and Chapter 11, we consider five dimensions of decision complexity: business units, objectives, alternatives, time, and uncertainty. All five of these dimensions apply to portfolio decision analysis. However, many of the dimensions can be more complex for portfolio resource allocation.

- **Business units.** Portfolio decision analysis can be performed within a business unit or across several business units. These could be business units within a corporation; oil fields in a geographic region; compounds in a drug company's pipeline; or military or government programs vying for organizational funding. We must evaluate both the individual projects and the aggregate portfolio for each business unit.
- **Objectives.** Instead of the business unit objectives for one alternative decision, we must consider the business unit objectives for the portfolio of projects (e.g., avoiding loss of key skills if a portfolio is too unbalanced across staffing in an organization). This may significantly expand the scope and number of objectives and involve more decision makers, stakeholders, and subject matter experts.
- **Alternatives.** As noted above, the number of alternatives can be very large and the number of possible portfolios can be extremely large. For example, suppose we have 10 projects and four funding levels for each project (none, 90%, 100%, and 110%), then the number of possible portfolios is 4^{10}. Of course, not all of the portfolios may be interesting (e.g., 0% funding for all) or satisfy the constraints.
- **Uncertainty.** As we consider large numbers of projects, more business units, and a broader set of objectives, the number of project and business uncertainties can become very large. Assessing all of the required joint probability distributions may become unwieldy.
- **Time.** The time period for the portfolio decision analysis usually varies from 1 year to many years. Private companies typically use a shorter planning horizon than public organizations. For example, resource allocation planning for government programs is done for at least 5 years into the future (and sometimes for 25 years, for example, for defense program planning).

However, project execution portfolio decision analysis might consider only 1 year.

In practice, in addition to the increase in complexity, there are additional social challenges to portfolio analysis:

- Identifying all of the projects can be challenging for organizations that do not have an existing systematic resource allocation process.
- Collecting consistent, credible information on the projects can involve many people who may not have incentives to provide timely information and can take a long time.
- Identifying the project dependencies and portfolio constraints is also difficult. Decision analysts should try to minimize the constraints, since adding a constraint can reduce the total value.

12.3 Single Objective Portfolio Analysis with Resource Constraints

Although we may find it compelling, the perfect capital markets point of view is rarely taken by clients. Instead, clients often view themselves as facing capital constraints.

12.3.1 CHARACTERISTICS OF PORTFOLIO OPTIMIZATION

If we have framed the decision, defined the portfolio alternatives, and obtained all the data, we can identify the capital-constrained global optimum portfolio by considering all possible portfolios, discarding those that are inconsistent with the capital constraint, and choosing the feasible portfolio with the best ENPV. If carefully done and properly explained, this global optimization approach can be useful. However, in practice, there are challenges to it:

- For large problems, it can take considerable computation (and sometimes special algorithms) to find feasible and, especially, optimal solutions.
- Clients sometimes state constraints that are not necessary. This destroys value. For instance, if an oil and gas manager arbitrarily requires that the aggregate success rate of an exploration portfolio exceed a certain threshold, this might rule out profitable high-risk high-reward exploration strategies.
- Small changes to inputs can require time-consuming reoptimization of the entire portfolio.
- The rationale of the recommended portfolio choices may not be clear, especially if a number of unnecessary constraints have been specified.

- Response to changing conditions is sometimes counterintuitive (e.g., if we increase the funding, a funded project may be removed from the portfolio and an unfunded project may be added). This can make for a very unstable solution.

12.3.2 GREEDY ALGORITHM USING PROFITABILITY INDEX AND THE EFFICIENT FRONTIER

Many of the challenges associated with global optimization can be met by developing a figure of merit for projects, rank-ordering them, and choosing from the top down until available funds are exhausted. In computer science, this approach is known as the "greedy" algorithm. Greedy algorithms are quick to calculate, their response to changing assumptions is easy to understand, and they often give results very close to the optimum. Furthermore, we do not have to re-evaluate the entire portfolio to ascertain the status of a new project; we merely compare its figure of merit with the prevailing threshold.

We can develop such a figure of merit by taking a brief look at the mathematics of constrained optimization. This discussion applies equally well to monetary or nonmonetary constraints (e.g., in oil and gas production, headcount or access to drilling equipment may be the constraint). If there are multiple constrained resources, this approach becomes noticeably more complicated. It can still be used as the basis of a computer algorithm, but it loses much of its attractiveness as a way of discussing a portfolio problem with stakeholders.

The derivation of the approach requires a brief reference to the mathematics of optimization. The pragmatic portfolio analyst with less mathematical background can skip the boxed discussion.

Optimization Theory

Let $\mathbf{x}$ be a set of portfolio choices, $h(\mathbf{x})$ be the amount of constrained resource it requires, c be the constraint, and $f(\mathbf{x})$ be the ENPV value of those choices. Our choice may be formulated as a constrained optimization:

$$\max f(\mathbf{x}), \text{ subject to } h(\mathbf{x}) \leq c. \tag{12.1}$$

The duality result for optimization (Luenberger, 1984) ensures that there will be a shadow price λ such that the constrained optimum is consistent with the solution of an unconstrained dual problem with the shadow price applied to any use of the constrained resource:

$$\max f(\mathbf{x}) - \lambda h(\mathbf{x}). \tag{12.2}$$

The process of solving the optimization increases the shadow price until the choices consistent with it employ a feasible amount of the constrained resource.

The limited resource constrains the value we can achieve. At the optimal solution, the shadow price is the value return we would get for one more dollar of budget constraint. A project's productivity index (PI)[1] is the ratio of the value returned per unit resource. We can use the PI to prioritize projects for funding, by choosing those with the best PI until the resource is exhausted, in greedy fashion (Reinsvold et al., 2008). This is equivalent to the PI equal to the shadow price.

The PI approach has often been used in cases where the constraint is financial, by splitting out a component of the cash flow that comprises investment, and treating its present value as the constrained resource. When formulated this way, the shadow price of capital specifies the minimal acceptable level of PI. While there is rarely an actual constraint on present-value investment *per se*, this formulation can still develop a figure of merit that is helpful. This approach was used in a prominent SmithKline Beecham portfolio review (Sharpe & Keelin, 1998), which is discussed in Chapter 1.

If we plot the total PV of investment dollars on the *x*-axis and the total ENPV of the portfolio on the *y*-axis, it is often informative to map out and display an "efficient frontier," which is the set of portfolios that provide more ENPV value than any other portfolio with equal or lower PV investment. Figure 12.1 gives an example. While this notion of efficiency may differ slightly from what is mathematically optimal,[2] the display can nonetheless be insightful.[3] The efficient frontier can always be identified by complete enumeration: by evaluating all possible portfolios, putting them all on a scatter plot, and noting those on the upper left extremes. If interproject dependencies are not too complicated, it can also be generated directly by rank ordering the opportunities using PI as we did above, and mapping out the curve that results from adding them one at a time to the portfolio, from best to worst. This chart is useful if the exact level of available investment capital is not yet known—once it becomes known, we look up the portfolio that uses it up and fund the opportunities comprising it. This chart is sometimes called the bang-for-buck curve, investment efficiency curve, or the Chief Financial Officer's (CFO) curve.

If some opportunities are prerequisites for the feasibility of others, or if the value of one project depends on whether another is chosen, the only com-

---

[1] To confirm the equivalence, interpret h as PV investment capital and λ as the PI, set $f(\mathbf{x}) - \lambda h(\mathbf{x}) = 0$, and solve for λ. Some firms and references (e.g., http://www.investopedia.com) define the numerator as future cash flow not considering investment, which adds 1 to all PIs. This does not affect interpretation or use.

[2] A portfolio made up of projects with a good ratio of value to PV investment may not be the same as the portfolio that has the highest value. This can be seen by noting that the former can be influenced by different definitions of which costs are considered to be "investment" while the latter is not.

[3] The differences from "optimal" are often small. Often it is better to let organizational momentum take its course than to try to force all decisions to honor small differences in value metrics.

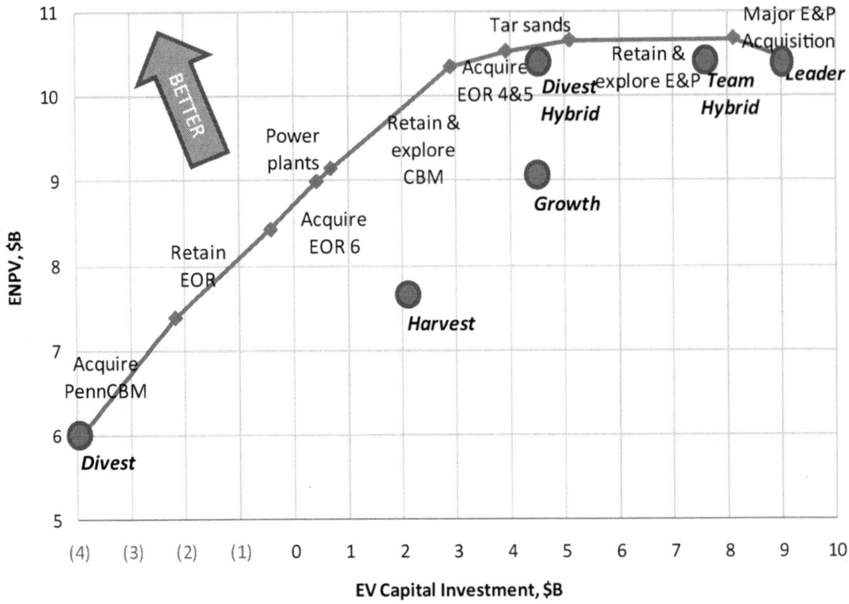

FIGURE 12.1 RNAS investment efficiency curve.

pletely general way to generate the efficient frontier is by complete enumeration. However, if there are only feasibility dependencies, and not too many of these, we may be able to construct an efficient frontier directly by manually combining prerequisite projects with their more-attractive follow-on projects and then treating these as single projects in constructing the bang for the buck curve.

This entire discussion addresses ENPV value for a given discount rate. Do we have a role in the specification of the discount rate? In the pure decision analyst role, we elicit alternatives, information, and preferences (including discount rate, which expresses time preference for money) from clients and develop and communicate insights from them. A decision analyst *per se* does not question preferences.

This can be contrasted with a broader role—the decision professional—in which we might also offer guidance on the formulation and articulation of preferences. Some preferences are easy to specify, and should not be questioned. Others, such as discount rate, are difficult to specify, due to their abstractness, and differing schools of thought on the nature and role of discount rate (e.g., the "risk adjusted" discount rates one sees in the finance literature). The following sidebar discusses choice of discount rate in the context of other preferences that are easier to specify, such as preferring more money to less.

Discount Rate

In a "perfect markets" view of portfolio management, an arbitrage argument shows that we must use our weighted average cost of capital (WACC) as our discount rate, and fund projects if and only if their ENPV (i.e., the ENPV of cash flow with the project minus cash flow without it) is positive. If we use a discount rate higher than WACC, projects whose rate of return is between WACC and the (higher) discount rate would have ENPV < 0, and would not be funded, even though there would be a net increase to wealth with no net incremental investment by acquiring capital at WACC and funding them. Likewise, if we use a discount rate lower than WACC, projects whose rate of return is between WACC and the (lower) discount rate would be funded, but would lose money relative to the capital they require. In other words, the WACC is the opportunity cost of capital, and projects should be funded only if they can bear this cost. Projects with ENPV = 0 using this as a discount rate are on the margin; small increments one way or the other could make them be funded or defunded.

If a firm is unable or unwilling to acquire debt and equity at the WACC to fund all attractive projects in their portfolios, the theoretical justification of using WACC as a discount rate is undercut. If funds are fully committed and a new attractive project arises, acquiring new capital at the WACC to fund it is not an option. Instead, the new project must be funded by defunding some other project (presumably a less attractive one). The opportunity cost of capital is now higher than WACC; it is the rate of return associated with the marginal project. In this situation, the decision rule to fund if and only if ENPV > 0 no longer works; capital is oversubscribed and unfortunately, we are forced to conclude that some projects with ENPV > 0 do not make the grade. We can reinstate the ENPV > 0 decision rule if we raise our discount rate enough that the marginal project has ENPV = 0. If we do this, the projects that are even more attractive than this one have ENPV > 0 and we fund them; others have ENPV < 0 and we do not fund them. With the new discount rate, using the ENPV > 0 decision rule gives us a portfolio whose capital requirement is consistent with our available capital. Finding a "market clearing" discount rate is usually straightforward, because in normal cases, increasing our discount rate makes each project less attractive, which reduces the number of projects with ENPV > 0 and the amount of capital demanded by them. We find the "market clearing" discount rate by simple iteration: make an initial guess, and increase it if capital demanded by ENPV-positive projects exceeds capital available, or decrease it otherwise.

These issues have practical consequences. Typical portfolio framing coupled with a discount rate lower than the opportunity cost of capital tends to undervalue projects that finish quickly. A typical framing of a portfolio analysis is to look at projects that start now or fairly soon, but not to model reinvestment of funds generated. In an expected net present value (ENPV) analysis, failure to reinvest is equivalent to reinvesting in projects whose ENPV = 0. Whenever a client is capital constrained, projects with ENPV = 0 are inferior to the projects that are viable. Analysis that essentially assumes capital is reinvested at the inferior rates of return depresses the return of each project, and depresses quick projects more because the reinvestment at inferior rate of return persists for a longer period of time.

We do not recommend that clients revise their stated discount rate every time they review their portfolio. However, if it is routinely the case that ENPV > 0 projects cannot be funded, this is evidence that the firm's discount rate is lower than their opportunity cost of capital. In this case, an arbitrage-style argument suggests that the firm would be better off to raise its discount rate. There is a simple diagnostic that indicates how far the discount rate is from the opportunity cost—the PI threshold below which projects are not funded. The further this is from zero, the further the discount rate is from opportunity cost. PI thresholds of 0.3 do not leave much value on the table, but if the threshold is above 1.0, senior management might be well advised to consider raising its discount rate.

12.3.3 APPLICATION TO ROUGHNECK NORTH AMERICAN STRATEGY PORTFOLIO

See Table 1.3 for further information on the illustrative examples. If we interpret the RNAS strategy table (see Chapter 8) as specifying different levels of investment at each of their business units (BUs), but discard the suggestion that high investment in one should be coupled with high investment in another, we can view the RNAS situation as a portfolio problem. There was no binding constraint other than capital, so we calculated the PIs of major RNAS investments (see Table 12.1). This was done by creating a set of strategies that included one opportunity from each business unit, then two, then three, and so on, and simulating the resulting set of strategies. Then looking at the delta between adjacent pairs of strategies and confining the display to the appropriate business unit (BU) gave the incremental cash flow, ENPV, and PV investment for each project, and allowed its PI to be calculated.

In this table, we see that merely retaining E&P assets and producing them was not worthwhile in itself, but it enabled profitable exploration and down-spacing,[4] so we merged these into one item "Retain & explore E&P" in the

[4]Down-spacing refers to additional drilling in an already developed area. This can increase and accelerate production, thereby increasing its value.

TABLE 12.1 **RNAS Portfolio Metrics ($M)**

Opportunity	ENPV	Invest	PI
CBM exploration	0.2	0.1	2.6
Explore at ENP 4	0.2	0.1	2.1
Acquire PennCBM	1.4	1.8	0.8
EOR_6 acquisitions	0.6	0.8	0.7
Build coal plant	0.1	0.2	0.6
Retain EOR assets	1.	1.8	0.6
Build cogen plant	0.1	0.1	0.5
Retain CBM assets	1.0	2.2	0.5
Explore at ENP 6	0.2	0.6	0.4
E&P Downspacing	0.1	0.2	0.4
Tar scands plant #2	0.1	0.4	0.2
Acquire EOR_4_5	0.2	1.0	0.2
Tar sands plant #1	0.1	0.8	0.1
Retain E&P assets	(0.5)	2.1	(0.2)
Major E&P acquisition	(0.2)	0.9	(0.2)

investment efficiency curve. Figure 12.1 shows the resulting RNAS investment efficiency curve, along with the strategies that were analyzed.

Often, the most effective use of a graphic like this is to identify projects that are clear winners and clear losers, and engender team consensus for funding and not funding, respectively. For RNAS, EOR and CBM were clear winners, while the Major E&P Acquisition was a clear loser. The projects in between have roughly zero value. The value of a portfolio is roughly the same if one of these projects is added or if it is removed. This means that little is lost by making such choices suboptimally. In a situation where the client has only a finite amount of time and patience to sort through fine points of valuation, it can be more helpful simply to identify winners, losers, and marginal projects, and allow the organization's momentum to carry the day on the latter, than to try to force decisions to honor every small difference in value metrics.

12.3.4 PORTFOLIO RISK MANAGEMENT

It may be possible to reduce portfolio risk by selecting opportunities whose values are influenced in opposite ways by uncertainty. A common example of this at the corporate level is vertical integration. For instance, when the price of oil rises, the value of an oil production company rises, while the value of a refinery falls. If an oil producer and refinery merge, the aggregate impact of oil price is largely neutralized, because the two business units naturally hedge one another. There can be opportunities like this within a portfolio, but none was seen at Roughneck.

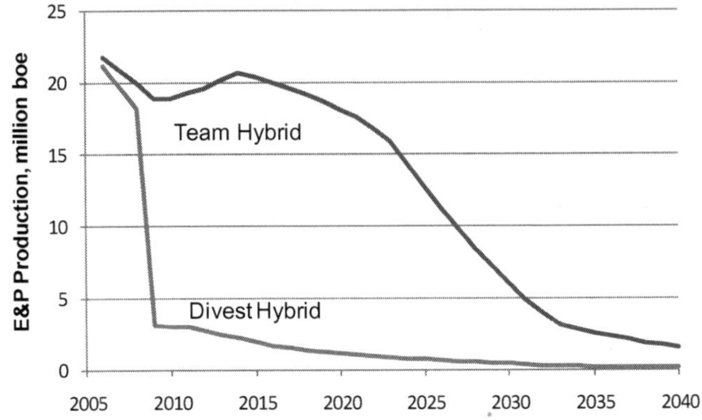

FIGURE 12.2 RNAS E&P production.

If an uncertainty influences all opportunities' value in the same direction, there are no synergies at the portfolio level. Of course, the sensitivities may suggest improvements at the level of individual opportunities.

12.3.5 TRADING OFF FINANCIAL GOALS WITH OTHER STRATEGIC GOALS

Some strategic goals are difficult to measure alongside financial goals. In principle, we could map out the possible outcomes and assess a utility function for all outcomes, and then we would be in position to optimize. In financial portfolio analyses, we often size up strategic goals qualitatively at the end, and adjust decisions to ensure greater alignment with those strategic goals.

For instance, the investment efficiency curve in Figure 12.1 suggests that Divest Hybrid might be the most attractive of the strategies considered. However, upon seeing this, Roughneck management reiterated that oil and gas production (measured in barrel-of-oil equivalents, or boe) was a fundamental objective (see Fig. 12.2), and divesting E&P created too large a drop in production to be accepted.

12.4 Multiobjective Portfolio Analysis with Resource Constraints

In the previous section, we describe an approach for achieving portfolio value that focuses on the use of ENPV and Profitability Index (PI) as a basis for selecting competing projects in a portfolio. In this section, we introduce and apply to the data center problem an additional approach that some of the authors have used in many highly successful portfolio analyses. The approach focuses on *incremental benefit/cost analysis using Pareto optimality criteria* (Buede & Bresnick, 2007; Phillips, 2007). This is entirely analogous to the PI approach, insofar as it treats cost as a constrained resource and develops a figure of merit by dividing benefit by cost. An alternative multiobjective decision analysis approach is to use a multiobjective value model to assess the value of the projects and an optimization model that determines the best value for the resource constraints (Burk & Parnell, 2011).

12.4.1 CHARACTERISTICS OF INCREMENTAL BENEFIT/COST PORTFOLIO ANALYSIS

The cost–benefit analysis approach to portfolio resource allocation focuses on determining the incremental benefit/cost ratios of competing program elements and on establishing the most cost-efficient way of spending additional resources, referred to as an "efficient order-of-buy." It can be used as a zero-based analysis tool, or can be used to examine allocations beyond some defined baseline. It can also be used in a budget-cutting mode to develop an "efficient order of sell." The solution approach explicitly identifies the set of feasible alternatives that are built to reflect the constraints, evaluates only the alternatives that provide the most "bang-for-the-buck" at any level of resource expenditure, and provides a clear audit trail for the results. This technique is best suited for problems:

- that have a very large number of alternative allocation schemes;
- where funding areas are independent (i.e., funding one area does not impact the benefit of funding another area);
- where programs can be divided into increments such that a piece of a program can be funded without funding the entire program;
- where budget constraints are expected to change greatly or are not known initially;
- where reasons for the resulting allocations need to be completely transparent; and
- where it is useful for participants to understand the underlying mathematical algorithms.

12.4.2 ALGORITHM FOR INCREMENTAL BENEFIT/COST PORTFOLIO ANALYSIS

12.4.2.1 Identify the Objective. For illustration purposes, assume that we are trying to maximize some notion of portfolio "benefit" subject to cost constraints (e.g., dollars, time, and people) and physical constraints (e.g., space, power, weight, and bandwidth). For convenience, we will call the areas to which we are allocating resources "programs," each of which can be funded at various "levels" of expenditure. Each program level that gets funded will contribute some "benefit" toward satisfying organizational needs. Our objective is to get the most "bang-for-the-buck" for any level of expenditure.

12.4.2.2 Generate Options. We assume that there are a limited, but perhaps large number of programs, i, to which we can allocate resources. We also assume that the programs are value independent, which implies that the level at which program A is funded should not affect the benefit associated with levels of program B.[5] For each program, there are at least two levels at which we can fund, which may range from none (do not fund) to "gold-plated" (with all the bells and whistles). A program is represented by a "row" in the matrix below, with the various funding alternatives being shown from left to right as levels for the row. A row may simply have two levels (for a Go/No-Go program that cannot be divided), or many levels for one with well-defined funding increments. Practically, the number of levels per row should be less than 10, although there is no analytical requirement for this. The levels may be independent of each other (called *substitute levels*, where we do level 1 *or* level 2 but not both), or inclusive (called *cumulative* levels where funding level 2 *includes* level 1). A set of alternative programs and funding levels is shown in Figure 12.3 in matrix form.

A specific portfolio is then defined by selecting one level in each row. One such portfolio is represented by the white dots in Figure 12.4. Each portfolio has an associated cost which is the sum of the costs of each selected program and level.

	1	2	3	4	5
Program T	Level 1	Level 2	Level 3	Level 4	Level 5
Program U	Level 1	Level 2	Level 3	Level 4	
Program V	Level 1	Level 2	Level 3	Level 4	Level 5
Program Y	Level 1	Level 2	Level 3		
Program Z	Level 1	Level 2	Level 3	Level 4	

FIGURE 12.3 Funding areas and levels.

[5]In reality, there is often a high degree of dependence among some programs, so we must deal with this problem by combining programs together and forming a larger program.

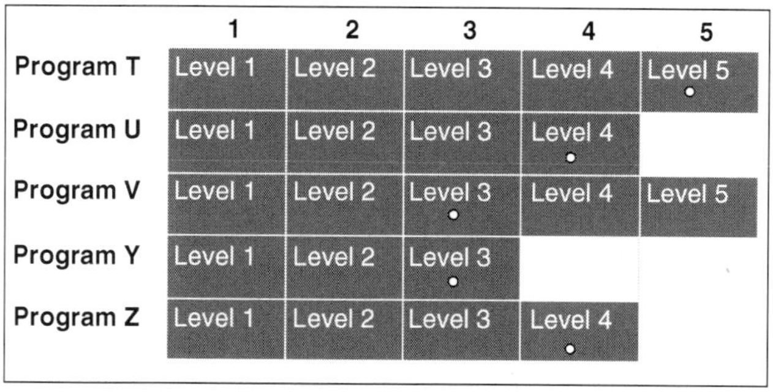

FIGURE 12.4 One possible portfolio.

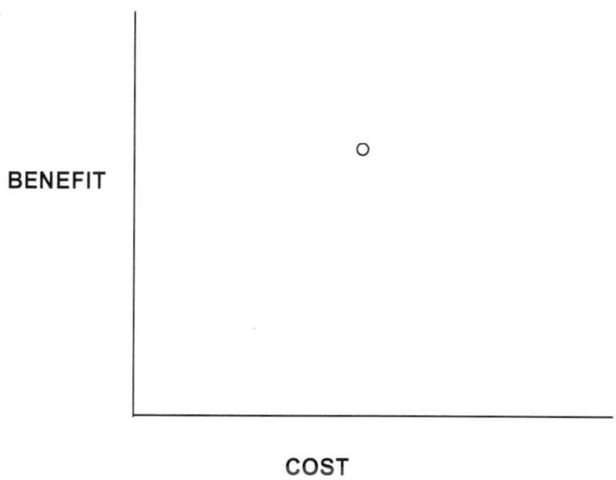

FIGURE 12.5 Benefit vs. cost plot of one possible portfolio.

As shown in the introduction, the number of possible funding options is the product of the number of funding levels of the programs.

Each such combination can be thought of as a *funding option* or *package*. Clearly, we do not want to evaluate every possible combination since this can easily reach the thousands or millions. The approach described here makes it unnecessary to do this; instead, it identifies and considers only those options that are determined to be "efficient."

Let us assume that we can calculate the "benefit" of each level using the techniques described earlier in the handbook, either through NPV, expected value, or MODA techniques. We can then plot the cost and benefit of any package, such as our "white dot" alternative on a benefit–cost graph as shown in Figure 12.5.

If we plot all possible portfolios (i.e., combinations of one level selected from each row), they have the "nice" mathematical property of falling within the football shaped region shown in Figure 12.6.

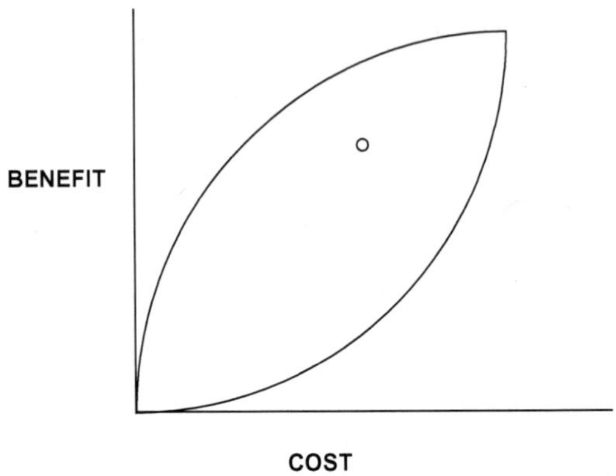

FIGURE 12.6 Trade space of portfolios.

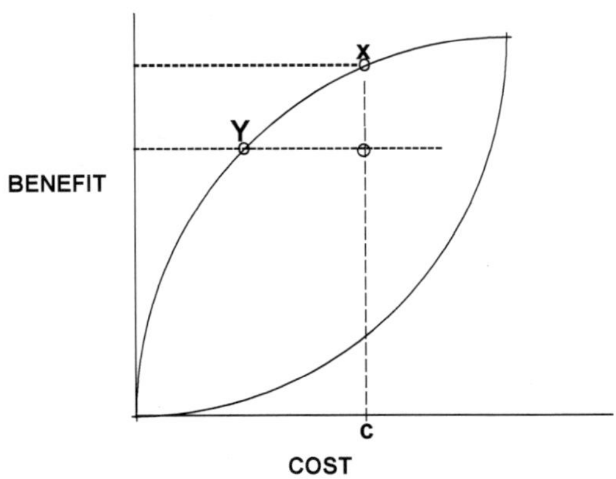

FIGURE 12.7 Selected portfolio for Cost C, better portfolio for same Cost (X), cheaper portfolio with same benefit (Y).

It is easy to see that for any cost *C*, we want to find the package that is as high on the benefit scale as possible; that is, we want to find the package *X* that falls on top of the football, known as the *efficient frontier* or the *Pareto optimal frontier*, as shown in Figure 12.7. Note that package X has the same cost as the white dot package, but gets more benefit. Also note that there is a package Y that gets the same benefit as the white dot package, but is cheaper. In fact, any package in the triangular wedge formed by *X*, *Y*, and the white dot is more "efficient" than the "white dot" solution.

The programs under consideration may come from a variety of internal and external sources. Programs may be couched in terms of the current baseline, additions to the baseline, deletions from the baseline, or other modifications to the baseline. It is usually wise to include the baseline (or "do nothing") option in the analysis to provide a benchmark for comparison with the other options.

12.4.2.3 Assess Costs. Assessing the costs associated with each program element is conceptually easy, yet may be very difficult to do in practice. At a minimum, the following issues should be addressed:

- Will life cycle cost estimates be used, and if not, which costs should be included?
- Can different "colors[6]" of money be combined or do we need to consider them separately?
- What assumptions will be made about inflation to enable comparison of constant dollars with current-year dollars?
- What assumptions will be made about discount factors to accommodate time-value-of-money considerations?
- Who provided the cost estimates, and was an independent cost estimate made as well?
- If costs were provided by different organizations, were the same "ground rules" followed?

The benefit/cost methodology allows for multiple dimensions of the constrained resource (such as various budget year dollars or different "colors" of money[2]), or even different types of resources (people vs. dollars vs. time). However, in order to combine costs, they all must have the same dimensionality. With dollars, this is no problem. With different types of resources, there are three choices. First, we can convert all resources to dollars. Second, we can allocate based on one resource at a time, and see where differences in the allocation schemes occur as we vary the resource considered. Third, we can establish a relative "value" scale on costs that allows us to treat one type of resource as more or less important than another by assigning weights to the different types of resources. Regardless of the method used, when calculating the benefit–cost ratios of the packages, a single resource number should be used.

12.4.2.4 Assess Benefits. This is typically the most difficult and most controversial part of the process. Often, there are insufficient "hard data" available to provide purely objective measures of benefit for the programs. Thus, it is necessary to rely upon more subjective and judgmental assessments from experts. Both the measures of performance used to evaluate the programs and the elicitation techniques themselves must be selected carefully and executed in a technically sound and defensible manner.

---

[6]In government funding, each budget category (research, development, procurement, operations, and so on) has restrictions on the types of programs that can be funded with that category.

In addition to the techniques for assessing value described earlier in the handbook, there are a variety of techniques that may be used for subjective benefit assessment (Brown et al., 1974; Barclay et al., 1977) and most use some measure of "relative value" to compare programs on one or more dimensions. Whether the measure is called "benefit," "pain," or simply "value" is not the important consideration, but rather, whether or not the measures are set up and defined in a way that facilitates assessment from the experts.

For the approach described here, we will use multiple attributes of benefit that relate to dimensions of performance (e.g., NPV, environmental impact, and fit with strategic direction). The benefit of a program at each level of funding can be assessed in terms of each of these needs separately, and then later combined across the dimensions. Benefits are assessed one row at a time, and one attribute of benefit at a time. For example, first consider Program X in terms of how well its elements contribute toward fit with strategic corporate goals. The level that provides the greatest satisfaction on "fit" arbitrarily scores 100; the level that provides the lowest level arbitrarily scores 0 (*Note*: A benefit score of 0 does not mean there is no benefit associated with the level; it means that this is the starting point against which additional benefit can be measured). There is no requirement that 0 be at the leftmost level or that 100 be at the rightmost level. The intermediate levels are scored with relation to 0 and 100 points. Thus, a level scoring 90 implies that funding to that level provides 90% of the total benefit that could be achieved in going from the lowest level to the highest. We then make similar assessments for the other attributes of benefit. When this is done, we move to the next row (program) and make the same assessments. When we are done with all programs, each row has its levels scored on a 0–100 basis for each attribute. At this point in time, it is not possible to compare one row with another or one attribute of benefit with another since they were all assessed on independent 0-to-100 scales. We therefore need a way of "calibrating" these scores to a common metric. This is done first by assigning "within criteria" weights, w_{ik}, which compare one row with another on each attribute, and second, by assigning "across criteria" weights, a_k, to the different attributes of benefit. We can then calculate the weighted average for each level of a program across all criteria. Thus, on the common scale, the benefit score B_{ijk} for level j of program i across k evaluation criteria would be calculated as:

$$\mathrm{B}_{ijk} = \sum_{k} a_k w_{ik} b_{ijk}, \qquad (12.3)$$

where w_{ik} is the "normalized" weight assigned to program i for criterion k, and a_k is the "normalized" weight assigned to criterion k.[7]

12.4.2.5 Specify Constraints. The above formulation does not explicitly include constraints that limit the resource allocation solutions, but instead builds

[7]To normalize, sum all of the components and divide each component by the sum.

them implicitly into the structure of the problem. For example, if for a given row, doing X or Y requires doing the other, we put them in the *same* level so both get funded together if at all. In the example below, X and Y never get selected without the other:

Level 1	W
Level 2	X,Y
Level 3	W,X,Y
Level 4	Z
Level 5	Z,X,Y

If we could do X or Y but not both, we build them into *separate* levels of the row where the levels are substitutes for each other. In the example below, X and Y never get selected together:

Level 1	W
Level 2	X
Level 3	Y
Level 4	W,X
Level 5	W,Y

If doing X requires Y, but Y does not require X, we put them in separate levels where the levels are *cumulative*, with Y in a lower level than X. In the example below, Y can be selected without X, but X cannot be selected without Y:

Level 1	W
Level 2	Y
Level 3	W,Y
Level 4	X,Y
Level 5	W,X,Y

Clearly, it is essential that the benefits and costs associated with each level reflect these relationships.

This formulation also allows the use of a single constraint in another way. The model can be queried to find the best solution for a specified resource amount (e.g., find the best portfolio for $250M), or to find the cheapest way to achieve a specified level of benefit (e.g., find the cheapest portfolio which will get us 90% of the potential benefit). In both cases, the specified amount acts as a constraint on the preferred portfolio.

12.4.2.6 Allocate Resources. The algorithm used is relatively straightforward. First, both the "across criteria" weights and the "within criteria" weights are applied to the assessed benefits for each level of each row. Second, the change in benefit (Δbenefit) and the change in cost (Δcost) are calculated for each increment of each row, where by increment, we mean going from one level in a row to the next. Third, we calculate the ratio of Δbenefit to Δcost for each increment. Finally, we order all of the increments in decreasing Δbenefit/Δcost order. This becomes the "order-of-buy," or the order in which the levels are funded in a purchasing exercise; the ordering can also be reversed to produce an "order-of-sell" for a budget-cutting exercise. Once a resource constraint such as a budget line is applied to the ordered listing, the resource allocation task becomes one of funding everything above the line, and not funding everything below the line. This results in a very stable program since changes in the budget do not require a total realignment of funding decisions; rather, changes simply result in moving the budget line up or down and determining which increments enter or leave the suggested mix at the margin.

An important complication can arise in this process. The algorithm assumes that we always want to fund in decreasing Δbenefit/Δcost order, and assumes that for each row, the levels are built to reflect such an order. If we look at a benefit/cost plot for such a row, this would result in a concave downward curve with an ever-decreasing slope (Fig. 12.8). In reality, this is not always possible to achieve, and in fact, where levels are substitutes for each other, is not likely to happen, and there will be "dips" in the curve (Fig. 12.9). The algorithm assumes that we do not want to fund less efficient increments before we fund more efficient increments, so it adjusts the curve to smooth out the dips. It does this by combining any levels where the subsequent slopes are increasing and then calculating the Δbenefit/Δ cost for the "combined" increment (Fig. 12.10). The combined increment is then placed in the order-of-buy in the appropriate place, and if it is funded at all, it will be funded completely. In a row with **cumulative** levels, this means that we never stop at the intervening level, but do it all. In a

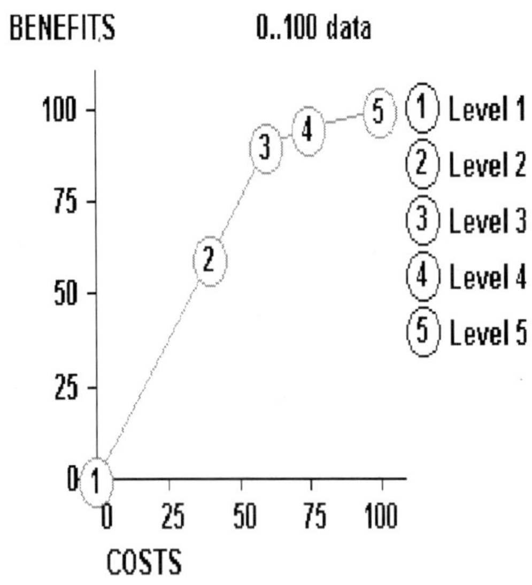

FIGURE 12.8 Curve with decreasing slope.

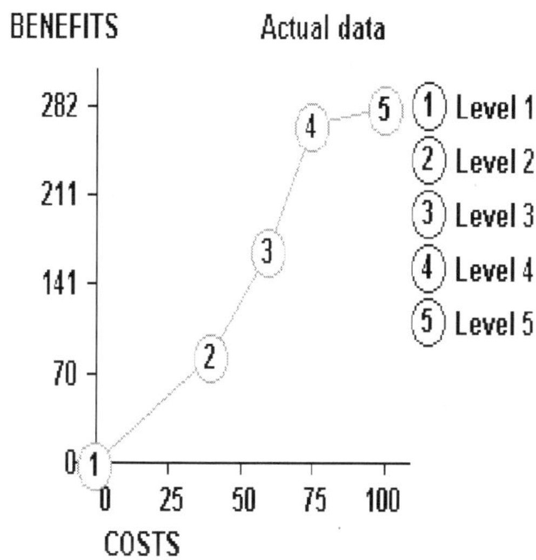

FIGURE 12.9 Curve with varying slope.

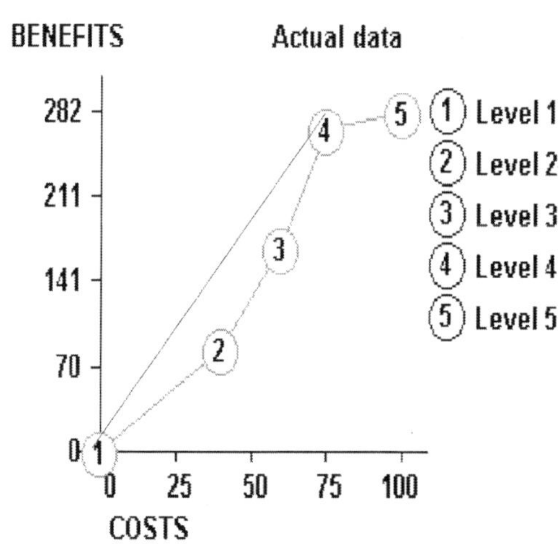

FIGURE 12.10 Combination of levels for curve with varying slope.

row with **substitute** levels, we completely skip over the less efficient level and fund the more efficient level. In the most commonly used COTS software packages for this approach, this happens automatically since the costs are built into the levels to reflect the nature of the row.

12.4.2.7 Perform Sensitivity Analysis. A critical part of the analysis is to challenge the initial assumptions by performing a sensitivity analysis. This can take various forms. By generating the order-of-buy, we can create a contingency table that reflects the best package arrayed against various budget targets. We can also do sensitivity analysis on the benefit measures either by varying the benefits associated with the programs, or by varying the "within criteria" or "across criteria" weights. We can also vary the assumptions on costs to include discount rates or costing models and see how the solution changes.

12.4.3 APPLICATION TO THE DATA CENTER PORTFOLIO

See Table 1.3 for further information on the illustrative examples. Assume that we have selected a single data center location and now must determine the portfolio of projects for the data center that we can fund. As is typical, the projects submitted for funding greatly exceed the resources available. There are four major business areas in the data center—Applications, Security, Platforms, and Connectivity, and they are all competing for the same funds. Three criteria have been identified to evaluate the programs: (1) maintenance of legacy systems (maintain); (2) advance strategic direction of the data center (strategic fit); and (3) ease of implementation, including integration issues. Life cycle NPV has been calculated for each proposed program as the cost dimension of the analysis. Table 12.2 summarizes the input data for the portfolio analysis. For the benefits, the pros and cons of each alternative were discussed in terms of each of the criteria, and the scales in Table 12.3 were developed:

Table 12.4 shows how each project was valued on the scales:

Using the swing weighting techniques described in earlier chapters and the weighting procedure described in Section 12.3.2, assume we assess weights of 10 for Maintain Legacy, 60 for Strategic Fit, and 30 for Ease of Implementation. For ease in analyzing results, the benefits in the model are normalized to sum to 1000 across all criteria. The trade space of projects is shown in Figure 12.11.

Note that an option is included to fund no projects in an area. Within each business area, the projects are ordered by decreasing incremental Benefit/Cost ratios. For example, in Applications, the incremental benefit divided by the incremental cost of going from project C to Project A is greater than that of going from Project A to Project B. In the example, we assume that all projects are value-independent, but this is not a requirement for the approach. The example is also built in a cumulative fashion in that funding Project A implies that Project C has been funded as well (again, not a requirement of the approach—projects can be cumulative or exclusive of each other).

We can examine the data for the model by looking at Figures 12.12 and 12.13, which show the structure of the Applications business area.

As described in Section 12.4.2, when we put all of the data for all business areas together, we produce the full trade space of portfolios along with the efficient frontier as shown in Figure 12.14. Each "dot" on the top of the football-shaped area (the "efficient frontier") represents adding the next best project based on incremental B/C ratio to the portfolio. For example, for a budget of \$450M, we would fund all projects represented by the black dots going from the origin to the \$450M point on the frontier.

TABLE 12.2 **Projects Requesting Funding**

	NPV ($M)	Helps to Maintain Legacy	Strategic Fit	Implementation Ease
Applications				
Project A	20	No support	Supports all strategic goals	Very easy
Project B	35	Supports critical systems only	Supports 50% of strategic goals	Moderate
Project C	15	Minimal support	Supports 100% of strategic goals	Easy
Project D	60	Fully support all	Supports all strategic goals	Very difficult
Project E	55	Full support all	Supports 25% of strategic goals	Moderate
Platforms				
Project F	50	Full support all	Supports all strategic goals	Difficult
Project G	80	Minimal support	Supports 50% of strategic goals	Moderate
Project H	40	Minimal support	Supports 50% of strategic goals	Easy
Project I	120	Supports critical systems only	Supports all strategic goals	Moderate
Security				
Project M	50	Supports critical systems only	Supports 50% of strategic goals	Difficult
Project N	120	Supports critical systems only	Supports all strategic goals	Moderate
Connectivity				
Project R	25	Full support all	Supports all strategic goals	Easy
Project S	55	Minimal support	Supports 50% of strategic goals	Moderate
Project T	75	Supports critical systems only	Supports all strategic goals	Difficult

We next explore the efficient frontier by looking at the order in which projects would be funded, or the "order of buy" as in Table 12.5.

Assuming that the budget is $450M, the funded portfolio across business areas would be as shown in Table 12.6.

As with all approaches, the next step would be sensitivity analysis on the weights, scores, criteria, and so on. Additionally, we would make some after-the-fact dependency checks, and revise the portfolio accordingly. If the portfolio that emerges from analysis of a multiobjective decision analysis seems to give

TABLE 12.3 **MODA Value Scales**

Value	Helps to Maintain Legacy	Strategic Fit	Implementation Ease
100	Full support all	Supports all strategic goals	Very easy
75	Supports critical systems only	Supports 75% of strategic goals	Easy
50	Moderate support	Supports 50% of strategic goals	Moderate
25	Minimal support	Supports 25% of strategic goals	Difficult
0	No support	No support	Very difficult

TABLE 12.4 **Values for Projects**

	Helps to Maintain Legacy	Strategic Fit	Implementation Ease
Applications			
Project A	0	100	100
Project B	75	50	50
Project C	25	100	25
Project D	100	100	0
Project E	100	25	50
Platforms			
Project F	100	100	25
Project G	25	50	50
Project H	25	50	75
Project I	75	100	50
Security			
Project M	75	50	25
Project N	75	100	50
Connectivity			
Project R	100	100	75
Project S	25	50	50
Project T	75	100	25

disproportionate weight to one value measure, this may serve as an opportunity to revisit the weights of the value measures in the value model.

12.4.4 COMPARISON WITH PORTFOLIO OPTIMIZATION

This approach, which is basically a "greedy hill climbing" algorithm, is very similar to the multiple objective decision analysis and optimization resource allocation approach used in Chapter 8 of Kirkwood (1997). In optimization, all constraints are added by equations instead of structuring the alternatives in rows.

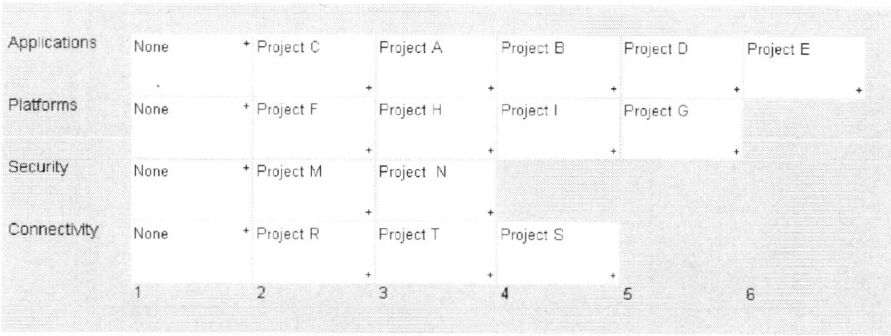

FIGURE 12.11 Trade space of data center projects.

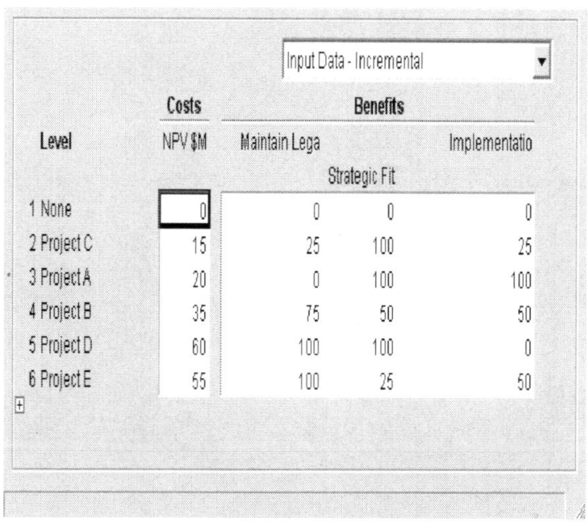

FIGURE 12.12 Data for applications projects.

A binary linear programming algorithm is then used to find the optimal portfolio for any budget level. The optimization approach has the strength that constraints can also be used to model the dependence of funding areas. However, the optimization approach does not always provide a stable allocation; as the budget increases, a program can enter the portfolio, leave the portfolio, and then reenter the portfolio. While mathematically correct, this property is hard to explain to program advocates when more money becomes available and their program drops from the selected set of programs.

12.4.5 STRENGTHS AND WEAKNESSES OF INCREMENTAL BENEFIT/COST PORTFOLIO ANALYSIS

The approach described above has the following strengths:

- Alternatives can be developed either as Go/No-Go or as individual funding increments of a program.

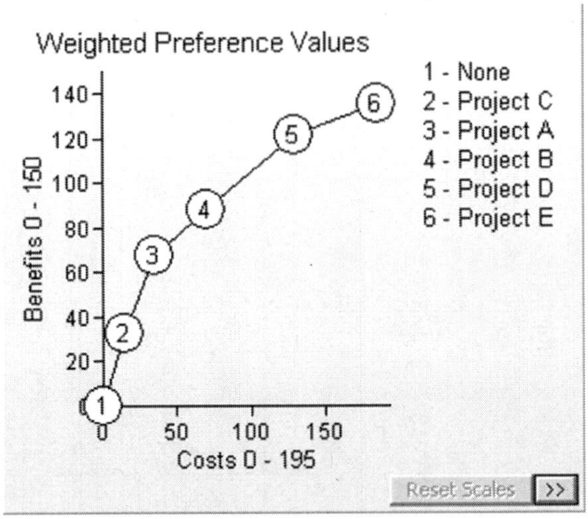

FIGURE 12.13 B/C for applications projects.

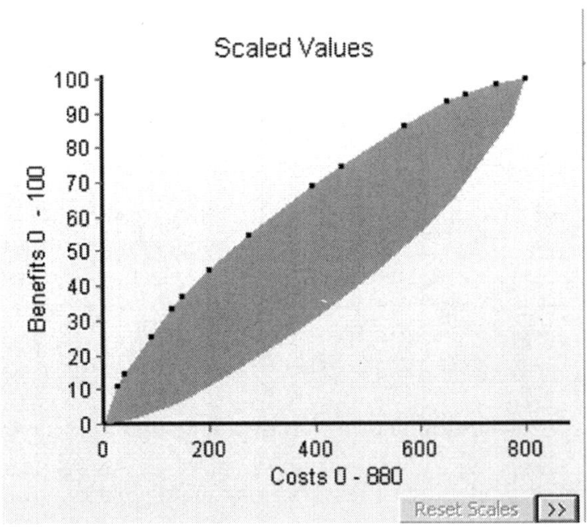

FIGURE 12.14 Efficient frontier for the data center.

- It provides a very stable allocation scheme; as the budget changes, funding increments enter or leave the mix, but the order of funding stays the same.
- The rationale for the "answer" is readily apparent in that the mathematics of the approach are straightforward and easy to understand.
- It provides an approach for quantifying costs and benefits (both objective and subjective).
- It is available in commercial off-the-shelf (COTS) software.

TABLE 12.5 **Project Order-of-Buy**

Order No.	Area		Costs	Cumulative costs	Benefits	Cumulative Benefits	ΔB/ΔC Ratio
1	Connectivity	Project R	25	25	114	114	4.55
2	Applications	Project C	15	40	32	146	2.15
3	Platforms	Project F	50	90	106	252	2.12
4	Platforms	Project H	40	130	84	336	2.1
5	Applications	Project A	20	150	35	371	1.75
6	Security	Project M	50	200	73	444	1.47
7	Connectivity	Project T	75	275	102	546	1.36
8	Security	Project N	120	395	143	689	1.19
9	Connectivity	Project S	55	450	57	746	1.03
10	Platforms	Project I	120	570	116	862	0.97
11	Platforms	Project G	80	650	70	932	0.87
12	Applications	Project B	35	685	20	952	0.58
13	Applications	Project D	60	745	34	986	0.56
14	Applications	Project E	55	800	14	1000	0.25

TABLE 12.6 **Best Portfolio for $450M**

Area	Funded Projects	Cost $M
Applications	C, A	35
Platforms	F,H	90
Security	M,N	170
Connectivity	R,T,S	155

- It accommodates multiple constraints and dependencies among levels of a funding area.
- It facilitates sensitivity analysis by supporting "what-if" analysis capability.
- It brings some structure and discipline to a difficult process, and provides an audit trail for the results of the analyses.

The approach has the following limitations:

- Since the funding increments are discrete, there may not be a package that falls *exactly* at the specified budget constraint; the analyst must decide offline whether to (1) "fill the gap" with a level further down in the funding order that stays within budget; (2) modify levels to get closer to the budget target; or (3) modify the budget target.
- It relies heavily on expert judgment.
- It can accommodate only a limited number of attributes of benefit, probably less than 10 before the assessments become overwhelming and the capabilities of available COTS packages are exceeded. This is not due to

mathematical constraints, but rather, due to the best practices in human factors that dictate how much should be displayed on a screen, how much scrolling is okay before it becomes distracting, and so on. Most of the COTS packages that implement this approach were designed to be used in group assessment sessions, so human factor concerns were built into the packages.

- The algorithm requires independence of funding areas (but there are various ways available to deal with dependencies if needed).

12.5 Summary

In many cases, the goal of a decision analysis is not to select the single best alternative, but rather, to select the combination, or portfolio of alternatives (e.g., projects), that best meets the articulated goals. In this chapter, we summarize several approaches that have been used successfully in government, petrochemicals, and pharmaceutical industries among others to make the portfolio problem manageable.

In "perfect capital markets," the weighted average cost of capital (WACC) is the opportunity cost of capital, and it is appropriate to use WACC as a discount rate and fund all projects with a positive expected net present value (ENPV). However, this "rule" becomes more difficult to use when capital is constrained and a firm is unwilling or unable to acquire debt or equity at the WACC in order to fund all acceptable projects.

One way around this problem is to state an acceptable discount rate and identify the constrained global optimum by considering all possible portfolios, discarding those that violate this constraint, and make feasible choices based on best ENPV.

The shadow price approach prioritizes projects for funding by choosing those with the best break-even shadow price until resources are exhausted. We can also use shadow prices by splitting out a component of the cash flow that comprises investment, and treating its present value as the constrained resource. This is known as using a Profitability Index.

Incremental benefit/cost analysis using Pareto optimality criteria has been used in many highly successful portfolio analyses. It establishes the most cost-efficient way of spending additional resources or accommodating budget cuts. The solution approach explicitly identifies the set of feasible alternatives that honors the problem constraints, and evaluates only the alternatives that provide the most "bang-for-the-buck" at any level of resource expenditure.

In this approach, programs that are competing for funding are described in terms of funding levels that could be as simple as Go/No-Go, or more complicated in terms of several funding increments. The increments can be dependent or independent, but the programs are assumed to be preferentially independent. For each funding level, incremental benefit and incremental costs are calculated, and program levels are then funded in order of decreasing Δbenefit/Δcost. This approach is similar to the MODA approach used in chapter 8 of Kirkwood

(Kirkwood, 1997), with one major difference. The optimization approach does not always provide a stable allocation; as the budget increases, a program can enter the portfolio, leave the portfolio, and then reenter the portfolio. With the approach described here, the order-of-buy remains constant, but the cut-off line for funding just moves up or down.

KEY TERMS

Across criteria weights Swing weights that are used to prioritize and calibrate the benefits from one criterion to another.
Bang for the buck chart A chart that plots the profitability and use of a constrained resource (e.g., investment capital) for a number of potential investment portfolios, allowing an efficient frontier to be identified
Benefit/cost analysis A technique for comparing programs or projects based upon the ratios of benefit to cost that each provides.
Business unit Segment of the entire enterprise that the decision team may want to view in isolation. This can apply not only to business units that are reported distinctly in an organization's financials, but also to opportunities or projects that the team wants to consider on their own.
Capital investment (capex) Money invested to initiate and enable a project.
Delta The difference of value or utility between two specified strategies.
Efficient frontier The set of portfolios that provide more value than any other portfolio with equal or lower cost.
Expected net present value (ENPV) Expected value of the NPV; typically applied to cash flow streams.
Expected value (EV) Sum (or integral) of outcomes multiplied by their probabilities (or probability densities).
Net present value (NPV) For a sequence of numbers through time, the sum of the numbers, each multiplied by a discount factor that represents the value today of one unit (e.g., dollar) in given time period.
Operating expense (opex) Ongoing expense associated with operating a project.
Order of buy The order in which programs or projects are added to the portfolio based upon decreasing incremental benefit to incremental cost ratios.
Pareto optimality A technique that selects the most efficient portfolios based upon funding in order of decreasing incremental benefit to incremental cost ratios.
Present value (PV) The amount of money today that is equally valuable as a stream of cash flow in the future. Typically calculated by weighting each period's cash flow by a discount factor that decreases exponentially through time at a specified rate.
Profitability index (PI) ENPV divided by PV investment.
Strategy A fixed set of choices from all pertinent decisions.

Trade space The exhaustive set of possible portfolios (combinations of projects) that are under consideration. Our goal is to find portfolios that are on the efficient frontier of the trade space.
Within criteria weights For a given criterion, swing weights that compare and calibrate benefits in one funding area to another
Utility The fundamental value measure to be maximized, regardless of whether it was generated via MODA or ENPV calculations. If there is no explicit risk preference adjustment, this term refers to the value measure or ENPV that is used for decisions under uncertainty.

REFERENCES

Barclay, S., Brown, R., Kelly, C., Peterson, C., Phillips, L., & Selvidge, J. (1977). *Handbook for Decision Analysis.* Arlington, VA: Defense Advanced Projects Agency.

Brown, R., Kahr, A., & Peterson, C. (1974). *Decision Analysis for the Manager.* New York: Holt, Rinehart, and Winston.

Buede, D.M. & Bresnick, T.A. (2007). Applications of decision analysis to the military systems acquisition process. In W. Edwards, R. Miles, & D. von Winterfeldt (eds.), *Advances in Decision Analysis,* pp. 539–563. Cambridge: Cambridge University Press.

Burk, R.C. & Parnell, G.S. (2011). Portfolio decision analysis: Lessons from military applications. In A. Salo, J. Keisler, & A. Morton (eds.), *Portfolio Decision Analysis: Improved Methods for Resource Allocation.* New York: Springer.

Kirkwood, C. (1997). *Strategic Decision Making.* Belmont, CA: Duxbury Press.

Luenberger, D. (1984). *Linear and Nonlinear Programming.* Reading, MA: Addison-Wesley Publishing Company.

Phillips, L. (2007). Decision conferencing. In R. Miles, D. Von Winterfelt, & W.E. Edwards (eds.), *Advances in Decision Analysis,* pp. 375–398. New York: Cambridge University Press.

Reinsvold, C., Johnson, E., & Menke, M. (2008). Seeing the Forest as Well as the Trees: Creating Value with Portfolio Optimization.

Salo, A., Keisler, J., & Morton, A. (2011). *Portfolio Decision Analysis: Improved Methods for Resource Allocation.* New York: Springer.

Sharpe, P. & Keelin, T. (1998). How SmithKline Beecham makes better resource-allocation decisions. *Harvard Business Review, 76*(2), 45–57.

CHAPTER THIRTEEN

Communicating with Decision Makers and Stakeholders

GREGORY S. PARNELL and TERRY A. BRESNICK

I know that you believe you understand what you think I said, but I am not sure you realize that what you heard is not what I meant.

—Author Unknown

The most important thing in communication is to hear what isn't being said.

—Peter Drucker

Handbook of Decision Analysis, First Edition. Gregory S. Parnell, Terry A. Bresnick, Steven N. Tani, and Eric R. Johnson.

13.1 Introduction

Decision analysis is a social-technical process. The art of communication is one of the most important soft skills of the decision analyst. Our opening quotes emphasize the three basic components of communications: the sender, the message, and the receiver. It is critical to understand that the best communication involves sending, receiving, and providing feedback to the sender that verifies the message is received. Figure 13.1 shows this basic communication diagram. The receiving (listening) skills are as important as the sending skills. One of the *Seven Habits of Highly Effective People*, "Seek first to understand, then to be understood" (Covey, 1991), emphasizes the importance of listening. In addition, the message and the mode of communication are also important. Some of us prefer to learn by reading, some by listening, some by doing, some by stories, some by pictures, and some by charts and graphs. The sender will be more effective if he or she knows and takes advantage of the receiver's preferred learning style and uses multiple modes when communicating to groups, decision makers, and stakeholders.

Chapter 5 introduces the Dialogue Decision Process, which focuses on the opportunities for formal communication (dialogue) between the decision analysis team and the decision team. However, decision analysis communication includes several other important communication paths. Figure 13.2 shows the major players in a decision analysis and the major communication paths. The major players are the internal decision team (the decision makers, the stakeholders, and the decision implementers), the study participants (the study champion, the internal stakeholder representatives, the subject matter experts, the decision

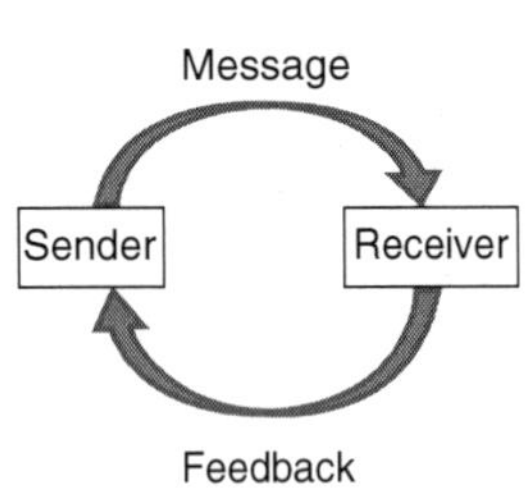

FIGURE 13.1 Communication.

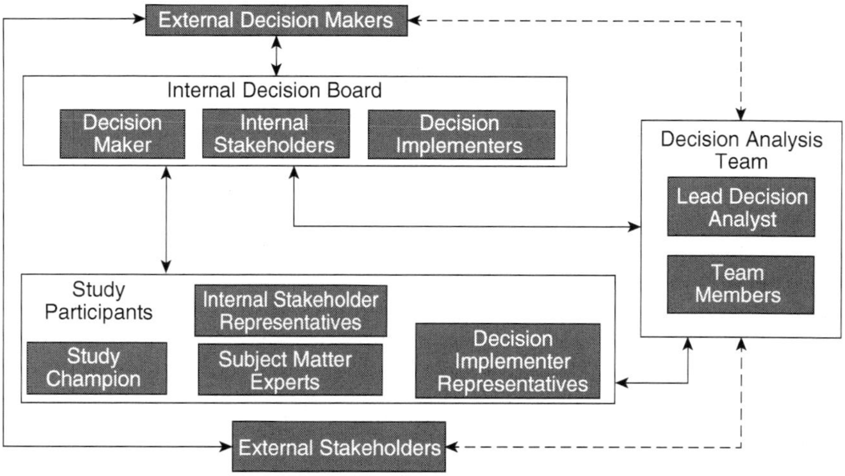

FIGURE 13.2 Decision analysis participants and communications paths.

implementer representatives[1]), the decision analysis team (lead decision analyst and team members), the external decision makers, and the external stakeholders. The two-way arrows emphasize that communication goes both ways. The solid arrows in the figure refer to the major formal communications that should occur. The dashed arrows refer to possible communication that may or may not occur depending on the access of the decision analysis team. As a simplification, the arrows go only to boxes rather than to individuals. Many formal communications go on between individuals in one box and individuals in another box. For example, the lead decision analyst and the study champion might communicate regularly, and one decision analysis team member might be assigned to obtain data from one subject matter expert. In addition to formal communications, informal communications can play an important role in a decision analysis study. Informal communications occur between individuals who have professional and/or personal relationships established before or during the study. Many times, decision makers, stakeholders, and decision analysts use informal communications to understand each others' issues and objectives.

Decision analysts should take advantage of interviews, facilitated meetings, focus groups, and decision briefings to expand their professional and personal networks. The individuals with whom an analyst meets in a decision analysis study even casually may prove to be very useful resources for the current or future studies.

[1]We could have included the decision implementers in the internal stakeholder box but we have singled them out due to the critical role they play in decision implementation, the subject of the next chapter.

The chapter is organized as follows. In Section 13.2, we discuss how the decision analyst can establish communication objectives for each communication opportunity with senior decision makers and stakeholders. In the Section 13.3, we discuss the need for decision analysts to understand the challenges faced by senior decision makers. In Section 13.4, we discuss the need to provide insights and not the details of the analyses, the challenges with presenting quantitative data, and best practices for communicating via presentations and technical reports. In Section 13.5, we describe how some of the key insights were presented in the illustrative examples. Section 13.6 provides a summary of the chapter.

13.2 Determining Communication Objectives

When planning for a major communication, it is critical for the senders (Decision Analysis Team) to assess not only their own objectives but also those of the message receivers—Decision Makers (DMs), Stakeholders (SHs), but also Subject Matter Experts (SMEs). In Chapter 7, we present several techniques for identifying decision-maker and stakeholder objectives for the decision analysis and recommend the use of the stakeholder issue identification matrix and the functional value hierarchy to document the key issues and the fundamental objectives for complex decisions. Our knowledge of these issues and objectives can be very useful background for any communications session. Table 13.1 provides a list of some typical communication objectives during each phase of the decision analysis process for the decision analysis team and the decision makers and stakeholders. The decision analysis team should tailor these objectives to the problem, the organization, the decision makers, the stakeholders, and the experts involved in the study.

We use these objectives as we prepare for communications in each of the major steps in the decision analysis process.

13.3 Communicating with Senior Leaders

Decision analysts need to understand senior leader decision making. Senior leaders are very busy and their time is one of their most important resources. As a result, they are difficult to see and have gatekeepers[2] who control access to them. One of the major roles of senior leaders is to establish the organization's vision and strategic objectives. Senior leaders always consider how a project supports their vision and objectives. Therefore, knowing their objectives is invaluable to the decision analyst. We usually obtain this information in the framing and objectives assessment phases. If this information is not readily available, a good secondary source is the decision analysis champion. Finally, another major

[2]Sometimes called Moat Dragons!

TABLE 13.1 Decision Team Communication Objectives and Stakeholder Objectives

Steps	Decision Analysis Team Objectives	DM, SH, SME Objectives
Establish the need for the study	Convince the DMs and SHs of the need for the decision analysis study	Achieve organizational, professional, and personal goals
Obtain approval for the study plan	Convey effective and efficient decision analysis plan Be able to perform the study in the resources and time available Obtain the necessary resources (dollars, people, and access to DMs/SHs/SMEs) to achieve the study objectives	Make the most effective and efficient use of study resources Allocate study resources
Scope the decision frame (Chapter 6)	Ensure the decision frame conveys the full scope of the decision Obtain approval of the decision frame	Protect organizational equities Include their important issues
Craft objectives and value measures (Chapter 7)	Show alignment of study objectives with organizational strategy Obtain approval of the list of decision objectives	Include their important objectives Understand the objectives of other DMs and SHs
Design creative alternatives (Chapter 8)	Describe high value, creative, doable alternatives Obtain approval of the alternative list	Include high value alternatives Include their preferred alternatives in the analysis
Perform deterministic analysis and develop insights (Chapter 9)	Demonstrate that a credible decision model has been used to assess the decision value and differentiate between alternatives Convey the sources of value and assess trade-offs Identify the most promising alternatives	Understand the decision model Understand how the information they provide will be used Be able to justify the information they provide (SMEs) Understand the most promising alternatives
Quantify uncertainty (Chapter 10)	Convey the key uncertainties and their impact on value	Understand sources of uncertainty that drive value and create risk
Perform probabilistic analysis and develop insights (Chapter 11)	Convey that a credible probabilistic decision model has been used Identify the value and risk drivers Create higher value alternatives Create risk management options	Understand the impact of the value and risk issues on their organization

TABLE 13.1 (*Continued*)

Steps	Decision Analysis Team Objectives	DM, SH, SME Objectives
Communicate insights (Chapter 13)	Convey the story and key insights Provide and justify recommendation(s)	Understand the impact of each alternative on their organization Select the best alterative(s) Have clear rationale for decisions
Enable decision implementation (Chapter 14)	Develop credible implementation plan and success measures Convey implementation risks Provide insights for managing implementation risks	Understand the impact of the implementation on themselves and their organization Approve implementation plan Manage implementation risk

role of senior leaders is to allocate resources to achieve organizational objectives. This is the role that usually results in the interaction of the decision analyst with the senior leader.

This book emphasizes both technical and soft skills of the decision analyst. Most of the academic and professional training of analysts focuses on the analytic techniques. However, decision making occurs in an organization with its own culture, decision-making processes, and political considerations (See Chapter 2). For example, a decision analysis presentation in an organization that has adopted a decision analysis process (e.g., Dialogue Decision Process) is very different than one in an organization using decision analysis for the first time. In addition, decision makers are individuals with their own decision-making experiences, knowledge of decision analysis methods, information-learning styles, decision-making preferences, and, perhaps, emotional considerations about decision opportunities. As an example, a presentation to a senior decision maker who has successfully used decision analysis for a major decision is very different than one to a decision maker who is seeing a decision analysis presentation for the first time, or a decision maker who has had a recent negative experience with another analytical technique. We believe that to be successful, decision analysts must provide sound analytical results in a manner consistent with the political and organizational factors and the individual decision maker(s)' knowledge and values. Figure 13.3 portrays this goal graphically.

We can also point to examples in which the major value added of the decision model was not the analysis itself but the ease with which the model allowed the stakeholders to communicate, exchange information and opinions, and

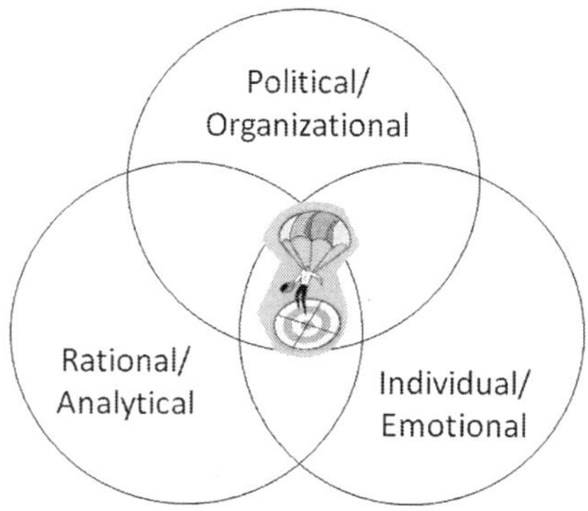

FIGURE 13.3 Communicating with senior leaders.

resolve disagreements. Frequently, decision analyses bring together diverse groups of stakeholders who will be differently affected by a decision. In an ideal world, the decision analyst can elicit values from all parties in a way that leads to consensus. In the real world, consensus is not always achievable—not because our techniques for achieving consensus are flawed, but because there may be no reason or incentive for consensus. There may be many issues about which the values stakeholders seem to be diametrically opposed and there is no reason to expect consensus. Yet a decision must be made that seems sure to displease some parties. Identifying the areas of agreement and disagreement both in probabilities and in fundamental objectives and using this knowledge as a springboard for reconciling differences is one of the most valuable ways that decision analysis models can be used. Emotions can mask what the debate is really about and can get in the way of decision making based on the decision rules; by providing a logical, separable framework for discussion, decision analysis can help us hone in on and communicate the "real" issues. An example of how this can work is presented in a sidebar.

Chicago Circulator: Using DA to Open Communications and Get Past the Emotions

The city of Chicago was considering alternatives for putting a light rail system, called the Circulator, in the heart of the Chicago "Loop." It was intended to increase shopper traffic to merchants while easing everyday vehicular traffic problems. There were several stakeholder groups, including the city managers, merchants, shoppers, commuters, and the Circulator developers and operators.

Meetings had been going on for months, and they often turned into very angry, emotional sessions involving name-calling and worse. Some felt that the Circulator would be unsafe, and there would be many accidents; others, while they liked the idea, felt that the city could not afford the Circulator given the tense economic times. Little was being resolved, and the project was making no progress.

The second author and his decision analysis team were brought in and built a MODA model to help understand and discuss the fundamental objectives of the stakeholders. We quickly discovered that there was little difference in fundamental objectives of the main stakeholder groups, but the perception of value of the alternatives was vastly different. Initially, to avoid the emotional outbursts, the same composite model of approximately 20 bottom-level criteria was evaluated separately by each of the two major constituencies, each of which assessed its own weights and values. When the smoke cleared, there were only two criteria on which there were significant differences—the safety of the system and the ability of the city to pay for the project. When the groups were brought together and told about these findings, they were able to agree on how to approach the areas of difference.

Regarding the safety issue, an independent safety firm, approved by both sides, would be hired do a thorough safety analysis and both sides agreed to abide by their analysis. Regarding the affordability issues, those who felt that the city could not afford the project learned that if the system was deemed to be safe, the merchants would be willing to underwrite a significant part of the cost, which would make the project financially viable. Thus, the MODA framework enabled the stakeholders to get past the emotions and logically discuss what really mattered—truly, a "socio-technical" process. (Note: The project was eventually canceled when it was learned that although the system could be made safe, the dollar costs and other impacts of making it acceptable were prohibitive.) (Bresnick, Circulator Project Notes, 1987).

13.4 Communicating Decision Analysis Results

Frequently, analysts communicate the wrong things. Although their medium, message, and modes of communication are understandable and meaningful to them, it is far more important that what they have to say is understandable and meaningful for the decision maker(s) and key stakeholders. Models such as influence diagrams, decision trees, NPV models, and MODA value hierarchies are often exactly what are needed for doing sound analysis, but the detailed models themselves may not be the best ways to communicate results and insights to the target audience. Sometimes, we need to simplify the model to focus on the key insights. See Section 13.5.1 for an example.

13.4.1 TELL THE DECISION MAKER THE KEY INSIGHTS AND NOT THE DETAILS

A natural tendency is to focus on highlighting the details of the analysis that support the soundness of the results, and to make sure that the decision maker follows the analysis every step of the way. The Coal Chute sidebar illustrates this point.[3]

The Coal Chute

Once upon a time, a young man inherited an old family vacation home. He remembered going to this majestic home to visit his grandparents during his youth but later moved away for college and had not visited the home in years. Since he had recently returned to the area to accept a new job, it was easy for him to inspect the home. Sadly, the lovely old home had fallen into disrepair over the past decade. All the doors and windows were boarded up. He made two trips around the house before he noticed an old coal chute that he was able to open. Since this was the only way into the house, he slid down the dirty, dusty, coal chute and found himself covered with coal soot from head to toe. As he inspected the house, he realized how much work was ahead of him to return the house to its former glory.

Over a period of a year, he spent many evenings, every weekend, and much of his vacation time repairing the house. First, he took all the boards off the windows and doors. Next, he cleaned, repaired, and repainted the house and bought new furniture. One of the special features of the house was a double door entry that opened to a lovely spiral staircase. As his work was coming to an end, he installed a beautiful chandelier over the spiral staircase and added silver door handles on each door. The double door entry opening on the spiral staircase lighted with the beautiful chandelier was a very nice way to introduce visitors to this grand home.

Finally the work was done and the young man invited all his friends and co-workers to see his restored vacation home. How do you think he took his visitors into the house? By the coal chute! Why? Because he wanted to impress them with his incredible diligence and hard work! After taking them into the house, he told them the details of how he had cleaned and repaired every item. How did his friends want to go into the house?

[3]The first author first heard this story on a video made by Professor William K. Linvill (1919–1980) of Stanford University. The tape was played at the Department of Engineering-Economic Systems 25-year anniversary celebration in 1992. The story has been embellished over the years, but the message is the same.

Through the double doors that opened on the spiral staircase with the beautiful chandelier!

The analogy is that the house is a difficult decision problem, the young man is the decision analyst, and his friends and coworkers are the decision makers. Just like the young man, the decision analyst wants to impress the decision makers with his or her incredible hard work and diligence in resolving the difficult decision problem. However, the decision makers do not want to make the same journey that the analyst took by going down the coal chute. Rather, they want to go straight to the decision insights by entering through the silver handled double doors and seeing the beautiful chandelier lighting the spiral staircase. Do not take the decision makers down the coal chute!

All too often, decision analysts believe that their job is mostly done when the analysis is completed. They have used the best methods, techniques, tools, and technologies, elicited the knowledge that is needed to perform credible, defensible analysis, and are ready to convey their recommendations to the decision maker. Yet it is at this point that many analyses fail: the analyst fails to recognize that communication of results is the most critical part of the analysis and that it is the most commonly deficient part of the analytical process. To emphasize this point, the first author always tells decision analysts "when the analysis is finished the job is 50% done." The analysts have the results, but they must identify the key insights from the results and determine how to best communicate those insights to decision makers and stakeholders. The communication of the analysis results is a critical soft skill. For some decision analysts, this part of the process is the most challenging.

While the Coal Shute is an important story, there are some situations where it is important to provide enough details on the decision process and models to convince the decision board that the team performed the detailed analysis required to support the story.

13.4.2 COMMUNICATING QUANTITATIVE INFORMATION

Decision analysts should communicate using principles of graphical excellence. One of the best sources of advice for excellence in the presentation of quantitative data is the series of books by Edward Tufte (Tufte, 1983). According to Tufte, graphical excellence is the well-designed presentation of interesting data that smartly combines substance, statistics, and design. It consists of complex ideas communicated with clarity, precision, and efficiency. Graphical excellence enables the recipient to view the greatest number of ideas in the shortest time with the least ink in the smallest space. Some basic principles of graphical excellence include: have a properly chosen format and design; use words, numbers, and drawings together; reflect a balance, proportion, and sense of relevant scale; display an accessible complexity of detail; have a story to tell; draw in a professional manner with careful reproduction; and avoid "chartjunk" that distracts from the message (Tufte, 1983). Tufte's books provide several examples of excellent graphical presentations as well as poor examples (Tufte, 1990; Tufte & Moeller, 1997). Although Tufte is recognized as one of the foremost experts in communicating quantitative information, some of his work is not without controversy, such as his disdain for PowerPoint presentations. Each decision analyst must personally decide what he or she likes of Tufte's philosophy and should tailor Tufte's communications ideas to their own situations and preferences.

13.4.3 DETERMINING AND TELLING THE STORY

In this chapter, we define the "story" as the major insights and key results of the analysis. We should strive to tell the story with one chart and a few sentences that capture the emotional aspects of the story. An important role of the story is to forge or undo beliefs. In complex decision analysis studies with conflicting objectives, many alternatives, and significant uncertainties, the standard decision analysis displays used in Chapter 9–11 may not be the best way to engage the decision maker in the story. Sometimes, we must do additional analysis just to determine the story and then think creatively about how to convey the story clearly and simply. The sidebar on the Only Robust Alternative provides an example of how the story was determined and presented in a complex decision analysis study for the Department of Energy. The chart in Figure 13.4 summarizes the results of 348(58 × 6) cumulative probability distributions and shows why one alternative, "2 Phase and DUS" was the preferred alternative.

Identifying and Telling the Story of the Only Robust Alternative

The first author worked on a research project to analyze the cleanup of a chemical spill at the Department of Energy nuclear production facility (Papatyi et al., 1997). The cleanup process was governed by the Comprehensive Environmental Cleanup Liability Act (CERCLA). The study used MODA to assess the value of each alternative. We used a Gold Standard value model (see Chapter 7) using the criteria specified in the CERCLA law and, therefore, called the value the CERCLA value. In addition, we used probabilistic analysis since there was significant uncertainty about the quantity of the spill. The initial spill estimate was 7500 gal. After new test data proved that the original estimate was too small, we performed a probability assessment that determined the spill could range from 50,000 to 500,000 gal with a mean of 100,000 gal. As a result, a probabilistic analysis was done of the 58 technology combinations that had the potential to remediate the spill. To assess the performance, 58 cumulative distributions were calculated—one for each technology alternative. Based on our experience assessing the probability distributions, we knew that our client and the stakeholders were not comfortable interpreting cumulative probability distributions. Clearly, we had two challenges: determining the story and telling the story.

In order to determine the story, we looked at the range of the expected value for all alternatives. See the heavy bars in Figure 13.4. Initially, using deterministic and stochastic dominance, we looked for a dominant alternative, but no alternative dominated. We looked at expected value. Several of the top alternatives had approximately the same expected value. Next, we looked for the alternative that was the most robust across the spill uncertainty range. The most robust alternative with a high expected value was 2 Phase and DUS.

We told the following story using Figure 13.4.

The graph displays the range of expected CERCLA value of all 58 alternatives. The range increases with the potential spill volume since there is more uncertainty about the alternatives' ability to remediate large spill volumes.

One alternative (2 Phase and DUS) stands out as the most robust alterative. This alternative is in the top four alternatives for each level of spill volume. The rank of 2 Phase and DUS improves at higher volumes.

It is the only alternative with this property. Therefore, we believe this is the most robust alternative.

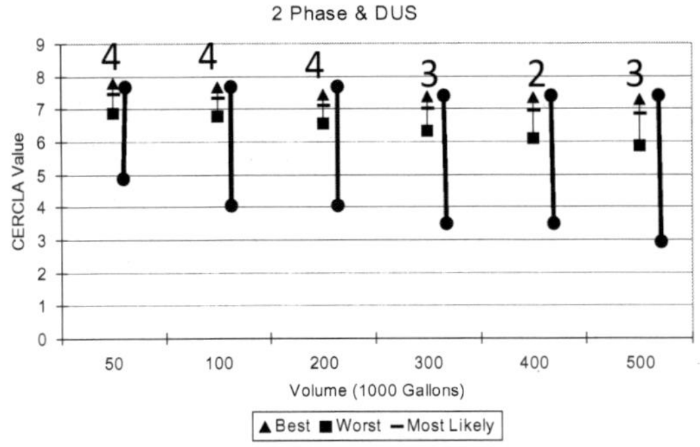

The number indicates the expected value ranking of 2 Phase and DUS out of the 58 alternatives assuming the specified spill volume. The arrows indicate the range of value uncertainty for 2 Phase and DUS at each spill level. The thermometers indicate the range of expected values of all 58 alternatives.

FIGURE 13.4 Chart used to tell the story of the best technology.

13.4.4 BEST PRACTICES FOR PRESENTING DECISION ANALYSIS RESULTS

The following is a list of the communications lessons learned from years of professional practice.

1. **Develop a communications plan.** As Covey says, "Start with the end in mind" (Covey, 1991). Understand where the analysis fits in the organization strategy, identify how the results will be used, and determine with which other studies or decisions the analysis must mesh. Develop a communication plan that includes who needs to be informed of the study results, who delivers the messages, what is to be communicated, how it is to be communicated, and when it is to be communicated. Each communication should be tailored to a specific purpose. The context and nature of what is to be communicated may play heavily in choosing communication mode(s). Informational discussions require different communications content and style than decision presentations. Presentations to a single decision maker may differ from presentations to multiple decision makers.
2. **Know the audience.** Knowing the decision makers and their objectives is very important. The communication should be tailored to the cognitive style of the decision maker. Some want only the bottom line, some like graphs, some are numbers oriented, and some want a lot of detail. The decision analyst should learn how the decision maker prefers to receive information. Sources of this information include the decision makers themselves, past experience communicating with them, and information from others who have interacted with the decision maker previously.
3. **Review the decision frame.** When communicating with senior managers, confirm that the problem is framed properly, and that the right problem is being solved (or the right opportunity is being exploited). Adding this to every presentation is a useful reminder and an important check to make sure that nothing significant has changed.
4. **Review the objectives of the decision team.** The decision objectives are identified early in the decision analysis. Understand how the decision analysis results support these objectives. Use the objectives to develop the communication plan.
5. **Determine the story.** After reviewing the decision frame, reviewing the decision team objectives, completing all the analysis, studying the analysis results, and identifying the important insights, determine the story to be communicated to the Decision Board. Stories are much easier to remember than endless charts with lists of bullets and they resonate on a personal level. For example, it is easy to remember the moral of the coal chute story (do not take decision makers down the coal chute), and the story of the robust environmental spill cleanup alternative. Sometimes it is useful to highlight the key features in the story. For example, in the

Robust Alternative story, the large potential spill was not accepted until test data became available that called into question the initial assumption of a small spill.

6. **Develop the presentation(s).** Remember that senior leaders are busy people and they may not have thought about the project since the last meeting. Make the presentation as short as possible while providing a clear story and convincing recommendation(s).

   a. **Allow time for the decision maker(s) to talk.** Use fewer and better slides. Use 10–15 slides for a 30-minute presentation and not more than 30! Do not just use the standard output slides of decision analysis software unless the decision maker is comfortable with this information. Develop new charts that better tell the story.

   b. **Use a bottom line up front (BLUF) chart.** Start the presentation by summarizing the story, the major analysis results, and the recommendation(s) in one chart. The BLUF chart will help the decision team understand the results and make the best use of their time during the presentation. They will be more likely to ask questions about the key elements of the analysis that are the foundation for the recommendations. In addition, this will help to ensure that the presentation supports the summary.

   c. **State the message on each chart.** Quantitative charts can have a lot of information. It is useful to write the message that the receiver should obtain from each chart. Many leading internal and external consulting organizations use this discipline. This helps the presenter stay on track and helps the listener summarize the information presented. It also is an excellent tool to make sure the analysis supports the message summary and that unnecessary analyses are not included in the presentation. It is important to have a colleague verify the clarity of the messages on the charts. Ideally, the message and the chart should be clear enough to stand alone.

   d. **Review the decision frame.** It is always a good idea to review the decision frame in every presentation. This is a useful reminder to busy senior leaders and, when they agree, provides confirmation that the decision analysis team is still focused on the right decision opportunity. If something important has changed based on new information, the decision frame may need to be adjusted.

   e. **Use one process chart.** One of the biggest mistakes the decision analysis presenter can make in a decision presentation is to have too much focus on process details. Senior leaders are usually more interested in results than in process. However, the use of *one* process chart can communicate the decision analysis methodology, the plan for the analysis, who has been involved, the current status of the analysis, and what remains to be done. The process chart may be essential in the study approval, and it may also be a useful reminder in the decision presentation.

f. **Provide summary information.** The presentation should provide summary results in a manner that allows the decision maker to "drill down" into the key assumptions and insights if desired. The audit trail of the logic and rationale can be an important part of the presentation. Backup charts can be very useful to anticipate questions.

g. **Identify the potential value added of the decision.** As we emphasize in Chapter 1 and Chapter 14, decision analysis identifies potential value. The presentation should clearly identify and quantify the potential value of the recommended decisions compared with the momentum or baseline alternative.

h. **Include decision implementation plans.** Decision execution provides the real value. The decision presentation should describe implementation plans, including responsibilities, schedule, success measures, risks, and risk mitigation plans. See Chapter 14 for more information.

i. **Assess the decision quality.** The decision briefing should explicitly assess the decision quality and the readiness to commit to the decision.

j. **Provide a clear summary.** The summary should reinforce the Bottom Line Up Front. A useful presentation guide is the old mantra: "Tell them what you are going to tell them, tell them, and tell them what you told them!" In addition, it is critical to make sure that you receive feedback that the message has been understood.

k. **Take care of the small things.** Several small things can improve the quality of a presentation: use consistent font, use similar chart formats, show the origins of scales, highlight scale changes between charts, show the most preferred region of the chart, and number the slides.

7. **Deliver the presentation(s).**

a. **Stay within the time allocation.** Time is a critical resource for senior leaders, so strive to complete the presentation during the allotted time, including allowing for questions and answers. Use summaries on each chart to help stay on message. A decision analyst who greatly exceeds the time limit may never have another chance to present to this senior leader.[4]

b. **Answer questions succinctly.** Answering questions should be a dialogue with the decision makers and not an opportunity for a monologue or a decision analysis tutorial. One of the keys to staying within the time limit is to answer questions clearly and concisely. Remember Covey's habit to "Seek first to understand and then to be understood" (Covey, 1991). Presenters should carefully listen to the full question before

[4]An analyst working for the first author greatly exceeded his presentation time limit for a senior leader in his organization. After the presentation, the senior leader told the first author "This analyst wasted my time, I never want him to give me another presentation."

responding and repeat the question if necessary to make sure they understand. The best answers to senior leaders' questions are in one or a few concise sentences. These short answers allow for a dialogue with the decision team.

13.4.5 BEST PRACTICES FOR WRITTEN DECISION ANALYSIS RESULTS

Some decision makers prefer to read a written report instead of, or in addition to, hearing a formal presentation. In this case, we recommend that the decision analysis team use the best practices for presentations in the previous section to prepare a written report of the decision analysis. Many organizations have a standard format for study reports that analysts are required to use. Some best practices for written reports are the following:

1. **Write an executive summary.** The executive summary should include a brief summary of the report. The usual topics are the decision frame, the objectives, the alternatives, the decision analysis method, the analysis, the potential value-added, the recommendation(s), and the plan for implementation. Like the BLUF chart, the executive summary may be the most important part of the report. It will have the most impact since senior leaders and key stakeholders are more likely to read the executive summary than the full report.
2. **Make the report readable to the audience.** The technical report should be readable by the intended audience. Technical terms should be defined. A glossary can be very helpful to the busy reader who is not familiar with the domain jargon or decision analysis terms.
3. **Use appendices for technical details.** Technical details about the models, detailed results of the analysis, and the full sensitivity analysis results should usually be placed in the appendices. This allows easy access by the interested readers.

13.5 Communicating Insights in the Illustrative Examples

See Table 1.3 for further information on the illustrative examples. Next, we return to the illustrative examples and provide examples of how important insights can be conveyed to decision makers.

13.5.1 ROUGHNECK NORTH AMERICA STRATEGY (by Eric R. Johnson)

We describe the Roughneck North America Strategy (RNAS) analysis in Chapter 9 and Chapter 11. Several RNAS charts were useful in telling the story of one particular analysis. We highlight the Tar Sands story here. In a complex decision

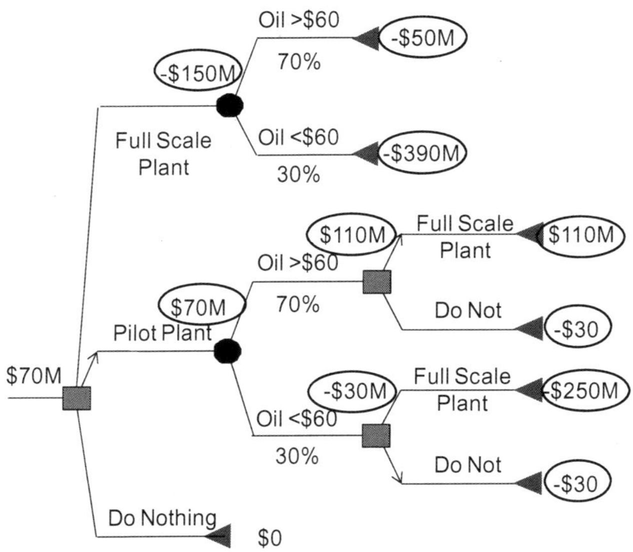

FIGURE 13.5 Tar sands decision tree.

situation like RNAS Tar Sands, there were too many uncertainties and decision opportunities to be displayed intelligibly in a decision tree. Once the analysis team had identified their decision recommendation, they wanted to show its logic to senior stakeholders. They identified the future scale-up decision and the then-current oil price as the most important things to communicate, and drew a simple decision tree indicating their relationship, shown in Figure 13.5. Values shown were Tar Sands EVs from the simulation, rounded to \$10M. Probabilities were inferred from the simulation. Immediate construction of the Full-Scale Plant failed even in favorable oil price scenarios, because operating expense reductions from experience with a pilot plant were not available. The pilot plant option has positive value because it generates the option to proceed only when oil price is above \$60.

13.5.2 GENEPTIN (by Sean Xinghua Hu)

The two graphics that were most meaningful to stakeholders in the Geneptin case were the flying bar chart (Fig. 11.20) and the waterfall chart (Fig. 11.21). The flying bar chart showed that the downside of the personalized strategy was the same as the conventional strategy, while its EV was better and its upside was a lot better. The waterfall showed why this is so: the smaller patient population addressed was a negative factor for the personalized strategy, but this was more than compensated for by higher market share within that segment, longer patient treatment duration (because they live longer!), and higher price (resulted from Geneptin's more compelling value propositions by targeting only the HER2-positive patient segment).

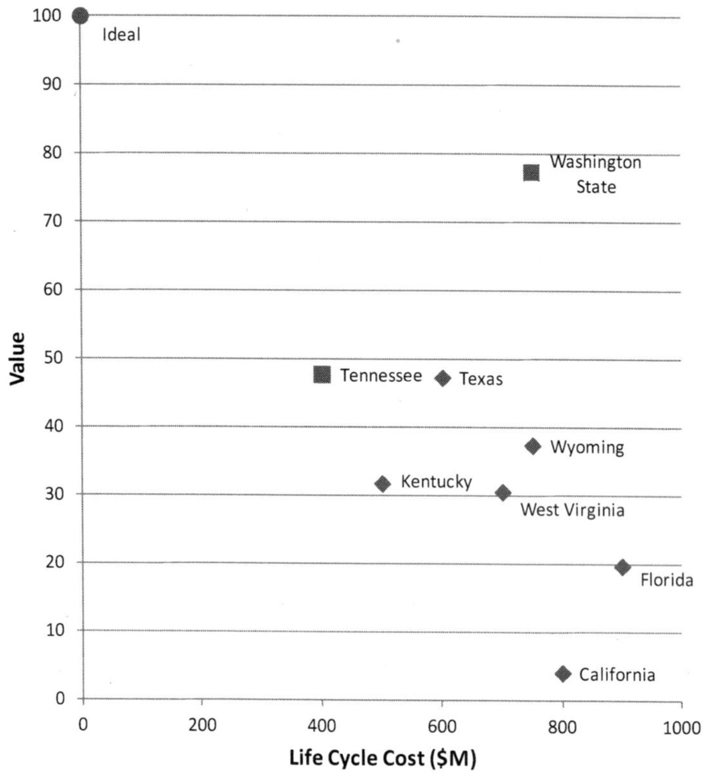

FIGURE 13.6 Data center cost versus value plot.

13.5.3 DATA CENTER LOCATION (by Gregory S. Parnell)

In the data center location decision, the chart that told the story of the analysis was the cost versus value plot in Figure 13.6. A plot of this data helped quickly identify the dominated alternatives and helped the decision makers understand the value added for the additional cost of the dominant alternatives. In our data center problem, we see that Tennessee and Washington states dominate the other alternatives. In Figure 13.6, we use squares to designate the nondominated alternatives and diamonds to designate the dominated alternatives. This chart focused the decision makers on the comparison of the two non-dominated alternatives using the value component and waterfall charts (see Chapter 9).

13.6 Summary

The purpose of this chapter is to identify the decision analysis communication challenges and provide best practices for communicating the results of decision analyses with the decision maker(s) and stakeholders on the decision team. Decision analysts must be aware of both the formal and informal communications pathways in an organization. The typical communication objectives of the decision analysis team vary in each step of the decision analysis process and vary when dealing with the decision maker, the stakeholders, and subject matter experts. Decision analysts must be able to communicate with senior leaders, a challenging task due to the constraints on their time. We believe that to be successful, the decision analyst must provide sound analytical results that consider the political/organizational factors and the cognitive preferences of the individual decision maker(s). One of the greatest values of the decision model itself can be its ability to communicate the story. A wonderful analytic study that misses an important political or organizational issue can be dead on arrival. A sound study that is not understood by the decision maker is not likely to be successful. We recommend providing a summary of the key insights (and not the details) and telling a compelling story of the analysis results. We present best practices for communicating decision analysis results built around all stages of the communications process: developing a communications plan, reviewing the decision frame, reviewing the objectives, knowing the audience, determining the story, developing the presentation, and delivering the presentation. We conclude with one example from each of our illustrative examples.

KEY TERMS

Bottom line up front (BLUF) The BLUF chart is an early chart in a presentation that summarizes the story, the major analysis results, and recommendation(s).
Coal chute The coal chute refers to a story in the text, the moral of which is that decision analysts should present insights of analysis rather than describing the process of analysis (bring decision makers through the front door and not through the coal chute).
Communications plan The decision analysis team's plan for communicating the decision analysis results to decision makers and stakeholders.
Executive summary The executive summary is a brief summary of a written report to provide senior leaders the essential results of the decision analysis.
Formal communications Formal communications are the vertical and horizontal hierarchical authority communications between decision analysts and others involved in the study.
Graphical excellence According to Edward Tufte, graphical excellence is the well-designed presentation of interesting data that smartly combines substance, statistics, and design.
Informal communications Informal communications are the communications that occur based on personal and professional relationship that are not authority based.
Process chart A process chart can communicate the methodology used in a decision analysis. The chart can tell the plan for the analysis, the participants, what has been done, and what remains to be done.

Story The "story" consists of the major insights and results of the decision analysis. We strive to tell the story with one chart and a few sentences.

REFERENCES

Covey, S. (1991). *The Seven Habits of Highly Effective People.* Provo, UT: Covey Leadership Center.

Papatyi, A.F., Deckro, R.F., Parnell, G.S., Jackson, J.A., & Klober, J.M. Jr. (1997). Screening technology trains for DNAPL remediation. *Remediation, Winter,* 87–105.

Tufte, E.R. (1983). *The Visual Display of Quantiative Information.* Cheshire, CT: Graphics Press.

Tufte, E.R. (1990). *Envisioning Information.* Cheshire, CT: Graphics Press.

Tufte, E.R., & Moeller, E.W. (1997). *Visual Explanations: Images and Quantities, Evidence and Narrative.* Cheshire, CT: Graphics Press.

CHAPTER FOURTEEN

Enable Decision Implementation

TERRY A. BRESNICK and GREGORY S. PARNELL

Decision analysts help create potential value. Project managers deliver value.

—Michael Menke

Handbook of Decision Analysis, First Edition. Gregory S. Parnell, Terry A. Bresnick, Steven N. Tani, and Eric R. Johnson.

14.1 Introduction

In Chapter 3 and Chapter 7, we discuss how to define value so that our decision models can effectively evaluate alternatives in terms of their relative worth. We continue our discussion about value here, but from a different perspective. When we perform a decision analysis, we want to make sure that the decision resulting from the analysis will achieve its intended benefit. All too often, a sound analysis is performed, only to sit on a shelf without the recommendations being implemented.

Commitment to action is a key link in the decision quality chain we present in Chapter 5. We emphasize the importance of involving decision makers and stakeholders in the decision process. It is easier to implement a decision when key stakeholders understand the need for the decision and the rationale for the alternative selected. When properly done, the decision analysis process can foster commitment to the decision from the start to the end.

In this chapter, we focus on a very important class of stakeholders, the decision implementers. The active participation of decision implementers is critical to effective and efficient decision implementation. If they are not included in the process right from the start, we may not consider important factors that could impact implementation performance, cost, and schedule. We discuss how we use standard decision analysis techniques and soft skills *during* the decision analysis effort to involve the decision implementers and *after* the decision to increase the chances the decision implementation will provide the intended value to the client.

> The active involvement of the decision implementers is very critical to effective and efficient decision implementation.

The chapter is organized as follows. In Section 14.2, we present some barriers to involving decision implementers. In Section 14.3, we describe how to explicitly involve decision implementers to consider decision implementation during each step in the decision analysis process up to the decision. Section 14.4 presents ways to use the decision analysis to manage decision implementation and the organizational changes that may be required after the decision. Section 14.5 discusses decision implementation considerations in each of the three illustrative examples. Section 14.6 presents a chapter summary.

14.2 Barriers to Involving Decision Implementers

There are five major barriers to involving decision implementers in the decision analysis process. The first barrier is organizational culture. Depending on the organizational culture, the decision implementers may have a strong voice in the decision process or they may have very limited participation. When needed, we try to overcome this barrier by educating the decision makers on the key role of the decision implementers in the process. The second barrier is difficulty in identifying the decision implementer(s). For example, for strategic decisions, the scope of the alternatives may be so broad that many decision implementers would be required. It is quite possible that for decisions with long-term implementation timelines, the implementer may not even be in the organization at the time of the decision analysis. This is particularly true of military or other government organizations, where rotation of personnel is frequent. In addition, we may have to go outside the organization to obtain decision implementation expertise for a new product and/or a new region. The third barrier is decision implementer workload. Decision implementers are usually fully employed, creating value for the company by implementing past decisions. The time they spend supporting decision analyses takes them away from their primary mission. The forth barrier is geography. For example, in an oil and gas company, the decision process may be performed in the corporate headquarters, and the decision implementers may be deployed in regions throughout the world. This can make coordination and clear communication more difficult. The fifth barrier may be the experience or inexperience of the decision analyst with decision implementation in this problem domain and/or organization. A decision analyst who has much past experience with implementation may think that it is adequate (and forget that it may be outdated!). A decision analyst who has little experience with implementation may not know what questions to ask during the decision process and may not know how decision analysis can be used during decision implementation.

The purpose of the chapter is to provide the key knowledge to decision analysts about the critical role that decision implementers can play during the decision process, and how decision analysis techniques can be used to support them. Hopefully, this information can be used to overcome these barriers related to both the analyst team and the decision maker.

14.3 Involving Decision Implementers in the Decision Process

Decision implementers should be involved in every phase of the process. Table 14.1 provides a summary of the roles of the decision implementers and the key decision implementation questions the decision analyst should ask at each step in the decision process. In this section, we discuss the rows of the table for Chapter 5 through Chapter 13.

TABLE 14.1 Decision Implementation Roles and Questions

Steps	Role of Decision Implementers	Key Questions
Select the appropriate decision process (Chapter 5)	Serve as decision team member(s) Serve as decision analysis team member(s)	Are the decision implementers included in the decision process?
Frame the decision (Chapter 6)	Identify decision implementation issues	Do the decision framing products (vision statement, decision hierarchy, and issue identification matrix) include these issues?
	Specify needed date for the decision	Does the analysis schedule meet the need of the decision implementers?
Craft decision objectives and value measures (Chapter 7)	Identify objectives and value measures	Are key implementation objectives and value measures included?
Design decision alternatives (Chapter 8)	Participate in alternative design	Are important implementation features included in the alternatives?
Perform deterministic analysis and develop insights (Chapter 9)	Identify implementation activities, benefits, and costs Validate insights	Are the benefits and costs of implementation included in the deterministic models?
Quantify uncertainty (Chapter 10)	Identify implementation uncertainties and risks Serve as SMEs	Are implementation uncertainties identified and assessed?
Perform probabilistic analysis and develop insights (Chapter 11)	Identify opportunities to increase value and/or reduce risk Validate implementation insights	Have we considered implementation opportunities to increase value (e.g., expand) or reduce risk (e.g., hedge)?
Optimize portfolio resource allocation (Chapter 12)	Identify implementation constraints Identify opportunities to relax implementation constraints	Have we considered implementation constraints and the opportunities to relax the constraints?
Communicate insights (Chapter 13)	Participate in decision briefings and validate implementation insights	Have decision implementers participated in development of the story?
Enable decision implementation (Chapter 14)	Lead development of decision implementation schedule Identify implementation success measures Identify implementation risks and risk mitigation plan	Can we use (with modification) our decision model to guide implementation? Can we use other decision analysis techniques to support implementation of decision results?

- **Select the appropriate decision process.** In Chapter 5, we discuss the importance of involving the decision makers and stakeholders in the decision process. The decision implementers are some of these key stakeholders, and as such, should be part of the decision team and the decision analysis team to ensure that they have input in the decision process. It is important to select individuals who have a corporate perspective in addition to the implementation perspective. If the analysis will be done over an extended period of time or in a series of decision conferences, we ask for management assurance that the same participants (primary and secondary) will be available for the duration, and that if the primary or secondary person is not available, it is better to send no one. One of the most destructive things affecting group commitment is to introduce new players late in the process who have not worked through the trials and tribulations of trying to reach informed consensus through open exchange of information. Finally, when applicable, we assign to specific people the responsibility for each task that comes out of the analysis, along with calendar milestones and exit criteria for completion of the tasks.
- **Frame the decision.** Commitment starts with decision maker and stakeholder participation in the problem definition and framing that we discuss in Chapter 6. The decision implementers can guide the framing process to ensure that decision implementation issues are included in the problem framing products, including the vision statement, stakeholder issue matrix, and the decision hierarchy. In our experience, the needed date for the decision is often driven by the schedules of the decision implementers.
- **Craft the decision objectives and value measures.** Involving the decision implementers in the development of the decision objectives and the value measures can help to ensure that the stated goals and objectives are feasible for and measurable by those who will have to execute the selected alternative. It also may be possible to reuse the value measures during the decision implementation.
- **Design the decision alternatives.** The decision implementers can have important ideas on how to create the initial set of alternatives and, after the preliminary analysis, how to continue to improve the alternatives.
- **Perform deterministic analysis and develop insights.** The deterministic model includes the benefits and costs of decision implementation. Decision implementers play an important role in providing knowledge to include in the model. In addition, they can help validate the major insights during the value dialogue described in Chapter 9.
- **Quantify uncertainty.** Decision implementers have the knowledge and experience to help identify the key uncertainties that will have significant impact on the value of the decision and may be SMEs for the assessment of these uncertainties.
- **Perform probabilistic analysis and develop insights.** Once we have done the probabilistic analysis, we need to identify opportunities to increase value (e.g., expand an alternative) or reduce risk (e.g., develop a hedge). Decision

implementers have the knowledge and experience to help identify these potential opportunities.

- **Optimize portfolio resource allocation.** Portfolio decision analysis may involve constraints that do not exist for individual decisions. For example, a critical resource (e.g., the number of key scientists for a pharma example) may limit the number of projects that can be implemented in the decision period. Decision implementers can help identify these constraints, as well as ways to relax the constraints to help us achieve more value.
- **Communicating insights.** Decision implementers on the decision analysis team can help identify the communication objectives for senior decision implementers and help communicate the analysis results and insights in formats that are useful for decision makers.

14.4 Using Decision Analysis for Decision and Strategy Implementation

Decision analysis models can be used to aid in decision implementation and to assess the implementation of strategies over time.

14.4.1 USING THE DECISION MODEL FOR DECISION IMPLEMENTATION

Decision analysis techniques can be used to manage decision implementation. We may be able to use the decision model directly or, more likely, we may have to modify the model to add additional implementation information. Probabilistic models are essential if we hope to understand the key risks that we face and to manage the consequences through risk avoidance, risk transfer, or risk mitigation. Decision trees and influence diagrams are excellent tools for representing the uncertainties and eliciting the associated probabilities and consequences. This is true whether we represent expected consequences using a MODA model or a model that converts all value measures to NPV. For the most part, decision analysis tools are often used for "static," nonrecurring analyses, but often, there is additional value in their ability to be used dynamically to enhance the risk management process. As new information is gleaned, probabilities get updated; as events unfold, consequences become conditional and change over time. If we can build our models to accommodate these dynamic effects, their value-added is increased significantly as they are used through all phases of implementation.

14.4.2 USING DECISION ANALYSIS MODELS TO SUPPORT DECISION IMPLEMENTATION

Decision analysis models can also be used to monitor decision implementation, to identify risks, and to develop risk mitigation approaches. Two examples of the use of decision analysis for decision implementation are provided.

14.4.2.1 Example 1: Gas Plant Implementation.

Oil and gas projects all over the world are often behind schedule and over budget. In 2009, a study of 230 large oil or gas projects (Goldman Sachs, 2009) found that "the average start-up delay has been 20 months, with a 135% cost increase." Consider the following decision analysis for an LNG regasification plant construction project. The plant was within a year of operation readiness. There was growing concern that time lost due to a hurricane could not be made up, leading to the possibility of $100,000 per day liquidated damages if the plant was not ready to operate on the contractually specified date, July 25. The company had developed a detailed timeline (>1000 tasks), with precedence information and point estimates of durations.

A decision analysis team converted the deterministic schedule to a probabilistic schedule and assessed activity costs as part of the economic analysis process. The team defined 18 activity classes across the tasks and assessed their duration factors and daily costs. This approach captured high-quality information from the experts quickly and effectively. A tornado diagram (see Chapter 9 and Chapter 11) of project duration (shown in Fig. 14.1) suggested interventions on activities in the areas of instrumentation and electrical and commissioning, aiming at reducing their duration. Interestingly, it turned out that most of these tasks were not on the deterministic critical path, but they were near it, and if they went wrong, their impact would be greater. The interventions were made, and the project was completed on time.

14.4.2.2 Example 2: Information Assurance Program Progress.

Parnell et al (Parnell et al., 2011) used MODA to assess the implementation status of the information assurance (IA) programs for the National Security Agency on Department of Defense (DoD) acquisition programs. First, they used the Systems Security Engineering Capabilities Maturity Model (International Systems Security Engineering Association, n.d.) to identify the IA base practices. Second, they determined three acquisition-oriented IA functions: *establish security needs*, *develop IA architecture*, and *certify security*, and categorized the practices under the appropriate function. Third, they used the

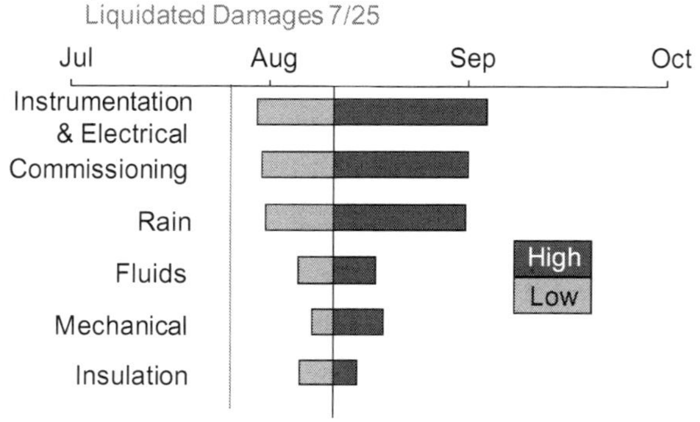

FIGURE 14.1 LNG plant completion date tornado diagram.

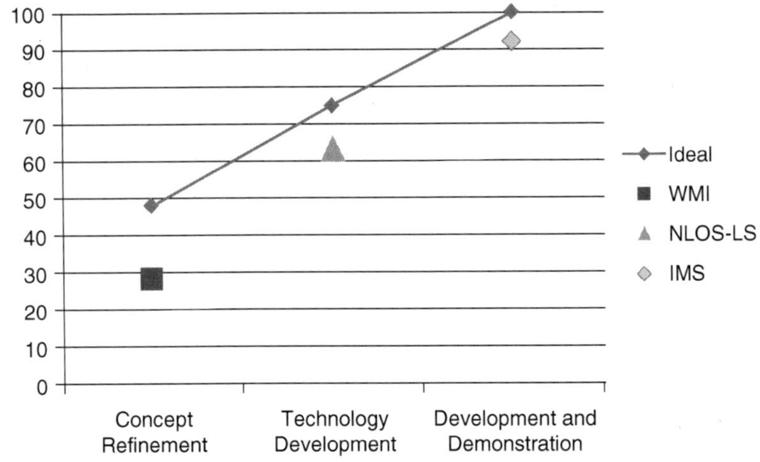

FIGURE 14.2 Plot of IA value versus life cycle phase.

DoD acquisition lifecycle stages to align the base practices with each acquisition life cycle phase to determine when each base practice should be completed at the end of each stage. They identified 33 base practices that should be completed by the end of the third life cycle stage. Fourth, they used MODA to convert the qualitative base practices to quantitative measures of the percent progress completing each IA base practice. A swing weight matrix as described in Chapter 9 was used to assess the relative importance of each of the base practices. Fifth, to demonstrate the model concept, they assessed three illustrative programs in each of the first three stages of the acquisition life cycle (concept refinement, technology development, and development and demonstration). Finally, they used the model to provide a program assessment, a prioritized list of the activities behind schedule, and a variety of sensitivity analyses.

Figure 14.2 shows the results of the analysis of three programs compared to their ideal progress at the end of each stage. The Warfighter Machine Interface (WMI), a program in the concept refinement stage, was dramatically behind schedule. The Nonline-of-Sight Launch System (NLOS-LS), a technology development stage project, was significantly behind schedule. The Integrated Munitions System (IMS), a program in development and demonstration stage, was slightly behind schedule.

The methodology highlighted the IA base practices that were behind schedule for each of the systems. For example, in Figure 14.3, we compare NLOS-LS with the ideal. The weighted completion is 11 points (top gray bar in the figure) out of 75 behind schedule. The figure also identifies the top 10 tasks (next 10 black bars), and each of these is behind schedule. If they were ahead of schedule, the bar would be to the left of the vertical line. The bottom black bar is the total time behind schedule of the remaining base practices. This means that there are many additional tasks behind schedule.

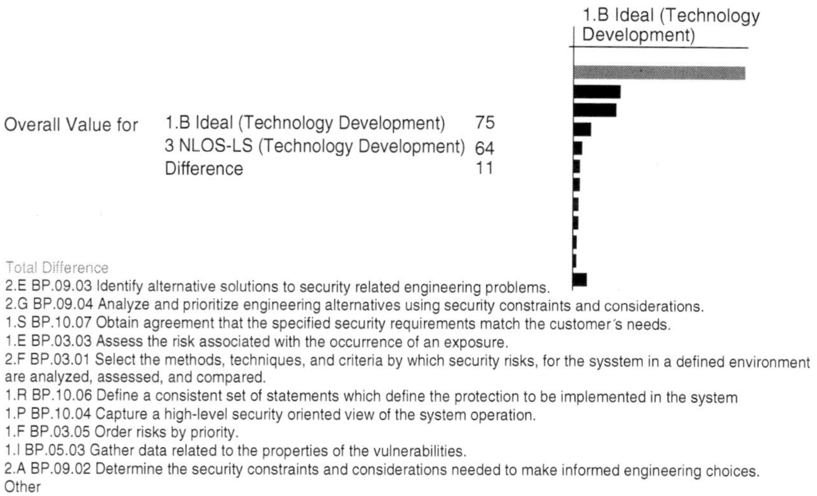

FIGURE 14.3 Base practices causing the IMS schedule delay.

14.4.3 USING DECISION ANALYSIS TO ASSESS STRATEGY IMPLEMENTATION

In government/military applications, we often use decision analytic models, particularly MODA and cost–benefit models, to assess the "state" of an organization from a strategic planning perspective. We can use such models to articulate in terms of fundamental objectives "Where are we today?", "Where do we want to be in the future?", and "How do we get there?" We can use strategic objectives and value measures to represent the first two questions, swing weights to articulate the relative importance of "closing the gap" in one area versus another, and portfolio models to allocate resources wisely to close the most significant gaps. This is a process that is frequently revisited on a 5-year basis, but the framework can provide value to the organization on a continual basis. We can use the value measures to show annual organizational performance and progress towards accomplishing goals, and we can use such an assessment as a report card on management's ability to execute what has previously been decided, and as a control mechanism on what needs to be changed. This dynamic use of the Multi-Objective Decision Analysis/Value-Focused Thinking (MODA/VFT) framework keeps the model "fresh" and the organization apprised on an ongoing basis of how well it is doing.

14.4.3.1 Example. The National Law Enforcement Telecommunications System (Nlets) is a nonprofit organization, primarily owned by the states of the United States, to maintain and operate the communications network that connects local, state, federal, and international law enforcement entities. For example, "AMBER alerts" are transmitted nationwide over the network. Nlets is governed by a board of directors made up of representatives from the states, and is managed by an executive director with a small staff. In 2000, Nlets developed its first strategic plan using a VFT/MODA framework built around the three questions mentioned above ("Where are we today?", "Where do we want to be in the future?", and "How do we get there?"). Every 2 years, Nlets management is evaluated in terms of its performance and progress on the goals and objectives laid out in the plan; this "report card" is presented to the board of directors and is used for performance reviews regarding the management staff. New "swing weights" are also assessed biannually, and areas of emphasis are reprioritized for the next few years. The evaluation session concludes with an action plan that identifies specifics that will be executed for each strategic goal over the next 2 years (Bresnick, 2000, 2011).

14.5 Illustrative Examples

See Table 1.3 for further information on the illustrative examples.

14.5.1 RNAS (by Eric R. Johnson)

The major implication of the analysis was that the large CBM acquisition was a $1B windfall at its anticipated price. The large CBM acquisition (paying $1B more than anticipated!) was implemented. However, the external decision analysis team was not involved in the unsuccessful attempt to implement the remaining Roughneck North American Strategy (RNAS) recommendations.

14.5.2 DATA CENTER (by Gregory S. Parnell)

Decision implementers played a major role in the data center decision analysis. The key decision implementers were the mission, IT, and logistics organizations. The senior leaders of these organizations were members of the decision team. The senior leader assigned the task of standing up the data center was in the IT organization and was a major participant in the study. His data center schedule determined the amount of time allocated to the decision analysis study. Key representatives from each of the organizations were members of the decision analysis team. The decision implementers played a central role in the development of the value model, the data used to score the alternatives, and the improvement of the alternatives. As a result, there was significant organizational consensus that the best alternative was selected and the decision implementation would be able to deliver the promised value to the organization. It turned out that this trust was well placed.

14.6 Summary

There is nothing more satisfying to a decision analyst than to see the results of an analysis implemented and value achieved as planned. Unfortunately, the decision analyst, particularly one who is external to the organization, is there for the client interaction, modeling, and insight generation, but not necessarily there to follow up on execution. One thing we can do, though, is to ensure that decision implementation is considered throughout the decision analysis process that will provide value long after the study effort has concluded. Value can be derived in many ways from a sound decision analysis. The analysis can provide ongoing value when used to:

- Involve the decision implementers in all phases of the decision analysis.
- Communicate the key factors and rationale for a decision, including identifying and reconciling key areas of stakeholder disagreement.
- Select the best alternative and generate better alternatives.
- Foster organizational commitment to a decision.

In addition, decision analysts should seek to use their decision analysis model (with modifications) to guide the decision implementation.

KEY TERM

Decision implementer An individual who is responsible for execution of the decision.

REFERENCES

Bresnick, T. (2000, 2011). Working notes for Nlets Strategic Plan.

Parnell, G.S., Buckley, M., Ernesto, A., McGrath, D., & Miller, M. (2011). Acquisition program information assurance assessment model. *Military Operations Research*, *16*(4), 46–50.

Goldman Sachs (2009). *Global Energy*. February 2009.

International Systems Security Engineering Association. (n.d.) System Security Engineering-Capability Model (SSE-CM). http://www.sse-cmm.org/index.html, accessed May 7, 2009.

CHAPTER FIFTEEN

Summary of Major Themes

GREGORY S. PARNELL

If a man will begin with certainties, he shall end in doubts: but if he will be content to begin with doubts, he shall end in certainties.

—Francis Bacon

(*Continued*)

Handbook of Decision Analysis, First Edition. Gregory S. Parnell, Terry A. Bresnick, Steven N. Tani, and Eric R. Johnson.

15.1 Overview

This chapter reviews and summarizes the most important themes of the handbook. It can be used either as a summary of the book or as a preview of the book.

Senior leaders face very difficult decision challenges that can have a critical impact on the future of their organizations. Decision analysis offers a useful framework for thinking about these decisions. We define a decision as an irrevocable allocation of resources. Decision analysis considers three elements: preferences (values), alternatives, and uncertainties. Decision analysis is an axiomatic method that helps create the best decision given the values we have, the alternatives we can do, and the uncertainties we perceive. Decision analysis has a rich history of over 45 years of development and practice. We should use decision analysis for our difficult decisions. As the Chevron Vice Chairman stated, "We use decision analysis because it works" (Chapter 1).

This chapter is organized as follows. Section 15.2 identifies the important professional decision-making questions that decision analysis helps answer and the chapter in the book that provide techniques to help answer the questions. Section 15.3 emphasizes that the purpose of decision analysis is to create value for decision makers and stakeholders and describes the two approaches to defining value. Section 15.4 focuses on the importance of understanding that decision analysis is a social-technical process. A successful decision analysis requires decision analysts who have both soft and hard skills. Section 15.5 lists the social skills (personal and interpersonal) that are critical to decision analysis success. Section 15.6 identifies two recommended decision processes and two that are not recommended. Section 15.7 summarizes the powerful decision analysis technical tools that can help us frame decisions, craft objectives, create alternatives, develop a composite decision model using the organizational expertise, analyze the alternatives, improve the alternatives, and provide the information to develop important insights for decision makers.

15.2 Decision Analysis Helps Answer Important Decision-Making Questions

Decision analysis provides the concepts, the mathematics, and the techniques to help answer some very important questions about decision making for important professional (and personal) decisions. The following is a list of the key questions and the corresponding chapter in this book where each of these questions is answered.

- What are the major challenges organizational decision makers face? (Chapter 2)
- What is a sound mathematical foundation for decision making for complex decisions? (Chapter 3)
- What are the soft skills that a decision analyst needs to deal with these decision challenges? (Chapter 4)
- How do we define a good decision versus a good outcome? What is decision quality for an organization? (Chapter 5)
- How can we design or tailor the organization's decision process to achieve decision quality? (Chapter 5)
- How do we frame the decision opportunity so we fully understand the vision for the decision or the opportunity before we start to build models? (Chapter 6)
- What do the decision makers and stakeholders value in the decision opportunity and how can we measure it? (Chapter 7)
- How do we develop a spanning set of creative, innovative alternatives from which to choose for the decision? (Chapter 8)
- How can we build a composite decision model to address the five dimensions of complexity often found in decision situations (multiple decision alternatives, ranges of uncertainty in key factors, multiple value components, multiple business units, and multiple time periods) using the expertise available to the organization? (Chapter 9)
- What are the major uncertainties that impact our decision values and how can we quantify and possibly reduce these uncertainties? (Chapter 10)
- How can we use the composite decision model to logically evaluate our strategies in light of the organization's values and the uncertainties they face? (Chapter 11)

- How can we use our analysis insights to develop improved strategies and reduce the risk of the best strategy? (Chapter 11)
- How can we help an organization make sound portfolio resource allocation decisions in the face of resource constraints? (Chapter 12)
- How can we build upon the data provided by models to best identify and communicate the analytical insights to decision makers and stakeholders? (Chapter 13)
- How can we increase decision maker and stakeholder commitment to decision implementation to achieve value? (Chapter 14)
- How can we use decision analysis to guide the decision implementation (Chapter 14)

The handbook provides decision analysis best practices on how to help an organization answer these questions.

15.3 The Purpose of Decision Analysis Is to Create Value for Shareholders and Stakeholders

We should never lose sight of the purpose of decision analysis: to create value for shareholders and decision makers. We believe that the philosophy of value-focused thinking is essential for every successful decision analysis. Decision analysis focuses on identifying potential future value and obtaining organizational commitment to achieve this future value. Decision analysis provides two important methods for defining and measuring value: single objective (typically shareholder) and multiple objective (typically stakeholder) decision analysis. We do not view these as separate fields. Both types of decision analysis are based on the same five rules. Both use value models that measure returns to scale and utility models that measure returns to scale and risk preference. Chapter 3 provides our taxonomy for thinking about the field of decision analysis applications and the different approaches used by decision analysts.

15.3.1 SINGLE OBJECTIVE VALUE

Single objective value usually focuses on shareholder value but it can be any measure. Net present value (NPV) is the most common single objective measure. When decision analysts use a single measure, like NPV, they strive to convert the other objectives to the single measure.

15.3.2 MULTIPLE OBJECTIVE VALUE

Multiple objective value models are developed for stakeholder values. The additive value model is the most common multiple objective model.

15.3.3 IT IS IMPORTANT TO DISTINGUISH POTENTIAL VALUE AND IMPLEMENTED VALUE

Decision analysts identify potential value; decision makers make decisions and provide resources; and decision implementers achieve value. Decision analysts are usually not the decision implementers. However, it is essential to involve the implementers in the decision process (see Chapter 14).

15.4 Decision Analysis Is a Socio-Technical Process

It is critical to consider both the social and the technical issues of the decision process. The social issues include the issues of people as individuals and in small and large groups in the organization or enterprise. The technical issues include understanding the foundational mathematical theory (Chapter 3), knowing how to develop credible decision models (Chapters 7–11), and knowing when and how to use decision analysis concepts and techniques to analyze alternatives.

15.4.1 SOCIAL

Decisions are made by leaders in organizations with many individuals, each having their own objectives. Stakeholders are individuals who represent organizations with additional objectives. Subject matter experts (SMEs) are individuals with outstanding experience and expertise in important areas relative to the decision. Decision analysts should understand the formal and the informal organization. Decision analysis requires the knowledge of decision makers (DMs), stakeholders (SHs), and SMEs about preferences, uncertainties, and alternatives. Decision analyses that are excellent technically will not be accepted if the social aspects of the analysis and the decision are not understood and considered in the decision process.

15.4.2 TECHNICAL

Most decision analysis academic courses focus on the concept and techniques of decision analysis. When used appropriately, these techniques can be very powerful. We summarize these tools and techniques in Section 15.7.

15.5 Decision Analysts Need Decision-Making Knowledge and Soft Skills

This section describes the important decision-making challenges and describes the soft skills that a decision analyst needs to successfully perform a decision analysis.

15.5.1 DECISION ANALYSTS NEED TO UNDERSTAND DECISION-MAKING CHALLENGES

Decision problems can be very complex; in Chapter 2, we characterize this complexity in three dimensions: content, analytic, and organizational. Content complexity ranges from few scenarios with little data and a relatively stable decision-making setting, to many scenarios with data overload, many subject matter experts (SMEs) involved, and a dynamic decision context. Analytic complexity ranges from deterministic problems with little uncertainty and few objectives, to problems with a complicated value hierarchy, many alternatives, and a high degree of uncertainty. Organizational complexity ranges from a single decision maker with a homogeneous set of stakeholders to multiple decision makers requiring consensus and a diverse set of stakeholders with conflicting objectives.

15.5.2 DECISION ANALYSTS MUST DEVELOP THEIR SOFT SKILLS

Soft skills include personal and interpersonal skills. In Chapter 1, we identify the soft skills by aggregating the soft skills required for each step in our decision process. In Chapter 4, we describe the best practices for each of the soft skills. We list them here again because we believe they are critical to the success of every decision analyst.

Based on our experience, we aggregate the soft skills into the following nine categories.

- **Thinking strategically** about the client organization, the problem domain, and the role of the decision analysis in achieving the current strategy or, when appropriate, developing a new strategy, and new decision opportunities
- **Leading teams**, including developing team goals, motivating individuals to achieve team goals, and guiding the client organization to achieve the most value from the study
- **Managing decision analysis projects**, including developing analysis plans; identifying and scheduling activities; and managing the completion of tasks
- **Researching the problem domain, modeling approaches, and data sources**
- **Interviewing individuals** (DMs, SHs, and SMEs) to frame the decision problem and elicit knowledge on preferences (value, time, and risk), probabilities, alternatives for modeling
- **Surveying stakeholders and experts** can be a efficient way to collect knowledge for geographically dispersed individuals
- **Facilitating groups** (and focus groups) of DMs, SHs, and SMEs to frame the decision opportunity, generate creative alternatives, and elicit knowledge on preferences (value, time, and risk), probabilities, and alternatives for modeling

- **Aggregating expertise** is needed to combine different views of SHs and SMEs
- **Communicating with DMs, SHs, and SMEs** to tell the story, the analytic results, and the key insights in ways that are understandable to the audience (see Chapter 13).

Each decision analysis provides an opportunity to apply and improve these soft skills.

15.6 The Decision Analysis Process Must Be Tailored to the Decision and the Organization

As we discuss in Chapter 2 and Chapter 5, intuition and "gut feel" are not sufficient for the complex, critical decisions of an organization. Since decision analysis is a socio-technical process, the interaction with decision maker(s), stakeholders, and SMEs is essential. In this section, we define decision quality and summarize two successful decision-making processes based on decision analysis.

15.6.1 DECISION QUALITY

One definition of a good decision that has been used successfully over many years of decision consulting is embodied in the Decision Quality Chain. A good decision is defined as one that has high quality in each of six essential elements:

1. Clear decision frame
2. Creative alternatives
3. Credible information
4. Clear values and value trade-offs
5. Logical reasoning
6. Commitment to action.

Two processes have been shown to be effective for interacting with decision makers and stakeholders.

15.6.2 DECISION CONFERENCING

Decision Conferencing has been successfully used within hundreds of organizations to achieve good decision making in the face of content and organizational complexities. Decision Conferencing brings together decision makers, stakeholders, and staff experts to provide substantive expertise, with experienced facilitators that guide the decision process. The result is a series of intensive meetings that seek to identify key issues, identify objectives, evaluate alternatives, reach decisions, and plan implementation. The expertise of the client organization is absolutely essential for success, and the level of expertise needed typically resides in the heads of participants. During the decision conference, computer-based models for multiple objective decision analysis, probability analysis, and resource allocation are often used as a focus for group discussion, and the structured conference process allows participants to debate issues constructively while encouraging the group to represent its collective judgments in a logically consistent and easily communicated fashion (see Appendix C).

15.6.3 DIALOGUE DECISION PROCESS

The Dialogue Decision Process has also been successfully applied within hundreds of organizations to achieve good decision making in the face of content, analytic, and organizational complexities. The process centers on a structured dialogue between two teams: (1) A team of decision makers, the Decision Board, and (2) a team created to assist the decision makers, the Project Team.

The Decision Board should comprise those people whose agreement is both necessary and sufficient for the decision to be made and successfully implemented. The Project Team consists of staff members from the organization who collectively possess the skills to perform the required analytics and conduct the Dialogue Decision Process, have access to key information sources required to make the decision, and are trusted by the Decision Board members. The Project Team may include decision professionals who are outside consultants. The Dialogue Decision Process centers on a sequence of three or four meetings between the Decision Board and Project Team, each focused on a particular subset of the six elements of Decision Quality.

In Chapter 5, we also describe two decision processes that we do not recommend: the analytical process and the advocacy process. The fatal flaw of the analytical process is the lack of contact and interaction with the decision makers and key stakeholders. The fatal flaw of the advocacy process is the focus on advocating an alternative instead of creating high-value alternatives. Decision makers are not well served by either process.

15.7 Decision Analysis Offers Powerful Analytic Tools to Support Decision Making

The foundation of decision analysis is the set of five rules presented in Chapter 3. These rules imply the existence of a utility function and the logical desire to make decisions that maximize expected utility. There are two important ways that decision analysis tools can be used to add value: choosing correctly among identified strategies and improving upon the selected strategy by increasing the value and/or reducing the risk of the downside consequences. Once we have developed a preliminary model of a composite perspective on what drives value, we undertake a value dialogue with the decision team aimed at verifying the authenticity of the composite perspective, and use it to choose correctly and find even more valuable and less risky strategies. The following are the major decision analysis techniques that are described in this handbook.

Framing the decision. The three major techniques are the vision statement, the issue identification matrix, and the decision hierarchy. A vision statement is an effective way to make sure that there is agreement on the purpose of the decision among the decision maker(s) and key stakeholders. The vision statement answers three questions: What are we going to do? Why are we doing this? How will we know that we have succeeded? The issue identification matrix (environmental factors and stakeholders) is a structured way to identify and document the major stakeholder issues. The decision hierarchy is a technique that defines the scope of the decision by identifying the decisions that have been made, the current decisions, and the future decisions.

Identifying objectives and value measures. The major technique is a value hierarchy (or functional value hierarchy) to structure and document the values of decision makers and stakeholders. A value hierarchy is a tree structure of the fundamental objectives and the value measures. For complex decisions, we also present the functional value hierarchy, which includes functions, objectives, and value measures. In Chapter 7, we describe four techniques to develop value hierarchies: the platinum standard (interviews with decision makers and stakeholders), the gold standard (strategic planning document), the silver standard (focus groups with stakeholder representatives), and the combined standard (interviews, focus groups, and documents).

Developing creative alternatives. The primary alternative-generation technique is the strategy generation table. A strategy table contains a number of columns, each of which represents a particular choice that is part of the overall decision. Selecting one option from each column defines a strategy. It is a powerful tool to develop and communicate decision strategies.

Building decision models. The purpose of the decision model is to develop a composite perspective of stakeholders and SMEs' knowledge of the factors that contribute to value. A deterministic decision model uses base values of decision model variables to calculate the value of the alternatives. The deterministic model may use single or multiple objective value. A probabilistic decision model incorporates uncertainty.

Identify uncertainties. The two major techniques to identify uncertainties are issue raising and the influence diagram. Issue raising brings to light all of the many stakeholder perspectives that can be used to view the decision situation and lays the groundwork for creating a good decision frame. Many of these issues may be modeled as uncertainties. An influence diagram is a compact graphical representation of a decision showing the interaction of decisions and uncertainties to produce value.

Performing deterministic analysis. We develop a composite decision model that codifies knowledge about the five dimensions of complexity often found in decision situations (multiple decision alternatives, ranges of uncertainty in key factors, multiple value components, multiple business units, and multiple time periods). The decision model can summarize any one or two of the five dimensions to help identify insights about the decision and the alternatives.

In addition, deterministic analysis uses tornado diagrams, difference tornado diagrams, and the value component diagram to produce insights to guide future probabilistic modeling and analysis.

- A tornado diagram is a deterministic sensitivity analysis plot that shows the impact on alternative value of changing parameters from their low to high values.
- A difference tornado diagram shows the impact that uncertainties have on the difference of a value metric for two decision alternatives. The difference tornado highlights those uncertainties that can change the optimal choice.
- The value component chart (or waterfall diagram) displays the contribution to overall value of various components of the decision.

Performing probabilistic analysis. The major techniques used for probabilistic analysis are influence diagrams, decision trees, and Monte Carlo simulation (Chapter 11). Each technique has advantages and disadvantages. Here are some important ways to present the results of probabilistic analyses:

- Value components chart, to show how much of each value component is created by each strategy, and spur thought on how to combine different strategies' successful approaches.
- Cash flow profiles of strategies, to highlight any time tradeoffs and spur thought on timing of costs or benefits.
- Direct tornado for a leading strategy, to identify opportunities for improvement by controlling a key uncertainty.
- Delta tornado of the best strategy versus another leading strategy, to ascertain whether the outcome of any uncertainty could change the decision, and to guide information gathering or construction of a contingent strategy.
- S-curves (cumulative probability curves) for strategies' NPVs, to show whether there is any risk-value tradeoff.

15.7.1 PORTFOLIO RESOURCE ALLOCATION

Portfolio resource allocation brings additional socio-technical challenges. We present two techniques and compare them with optimization. First, we present portfolio decision analysis with a single financial objective and capital constraints using the greedy algorithm approach. In this approach, we order the project by the Profitability Index (PI = NPV divided by dollar cost) and fund all projects until we run out of budget. Second, we present a multiple-objective decision analysis with resource constraint approach using an incremental benefit/cost portfolio analysis. This approach is easily modified to handle a diverse type and number of constraints. The major benefits compared with optimization are a stable order of buy and more understandable decision rationale.

15.8 Conclusion

The purpose of the handbook is to provide the art and the science of decision analysis. We believe that decision analysis is more than an operations research technique with a sound mathematical foundation. We believe that decision analysis is a way of thinking that can be used to help decision makers and stakeholder create great value in very complex and challenging decision opportunities.

APPENDIX A

Probability Theory

STEVEN N. TANI

A reasonable probability is the only certainty.

—E.W. Howe

Handbook of Decision Analysis, First Edition. Gregory S. Parnell, Terry A. Bresnick, Steven N. Tani, and Eric R. Johnson.

A.1 Introduction

This appendix is not intended to be a comprehensive discourse on probability theory. For that, please refer to any good textbook on probability, such as those by William Feller (1968) and by K.L. Chung and Farid AitSahlia (2003). Rather, we here present a number of key ideas from probability theory that every decision practitioner should know.

A.2 Distinctions and the Clarity Test

A fundamental requirement for addressing the important uncertainties in a decision is to create *distinctions* that are both clear and useful. A distinction is a separation of the universe of possibilities (outcomes) into two or more subsets. By definition, these subsets will be mutually exclusive (i.e., a possibility cannot be in more than one subset) and collectively exhaustive (i.e., every possibility must be in a subset). An example of a distinction is whether or not a new product achieves high sales volume. This distinction defines two events that are mutually exclusive (the product cannot both have high sales and not have high sales) and collectively exhaustive (either the product will have high sales or it will not, which includes the possibility that it is not launched and therefore has zero sales.)

A distinction is clear when there is no ambiguity in its definition. To determine if a distinction is clear, we apply the clarity test (Howard, 2007), which employs the concept of a clairvoyant—someone who perfectly sees and truthfully reports any observable event or quantity, either in the future or in the past.[1] A distinction is deemed to be clear if the clairvoyant, without exercising any judgment, can say whether or not an event that is defined by the distinction will occur. The distinction in the example above would fail the clarity test. The clairvoyant would not be able to say whether the product will achieve high sales volume without judging what level of sales volume qualifies as "high." A much clearer distinction in this example would be whether or not the new product achieves worldwide sales volume of at least 1 million units during the next calendar year. The clairvoyant might still wonder whether or not free samples given to prospective customers should be included in "sales volume," so the distinction might need to be refined to make that clear. Creating distinctions that are clear is important in decision making because it avoids a situation in which subject

[1]This definition is similar to the definition of an ideal observer used in Chapter 10, insofar as it highlights the application of a criterion to events without use of judgment. This definition differs slightly from the definition of clairvoyant used in Chapter 11. The definition used in Chapter 11 also requires that the event in question be unaffected by the asker's actions. This additional requirement is necessary for value of information to be formulated properly.

matter experts give different probabilities for the same event because they have different interpretations of its definition.

A distinction is useful when it contributes to a full understanding of why one decision alternative is preferred to others. The distinction can be a direct measure of value or it can be an important parameter in the calculation of value. For example, the level of sales volume is a useful distinction in the calculation of net present value. Creating distinctions that are both useful and clear is an important skill of a decision practitioner.

A.3 Possibility Tree Representation of a Distinction

A highly useful diagrammatic representation of a distinction is a possibility tree. The distinction is shown as a tree *node* from which *branches* emerge representing the possibilities. The number of branches at a node is called the *degree* of the distinction. Figure A.1 shows the possibility tree for the example distinction of high sales or not high sales (two degrees), suitably defined to pass the clarity test. It is easy to imagine a distinction with three or more degrees. For example, a distinction defined by a student's grade in a course could have six degrees: A, B C, D, E, or F.

A.4 Probability as an Expression of Degree of Belief

A probability is the quantitative expression of someone's uncertainty about a distinction based on his or her state of information. More precisely, it expresses the person's *degree of belief* that an event will occur, ranging from 0 if the person is certain that the event *will not* occur to 1 if the person is certain that it *will* occur. A person's belief about the occurrence of any event, such as high sales volume, depends, of course, on the information that the person possesses. If the person's information changes (by observing the results of a test market, for example), then the probability may change.

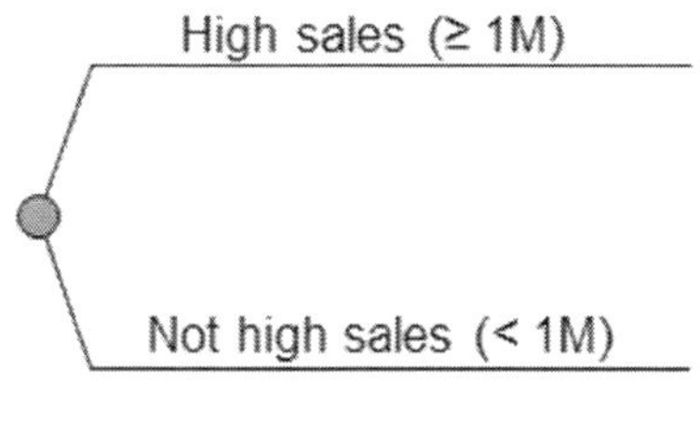

FIGURE A.1 Possibility tree.

A.5 Inferential Notation

Using inferential notation (Howard, 1966) emphasizes that every probability is based on a particular state of information. The probability of occurrence of an event A is notated as follows:

$$\Pr(A|\&),$$

where the event is represented to the left of the vertical bar and the information on which the probability is based is represented to the right of the bar. The ampersand character "&" stands for all of the background information possessed by the person making the probability statement.

A.6 Multiple Distinctions

In most decision-making situations, we must deal with multiple distinctions. Suppose, for example, that in addition to the previous distinction of whether or not the product achieves high sales volume (as defined to meet the Clarity Test), we have a second distinction of whether or not our major competitor launches a similar product (again, suitably defined to pass the Clarity Test). In the discussion that follows, we use the following notation:

SY = High sales (≥1 M)

SN = Not high sales (<1 M)

CY = Competitive product

CN = No competitive product.

As shown in the possibility tree in Figure A.2, these two distinctions, each with two degrees, together define four *elemental* possibilities—(1) high sales, competitive product (SY,CY), (2) high sales, no competitive product (SY, CN), (3) not high sales, competitive product (SN, CY), and (4) not high sales, no competitive product (SN, CN).

A.7 Joint, Conditional, and Marginal Probabilities

There are three types of probabilities that we use to express uncertainty about multiple distinctions—joint, marginal, and conditional.

A.7.1 JOINT PROBABILITY

The probability of an elemental possibility with multiple events is called a joint probability. In the example, the probability of high sales combined with the competitive product is a joint probability, notated as follows:

$$\Pr(\text{SY}, \text{CY}|\&).$$

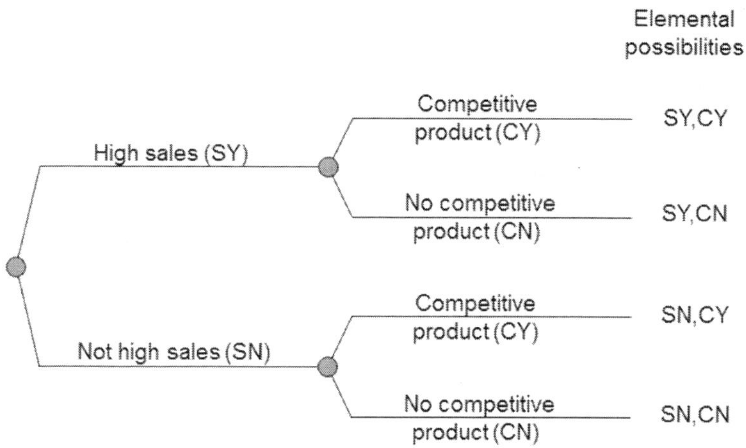

FIGURE A.2 Possibility tree with two distinctions.

A.7.2 MARGINAL PROBABILITY

The probability of an event defined by just one of the multiple distinctions is called a marginal probability. In the example, the probability of high sales is a marginal probability, notated as follows:

$$\Pr(\text{SY}|\&).$$

A.7.3 CONDITIONAL PROBABILITY

The probability of an event defined by one distinction given that we know of the occurrence of an event defined by a second distinction is called a conditional probability. In the example, the probability of competitive product given high sales is a conditional probability, notated as follows:

$$\Pr(\text{CY}|\text{SY}, \&).$$

We can illustrate these types of probabilities in a tree diagram as shown in Figure A.3.

Since the probabilities of all events given the state of information must sum to 100%, we know that

$$\Pr(\text{CY}|\text{SY}, \&) + \Pr(\text{CN}|\text{SY}, \&) = 20\% + 80\% = 100\%.$$

It is clear from the tree diagram that a marginal probability is the sum of the appropriate joint probabilities.

$$\Pr(\text{SY}|\&) = \Pr(\text{SY}, \text{CY}|\&) + \Pr(\text{SY}, \text{CN}|\&) = 12\% + 48\% = 60\%.$$

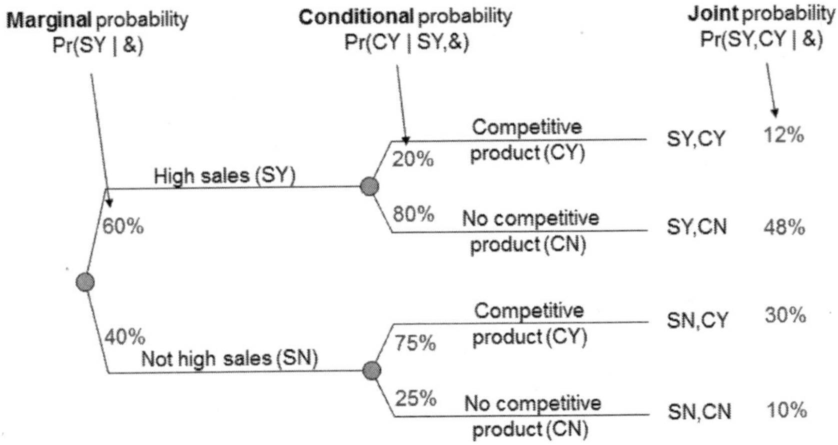

FIGURE A.3 Probability tree with two distinctions.

A.8 Calculating Joint Probabilities

One of the fundamental results from probability theory is that a joint probability can be calculated as the product of a conditional probability and a marginal probability:

$$\Pr(SY, CN|\&) = \Pr(CN|SY, \&) \times \Pr(SY|\&).$$

When applied to the tree diagram in Figure A.3, this result means that each joint probability at the right-hand end of a path through the tree is equal to the product of the marginal and conditional probabilities of the branches comprising that path. For example,

$$\Pr(SY, CN|\&) = 60\% \times 80\% = 48\%.$$

Note that since

$$\Pr(SY, CN|\&) = \Pr(CN, SY|\&).$$

we can reverse the conditionality and get the same result:

$$\Pr(SY, CN|\&) = \Pr(SY|CN, \&) \times \Pr(CN|\&).$$

Also note in the diagram that if we know the four joint probabilities at the end of the tree, we can calculate all of the other probabilities on the tree. For example, the marginal probability of high sales is the sum of two joint probabilities (12% + 48% = 60%). The conditional probability of a competitive product given high sales is a joint probability divided by a marginal probability (12% / 60% = 20%).

A.9 Dependent and Independent Probabilities

The probabilities of two distinctions are either *dependent* or *independent.* The test is whether knowing the outcome of one distinction affects the probabilities of the other. If the probabilities for one distinction are different depending on the outcome of the other distinction, the probabilities are dependent. If the probabilities for one distinction are unaffected by the outcome of the other, the probabilities are independent. Stated mathematically, events A and B are probabilistically independent if

$$\Pr(A|\&) = \Pr(A|B, \&),$$

which implies that

$$\Pr(A, B|\&) = \Pr(A|\&) \times \Pr(B|\&).$$

It is readily apparent that the probabilities of the two distinctions shown in Figure A.3 are dependent because the probability of a competitive product given high sales (20%) differs from the probability of a competitive product given not high sales (75%).

Probabilistic dependence is a mutual property. If the probabilities of one distinction are dependent on a second distinction, then the reverse must also be true.

A.10 Reversing Conditional Probabilities: Bayes' Rule

As indicated in Section A.8, when dealing with two distinctions with mutually dependent probabilities, it is often useful to be able to reverse the order of conditioning. That is, if we know the conditional probability of event A given that event B has occurred, we would like to calculate the conditional probability of B given A.

This operation, which is known as Bayes' rule, is done quite easily using tree diagrams, as illustrated in Figure A.4. The original order of the distinctions is shown in the tree on the left.

The following steps are taken to create the tree on the right with the order reversed.

Step 1. Draw the tree structure with the order reversed

Step 2. Copy the four joint probabilities from the original tree, taking care to put them in the correct position in the reversed tree.

Step 3. Calculate the marginal probabilities in the reversed tree as the sum of the appropriate joint probabilities. For example, Pr(CY|&) = 12% + 30% = 42%.

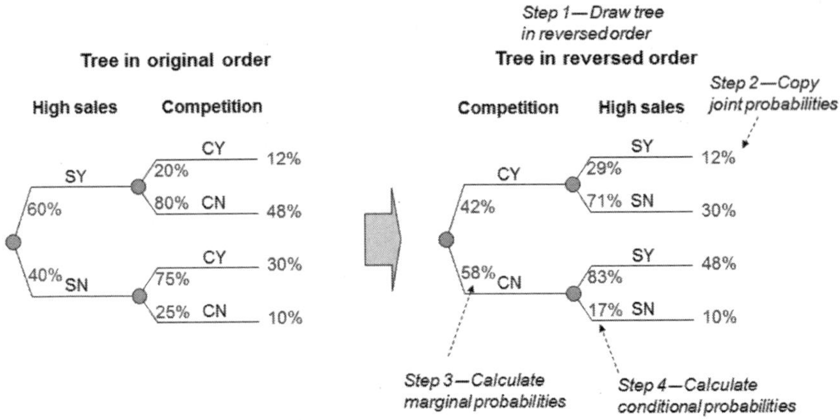

FIGURE A.4 Reversing the order of a tree.

Step 4. Calculate the conditional probabilities in the reversed tree as the ratio of a joint probability to a marginal probability. For example, Pr(SY|CY,&) = 12%/42% = 29%.

We can express Bayes' rule mathematically for events A and B as

$$\Pr(A|B, \&) = \Pr(B|A, \&) \times \Pr(A|\&) / \Pr(B|\&).$$

Bayes' rule plays a central role in decision analysis since it provides a sound mathematical way to update our probabilities based on new information (See Chapter 11).

A.11 Probability Distributions

Often, the distinctions that we create in a decision situation are described by quantitative measures. For example, for a new product introduction decision, the distinction of the average unit cost of manufacturing the product might be important. In such cases, we characterize our uncertainty about the quantity as a probability *distribution*, which specifies the probability that the quantity is in any given interval. One way to visualize a probability distribution is as a histogram, as shown in Figure A.5. We divide the possible range for the quantity into intervals and draw a bar for each interval whose height is proportional to the probability that the quantity will be in that interval. If the intervals are suitably small enough, the histogram in some cases is the familiar "bell-shaped" curve.

An alternate way to portray a probability distribution is in its cumulative form, which is sometimes called an "S-shaped" curve. The cumulative form makes it especially easy to gauge the probability that the quantity will be in a given range. For example, the cumulative curve in Figure A.6 shows that the

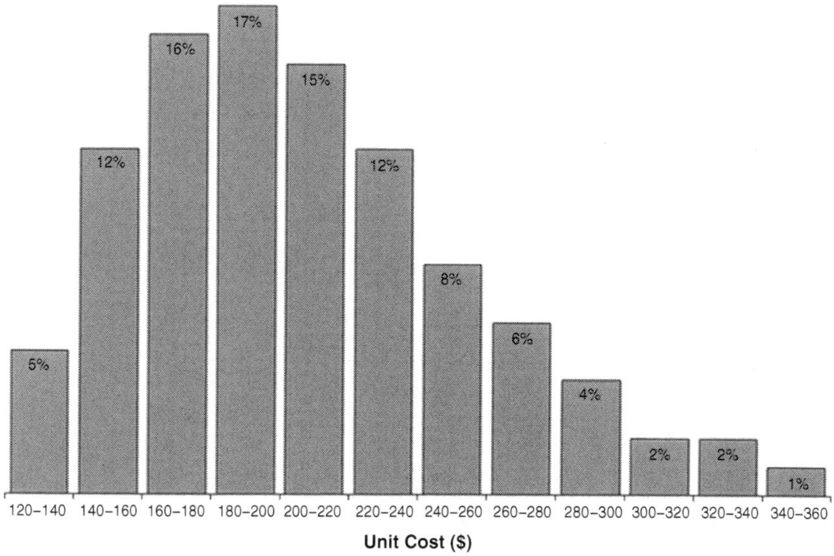

FIGURE A.5 Probability distribution as a histogram.

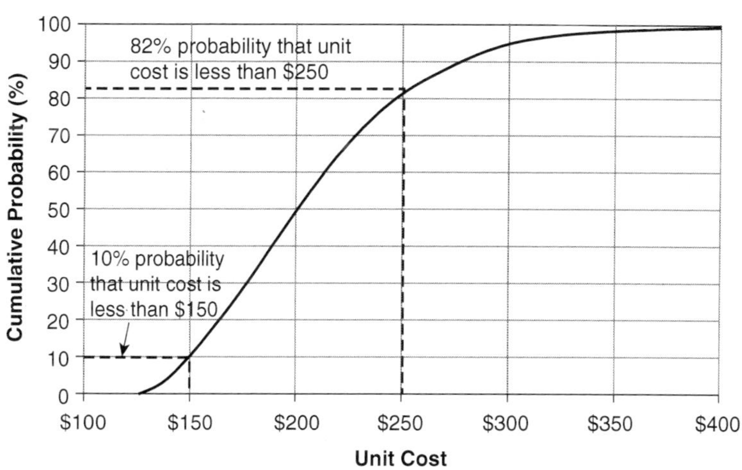

FIGURE A.6 Probability distribution in cumulative form.

probability that cost will be less than $250 is 82% while the probability that it will be less than $150 is 10%. We can therefore deduce that the probability that cost will be between $150 and $250 is 82% – 10% = 72%.

The quantitative measure on which a probability distribution is defined can be either *continuous* or *discrete*. A continuous measure is one which can take any

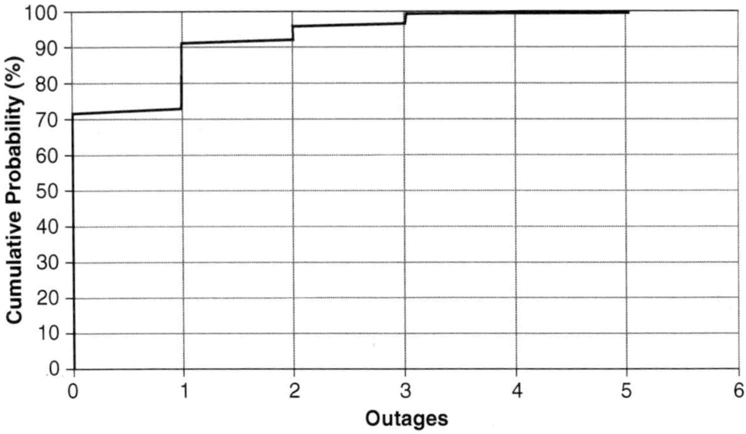

FIGURE A.7 Cumulative probability distribution of a discrete measure.

value in a specified range while a discrete measure is restricted to a countable number of possible values. The unit cost example above exemplifies a continuous measure. An example of a discrete measure is the number of unplanned outages experienced by a power plant in a year. The cumulative probability distribution of a discrete measure has a stairstep shape (see Fig. A.7).

A.11.1 SUMMARY STATISTICS FOR A PROBABILITY DISTRIBUTION

A number of key statistics are often used to summarize a probability distribution. The most widely used is the probability-weighted average, commonly called the *mean* or the *expected value*. Two other summary statistics of interest are the *variance*, which measures the dispersion about the mean and the *skewness*, which measures the degree of asymmetry of the distribution. *Percentiles* are another set of useful summary statistics for a distribution. The P-th percentile is the value of the quantity such that there is a probability of P that the quantity will not exceed that value. The cumulative curve is a display of the percentiles of the distribution (read in reverse—the x-axis value of a point on the curve is the P-th percentile, where P is the y-axis value of the point). The frequently used 50th percentile is given the special name *median*.

A.12 Combining Uncertain Quantities

When uncertain quantities are combined via mathematical operations, the combined quantity is also uncertain and therefore has a probability distribution. Indeed, the primary goal of analysis in a decision situation is to determine the

TABLE A.1 **Equalities When Combining Uncertain Quantities**

	Quantities Are Mutually Independent	Quantities Are *Not* Mutually Independent
Mean of sum = ? Sum of means	True	True
Variance of sum = ? Sum of variances	True	False
Skewness of sum = ? Sum of skewnesses	True	False
Percentile of sum = ? Sum of percentiles	False	False
Mean of product = ? Product of means	True	False
Variance of product = ? Product of variances	False	False
Percentile of product = ? Product of percentiles	False	False

probability distribution of the value measure of interest, which is a (usually complicated) combination of the decision choices taken and external uncertain factors.

Probability theory tells us some very useful facts about simple combinations of uncertain quantities—sums and products. The most important of these facts is that the mean of the sum of uncertain quantities is *always* equal to the sum of the means of the individual quantities. For example, if a corporate portfolio comprises a number of business units, each of which has uncertain profits, then the mean of corporate profit is always equal to the sum of the means of the business unit profits. The word "always" means that the equality holds even if the business unit profits are probabilistically dependent on each other.

Table A.1 lists the results of probability theory regarding the sums and products of uncertain quantities. It is important to note that the percentiles of sums and products are, in general, not equal to the sums and products of corresponding percentiles. That is, one should not attempt to calculate the median, for example, of the sum of uncertain quantities by summing the individual medians.

Another property of note in the table is that the mean, variance, and skewness of the sum of independent quantities can be found by summing the corresponding measures of the individual quantities. So, a very quick way to get a good approximation of the probability distribution for the value of a portfolio of independent assets is to calculate the mean, variance, and skewness of the portfolio by summing across the individual assets and fitting a probability distribution to those three statistics.

REFERENCES

Chung, K.L. & AitSahlia, F. (2003). *Elementary Probability Theory*, 4th ed. Springer.

Feller, W. (1968). *An Introduction to Probability Theory and Its Applications*, Vol. 1, 3rd ed. John Wiley & Sons.

Howard, R. (1966). Information value theory. *IEEE Transactions on Systems Science and Cybernetics*, *2*(1), pp. 22–26, reprinted in R. Howard, & J. Matheson, *The Principles and Applications of Decision Analysis* (pp. 779–783). Menlo Park, CA: Strategic Decisions Group.

Howard, R.A. (2007). The foundations of decision analysis revisited. In W. Edwards, R.F. Miles, & D. von Winterfeldt (eds.), *Advances in Decision Analysis: From Foundations to Applications*, pp. 32–56. Cambridge University Press.

APPENDIX B

Influence Diagrams

STEVEN N. TANI and GREGORY S. PARNELL

B.1 Introduction

The purpose of this appendix is to provide an introduction to influence diagrams (ID) and pointers to some of the foundational references. An influence diagram

Handbook of Decision Analysis, First Edition. Gregory S. Parnell, Terry A. Bresnick, Steven N. Tani, and Eric R. Johnson.

is a compact graphical representation of a decision. Howard and Matheson published the seminal paper on influence diagrams in 1980. The paper was republished in *Decision Analysis* to be more broadly available (Howard & Matheson, 2005). They developed influence diagrams as a decision problem representation that could be understood by both computers and people. Currently, Howard (Howard, 2004) uses the term "relevance diagram" to describe an influence diagram that contains only uncertainties and "decision diagram" for one that also includes decisions and values. Howard prefers "relevance" to "influence" since an arrow in an ID means that knowledge of one uncertain event is relevant to knowledge of a second uncertain event and *not* that one event "influences" the outcome of another event.

Shachter developed computation algorithms for influence diagrams that are equivalent to the decision tree algorithm (Shachter, 1986, and Shachter, 1988). This research put influence diagrams on a sound theoretical foundation. Influence diagrams were originally developed for decisions having a single value measure. Merkhofer (Merkhofer, 1990) demonstrated the use of influence diagrams for multiple objective decision analysis.

Influence diagrams are often used in decision analysis textbooks (e.g., Clemen & Reilly, 2001) and decision analysis software is available to solve influence diagrams, producing the same results as the decision tree algorithm (Buckshaw, 2010). Influence diagrams are an important tool for DA practitioners to define the decision frame, identify the variables and decisions that should be included in the decision model, and communicate the structure of the decision model to decision makers and stakeholders (Buede, 2005).

B.2 Influence Diagram Elements

Although an influence diagram may appear to be just an informal "boxes and arrows" drawing, there are precise rules for constructing influence diagrams, which represent N-dimensional probability distributions. The shapes in an influence diagram have specific meanings, and there are a few rules that must be obeyed when drawing an influence diagram. The influence diagram elements are shown in Figure B.1 using common symbols.[1]

A *rectangle* represents a decision, which is specified by a set of alternatives. Depending on its purpose, an influence diagram may contain only one rectangle representing a high-level strategic decision, which comprises a set of choices of lower-level decisions, or it may contain several different rectangles each representing a sequence of lower-level decisions. If some decisions are made initially and others are made after some uncertainties are resolved, it is essential to show those decisions as separate nodes in an influence diagram.

An *oval* represents an uncertainty. Uncertainties differ in two dimensions: (1) continuous versus discrete and (2) scalar versus vector.

[1]Different software uses different symbols.

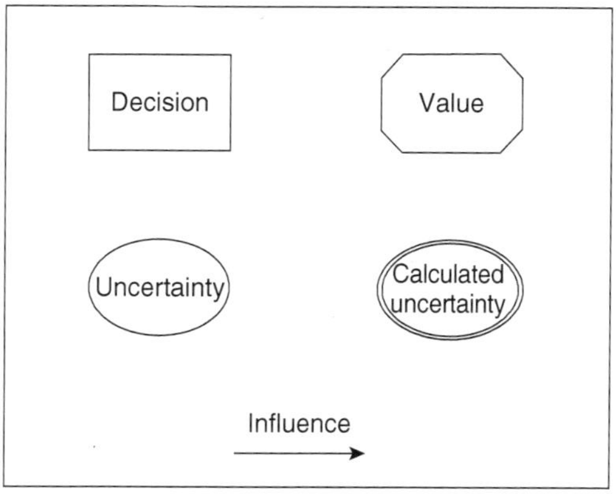

FIGURE B.1 **Elements of an influence diagram.**

An uncertainty may be either continuous or discrete. A continuous uncertainty is a random variable whose value can be any real number in a specified range of outcomes. An example of a continuous uncertainty would be the weight of an object. A continuous uncertainty is specified completely by a continuous probability distribution, such as the normal distribution, but is often characterized by a discrete approximation, usually with only three possible values—low, base, and high.

A discrete uncertainty is a random variable whose value can be any of a specified countable subset of the real numbers, such as the set of integers in a given range. An example of a discrete uncertainty is the number of successful product launches in a given time period. An important type of discrete uncertainty is a binary event, which may either occur or not occur.

An uncertainty may be either scalar or vector. A scalar uncertainty is one described by just a single number. All three examples given above are scalar uncertainties—the weight of an object, the number of product launches, and the occurrence of an event. Often, however, a vector of numbers can also describe an uncertain factor, usually as a time series of values, such as the market share of a product as it changes over time. Each element of the vector might be either continuous or discrete. Uncertainty about the entire vector can be represented by a single oval in the diagram, unless some elements of the vector are known at the time of a decision while others remain uncertain.

A *double oval* represents a special case of an uncertainty called a calculated or determined uncertainty. This is a factor that is uncertain only because it is calculated from other factors that are uncertain (e.g., the sum of uncertain factors). If each of those other factors were known for sure, the factor represented by the double oval would also be known for sure.

An *octagon* represents a value measure. The value measure is a decision criterion—a quantity to be maximized (or minimized) by the choice of decision alternative. For example, in most business decisions, the net present value of future cash flows is a value measure.

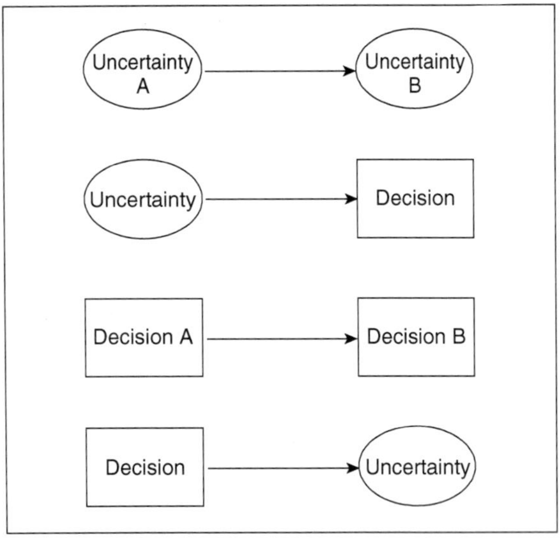

FIGURE B.2 Types of influences.

An *arrow* represents a relationship between two elements of the decision. The nature of the relationship depends on the types of elements connected by the arrow

Figure B.2 shows the four types of relationships. An arrow between two uncertainties means that there is probabilistic dependence between the two. Note that the *absence* of an arrow between two uncertainties is actually a stronger assertion—that there is no probabilistic dependence between the two. In other words, knowledge of the outcome of uncertainty A does not provide information about the probability of the outcome of B. The direction of the arrow indicates the order of conditionality. The probabilities of the uncertainty at the head of the arrow are conditional on the outcome of the other uncertainty. The arrow represents an informational relationship and does not necessarily imply causality. For example, an arrow might be drawn from an uncertainty about whether people are carrying umbrellas to an uncertainty about whether it is raining. Such a diagram indicates that observing people carrying umbrellas might affect the probability we assign to rain, but it would not be valid to say that rain is caused by people carrying umbrellas.

An arrow from an uncertainty to a decision means that the outcome of the uncertainty is known when the decision is made. This arrow represents a strong assertion about timing—that the resolution of the uncertainty occurs before the decision is made.

An arrow between two decisions means that when the decision at the head of the arrow is made, the choice taken in the other decision is remembered perfectly. This "no forgetting" arrow makes a strong assertion about the relative timing of the two decisions.

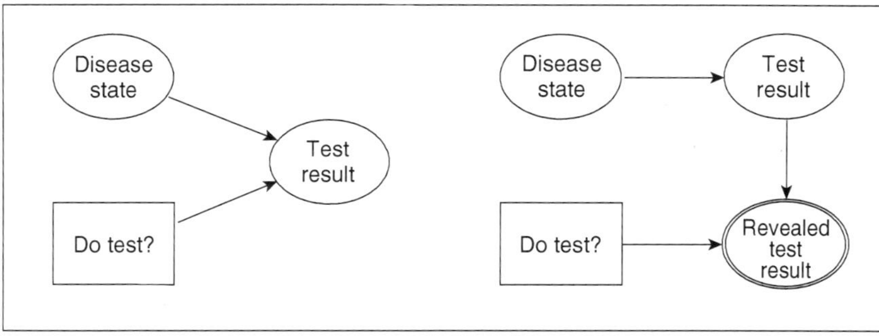

FIGURE B.3 Probabilities conditional on a decision (left) and Howard canonical form (right).

Finally, an arrow from a decision to an uncertainty means that the probabilities for the uncertainty depend on the choice made in the decision. Such an arrow is often used to describe a situation in which an uncertain outcome is revealed only if a particular action is taken. For example, the decision might be whether or not to do a diagnostic test and the associated uncertainty would be the result of that test with three possible outcomes—good result, bad result, or no result. Clearly, the probabilities for these outcomes depend on the choice made in the decision, because the probability of no result is 100% if the test is not done, whereas it is 0% if the test is done.

Drawing an influence diagram with an arrow from a decision to an uncertainty places a severe restriction on the decision situation so described—it may be impossible to calculate the value of clairvoyance (perfect information) on that uncertainty. It is always possible to modify such an influence diagram by adding calculated uncertainties so that the diagram has no arrows from decisions to uncertainties, as illustrated in Figure B.3. An influence diagram in which no arrow exists from a decision to an uncertainty is said to be in Howard canonical form.

B.3 Influence Diagram Rules

There are three important influence diagram rules, one of which (Rule 1) must be obeyed for all influence diagrams.

B.3.1 RULE 1: NO LOOPS

There must not be any path in the diagram that forms a directed loop. That is, it must always be impossible to return to any element in the diagram by following arrows in the direction that they point. Violation of this rule invalidates the underlying mathematical foundation of the diagram and creates a situation in which it could be impossible to assess probabilities for all of the uncertainties.

Two additional rules must be obeyed if the influence diagram is to describe a situation that can also be represented by a decision tree.

B.3.2 RULE 2: ONE VALUE MEASURE

All decisions represented in the diagram must be made to optimize the same value measure. However, multiple value nodes can be used to calculate one value measure.

B.3.3 RULE 3: NO FORGETTING

Any information known when a decision is made must be remembered perfectly when any subsequent decision is made. This implies that between every pair of decisions in an influence diagram, there must be a directed path that includes only decision nodes. And, if there is an arrow from an uncertainty to a decision, there must also be an arrow from that uncertainty to any subsequent decision.[2]

B.4 SUMMARY

This appendix provides an introduction to influence diagrams and pointers to some of the foundational references. Influence diagrams are an important tool for DA practitioners to define the decision frame, identify the variables and decisions that should be included in the decision model, and communicate the structure of the decision model to decision makers and stakeholders. Decision analysis software is available to solve influence diagrams that provide the same results as the decision tree algorithm.

REFERENCES

Buckshaw, D. (2010). *ORMS Today*. Retrieved December 17, 2011, from Decision Analysis Survey, http://www.orms-today.org/surveys/das/das.html.

Buede, D.M. (2005). Influence diagrams: A practitioners perspective. *Decision Analysis*, *2*(4), 235–237.

Clemen, R.T. & Reilly, T. (2001). *Making Hard Choices with Decision Tools*. Duxbury.

Howard, R.A. (2004). Speaking of decisions: Precise decision language. *Decision Analysis*, *1*(2), 71–78.

Howard, R.A. & Matheson, J.A. (2005). Influence diagrams. *Decision Analysis*, *2*(3), 127–143.

[2]Some influence diagram software, for example, DPL, does not require these arrows to be in the influence diagram, since the sequence is specified in the decision tree window.

Merkhofer, M.M. (1990). Using influence diagram in multiattribute utility analysis: Improving effectiveness through improved communications. In R.M. Oliver & J.Q. Smith (eds.), *Influence Diagrams, Belief Nets, and Decision Analysis*, pp. 297–316. John Wiley & Sons.

Shachter, R.D. (1986). Evaluating influence diagrams. *Operations Research*, *34*, 871–882.

Shachter, R.D. (1988). Probabalistic inference and influence diagrams. *Operations Research*, *36*, 589–605.

APPENDIX C

Decision Conferencing

TERRY A. BRESNICK

Decisions of the kind the executive has to make . . . are well made only if based on the clash of existing views, the dialogue between different points of view, the choice between different judgments. The first rule in decision making is that one does not make a decision unless there is a disagreement.

—Peter Drucker, *Management: Tasks, Responsibilities, Practices*

None of us is smarter than all of us.

—Japanese proverb

Handbook of Decision Analysis, First Edition. Gregory S. Parnell, Terry A. Bresnick, Steven N. Tani, and Eric R. Johnson.

C.1 Introduction

The purpose of this chapter is to provide an overview of a specialized group facilitation approach that has been in use since the late 1970s known as *decision conferencing*. According to the website of the International Decision Conferencing Forum, a professional association of decision conference practitioners, the formal definition of a decision conference is as follows: (IDCF, 2012)

> Decision Conferencing is a scientifically-grounded methodology that managers of an organisation can adopt to take decisions as a group. It is basically a series of intense day-long meetings, normally stretching for no more than 2-3 days, attended by all the decision makers that are in one way or the other involved, impacted, or interested in a particular issue requiring a decision. Unique features of Decision Conferencing are the dynamic creation of a mathematical computerized model, normally associated with a few proven Decision Analysis (management sciences) techniques, the absence of any pre-configured schedule and the use of a professional facilitator. The results (decision) are shaped by the group of participants, dynamically and in a way that allows them to see the effects of any individual preference; this reduces conflict and channels positively any individual concerns. The ultimate effect is that the group creates decisions that last, learning along the way from each other.

This approach attempts to combine the best of staff experts and managers from the field that provide substantive expertise, with external facilitators who provide process expertise. The result is a series of intensive meetings that seek to identify key issues, evaluate alternatives, and introduce an implementation mechanism. The overall goal of the conference is to develop informed consensus among key players.

Decision conferencing was introduced by Dr. Cameron Peterson, Director of Decision Analysis at Decisions and Designs, Inc. (DDI) in 1979. According to Dr. Larry Phillips, who brought decision conferencing to the United Kingdom, (Phillips, 2007), Dr. Peterson wanted to change the traditional "doctor-patient" model of consultancy to one in which key people who knew the problem were brought together for an open exchange of information via discussion, to provide relevant data and judgments, and to make decisions. This new model ensured that the customer owned the problem and the solution, while the decision analysts managed the process for the client's problem solving. Decision conferencing recognized that: (Kuskey, 2004, Overview of Decision Facilitation)

- Most significant decisions require decision-maker judgment regarding options, consequences and value of the consequences, and uncertainty.
- Most significant decisions are made collaboratively, some people bringing technical or policy expertise, some decision-making authority.

- A face-to-face environment provides the best forum for successful collaboration.
- Without effective facilitation, meetings can go on and on without closure

The decision conferencing approach has evolved over years of trying to assist decision makers with time-urgent decision problems. The approach is not one of an external analyst coming in to an agency, gathering data on the facts, taking the problem away for study, and later returning with a recommended decision. Instead, it can be viewed as facilitation of the decision process (O'Connor, 1984), but with the major effort being accomplished in several days through a series of intense group meetings in which key players interact to explore the decision process as well as the decision itself. The expertise of the client organization is absolutely essential for success, and the level of expertise needed is typically that which resides in the heads of business and technical managers. While supporting information is important, it is supplemental to the process rather than being its focus. During the decision conference, computer-based models for multiobjective decision analysis (MODA), for probability analysis, and for resource allocation often are used as a focus for group discussion. The structured conference process allows participants to debate issues constructively while encouraging the group to represent its collective judgments in a logically consistent and easily communicated fashion (Kuskey, 1983). One of the strongest features of a decision conference is that it allows participants to work towards consensus regardless of the individual decision making processes being used by the participants (Kuskey, 2004).

One point needs to be emphasized here. In the subsequent sections, we discuss typical processes and formats for decision conferences, and the information can provide good background for practitioners. That said, the authors have found that there truly is no such thing as a "typical" decision conference. Every conference is unique, and there is no "cookbook" approach to conducting them. The facilitator must be agile enough to react to the unfolding events, and must be prepared to make major changes to the "plan" on the fly. Not only that, each facilitator must work to his or her own strengths—what works for one facilitator may not work for another.

The remainder of the Appendix is organized as follows. Section C.2 and Section C.3 discuss typical formats and facilities and equipment used in decision conferences. Section C.4 introduces group processes. Section C.5 presents advantages and disadvantages of decision conferences. Section C.6 presents best practices. Section C.7 offers a summary.

C.2 Conference Process and Format

Decision conferences can be used for a variety of decisions. First are institutional decision processes that may involve many people over a series of meetings that extend over weeks or even months. Second are routine, one-time responses aimed at a specific topic, such as resource allocation, evaluation of alternatives, or strategic planning that extend over a 1- to 4- week timeframe. Third are quick-response, quick turn-around sessions that leave little time for planning and are completed in hours or days (Kuskey, 2004, Overview of Decision Facilitation).

A typical decision conference consists of a 2- to 3-day session with key players and facilitators, followed by further analysis and reporting. The key players are the client organizational planners, decision makers, and decision implementers who are responsible collectively for the issues under discussion, supported by the facilitation team. The facilitators are professionals who specialize in leading, moderating, model building, and documenting the working sessions. Normally, each facilitator plays a distinct role in the conference. The group leader moderates and controls the sessions, elicits information, asks questions, channels responses, and builds analytical models (often, evaluation and resource allocation models) in response to group input. A second team member implements in real time the computer models[1] developed by the group leader. The third team member acts as a conference recorder, documenting all major decisions and providing an audit trail of the decision rationale for the session. Ideally, all team members are qualified to assume any of the three roles. Facilitator teams are also structured to bring to bear a broad base of experience, with a variety of disciplines represented to include decision analysis, business administration, computer science, mathematical and cognitive psychology, engineering, economics, and operations research. Additional desirable characteristics of facilitators include the ability to think quickly and clearly on their feet, strong leadership skills, a results-oriented philosophy, and self-confidence (Ring, 1980).

Larry Phillips (Phillips, 2007) developed a schematic of the decision conferencing process (Fig. C.1):

C.3 Location, Facilities, and Equipment

Decision conferences can be and have been conducted at the client's location, an off-site location, or at the facilitator's location. With the advent of technology, they have also been held in virtual environments using same time, different place teleconferencing systems, but this approach poses its own challenges in maintaining the interpersonal nature of face-to-face discussions.

The client's location offers the advantage of maximum exposure to the broadest number of in-house experts. Little time is lost due to travel, and personnel can be brought into the sessions on an as-needed basis. However, this colocation with the normal workplace can be a severe disadvantage. There is a strong tendency for key personnel to be distracted by phones, messages, and other business. Such interruptions can have a debilitating effect on the intensity of the session. Off-site locations help to avoid these distractions but have other

[1]For evaluation, the most common software used includes HIVIEW, Logical Decisions, or Excel with add-ons. For resource allocation, the list includes Equity and Logical Decisions.

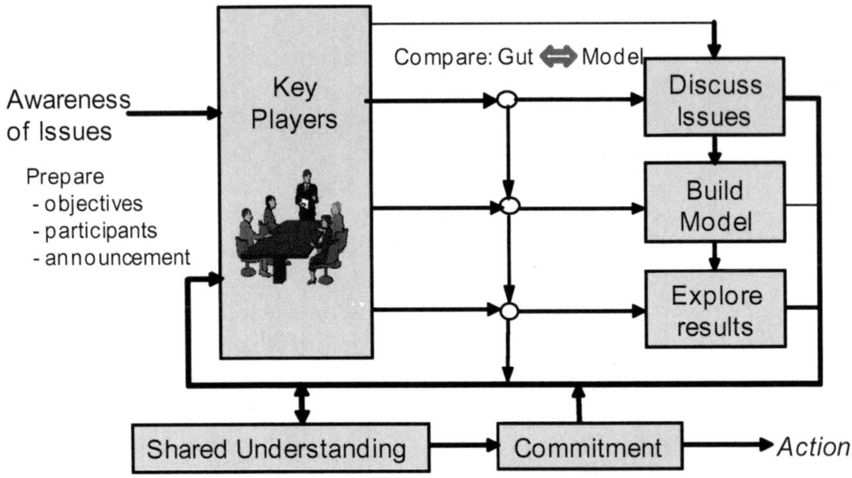

FIGURE C.1 The decision conferencing process. (Modified from Phillips, L.D. [2007] "Decision conferencing," in Edwards, W., Miles, R., & von Winterfeldt, D. (eds.), *Advances in Decision Analysis from Foundations to Applications*, Cambridge University Press.)

limitations. They tend to be more costly, the number of participants is reduced, and critical facilities may be lacking (e.g., sufficient blackboard space and computer hookups). The facilitator's location usually is the most efficient site. The number of participants may be limited (usually due to travel constraints), but the number of distractions is minimized and the facilitators can control more easily the flow of the agenda.

Decision conference facilities range from ordinary conference rooms to very sophisticated and specially designed facilities. White boards can be used with numerous colored markers to enhance visual displays, and computers are used to develop interactive decision models that can be projected onto large screens for all to see (Adelman, 1982). This arrangement allows the computer to be available for computational purposes, while, at the same time, being configured so as not to intimidate the participants (Ring, 1980). Coffee and other beverages can be available right in the conference rooms, and lunches are brought in at an appropriate time. Sessions typically last 8–10 hours, and participants often feel quite emotionally drained at the end of each day.

C.4 Use of Group Processes

Many of the inputs used to develop analytical models in the decision conference are elicited using well-documented group processes. Various methods are used to take advantage of the positive aspects of group behavior and to minimize the negative effects. These techniques can be categorized into three approaches:

Delphi techniques, nominal group techniques, and consensus group techniques (Ulvila, 1984).

Delphi can be described as a mechanical method for controlling group processes. Each participant gives an individual opinion in an anonymous fashion, and is then shown all responses but not told who provided which response. Participants then revise opinions, and the assessment process goes through additional iterations. Delphi stresses anonymity, and works well to minimize the influence of strongly dominant personalities in the group. However, it is time consuming, and participants often lose interest after a few iterations (O'Connor, 1984).

Nominal group techniques call for all individuals to provide assessments to the group without discussion. Once these comments are made, and the contributors are identified, the group then discusses *all* judgments for clarification and evaluation. Individuals may reconsider their assessments at this point, and any remaining differences are resolved mathematically through averaging. The approach is appropriate for combative groups with wide variances in their sources of data (Ulvila, 1984).

Group consensus techniques involve open group discussions aimed at producing a consensus view. They stress face-to-face exchanges of information, and direct interactions are encouraged. They are most effective with cooperative groups, but skilled facilitators usually can overcome problems introduced by combative participants. The facilitator focuses discussion and keeps any individual from dominating the group. Most of the documented decision conferences have used group consensus techniques rather than the other approaches (Kramer & King, 1983).

From the perspectives of the authors, the "best practice" is to use consensus group techniques. While Delphi and nominal group process may be useful in preliminary discussions to get the group more quickly to a useful starting point for the consensus group process, we do not recommend Delphi or nominal group process as the primary technique for conducting decision conferences.

C.5 Advantages and Disadvantages

The decision conference approach offers several advantages when compared with less intensive, less structured methods, such as the dialogue decision process and other analytical approaches discussed in Chapter 5 (Table C.1).

In summary, the decision conference has the following positive features:

- The process is highly focused and productive; results that usually take weeks or months to achieve can be accomplished in a few days of concentrated effort.
- The process involves a broad base of technical expertise and management support throughout the organization; this widespread participation leads to greater commitment and more successful implementation.

TABLE C.1 **Advantages and Disadvantages**

	Analytical Approach	Decision Conference	Dialogue Decision Process/Systems Decision Process
Analytical role	Analyst Team	Facilitators, Client Analyst Team	Decision Team/Design Team
Advantages	May be appropriate for well framed problems Detailed analytical models developed Least time demand on decision makers and stakeholders	High confidence you are solving the right problem, achieved by discussions with decision makers in advance All participants develop common understanding and shared commitment Develop and use requisite decision models that use the essential information to distinguish between the alternatives Multiple conferences can support hierarchical decision making Reaches a decision very quickly	High confidence you are solving the right problem Planned involvement of key decision makers and stakeholders at major decision points More analytical models of values and uncertainty Less time demand than analytical approach, but more than decision conferences (several short meetings)
Disadvantages	Analysts can lose focus on the evolving views of decision makers and stakeholders Models may become overly complex Lack of stakeholder participation and data	Must schedule all key players for same time Time commitment of managers and stakeholders (days) If needed, complex models must be developed before the conference	Requires scheduling periodic meetings with key players Key stakeholder availability and data collection challenges between dialog points

- The approach specifically embodies and capitalizes on the substantive expertise of the client organization; the plan is produced *with* the client instead of *for* the client.
- Because everyone participates in a single directed discussion focused specifically on the decision at hand, communication among the participants is reliable and efficient; thorough documentation provides a permanent record of the conference proceedings and results, thus maintaining a reliable organizational memory of why certain decisions were made.
- The use of explicit computer-based decision analysis models helps the participants to understand and focus on the most critical issues, to debate those issues logically and clearly, and to apply their priorities consistently.
- The approach pays significant attention to the implementation of results; that attention ensures that the conclusions and recommendations are not only analytically sound, but also are directed toward immediate implementation.
- Finally, the approach ensures that participants have a vested interest in the decision they have analyzed; all participants are more likely to support the results if they have understood the approach and had a fair opportunity to present and debate their own viewpoints.

At the same time, the decision conference has the following limitations:

- It is sometimes difficult to get key personnel together for extended periods of time;
- The costs in terms of man-hours devoted to the decision conference can be high.
- Some people believe that there is too much reliance on expert subjective judgments and not enough on hard verifiable data.
- Facilitators are not experts in the substantive organizational matters, and may not understand the implications of the elicited judgments.
- If the wrong participants are selected for the conference, the results can turn out to be ineffective.

On balance, the decision conference is an innovation that enables managers to collaborate very efficiently and effectively to plan for and address difficult decisions. The product of a decision conference is a well-supported "way ahead" for the client organization, backed by a thorough documentation of the key players' collective analysis and judgments.

C.6 Best Practices

The following paragraphs comprise best practices as per the author's experience:

- Face-to-face conferences (same-time, same place) work best. The ability to observe body language, gestures, eye rolling, and so on is very important, and often allows the facilitator to pick up on messages that might otherwise have been missed.
- Make use of consensus group techniques when facilitating. While Delphi and nominal group process may be useful in preliminary discussions to get the group more quickly to a useful starting point for the consensus group process, we do not recommend Delphi or nominal group process as the primary technique for conducting decision conferences.
- It should be noted that, although we talk about consensus often in this Handbook, disagreement is not only okay during deliberations, it is essential. A decision conference is not likely to succeed without it, and some of the more successful facilitators such as Cam Peterson and Roy Gulick,[2] used this message to great advantage, even to the point of forcing disagreement when there appeared to be none. A basic truth is that we learn more when we disagree than when there is no dissent.
- Although most of the discussions and knowledge elicitation happens within the formal conference schedule itself, we should not underestimate the importance of lulls in the action, such as coffee breaks, nighttime social gatherings between working session days, breakouts, and so on. The author has found that inevitably, someone will want to talk during these "breaks," and frequently, will bring to the fore some critical information that they were reluctant to raise in the formal group. Building these opportunities into the decision conference schedule is essential.
- While technology such as large-screen computer displays can be of great assistance to the process, they can also get in the way. Trying to take notes on a computer visible to the participants in real time can be a great distraction, and takes their attention away from the issue at hand. Computer displays typically allow for only one screen to be seen at a time. It is far more productive to make use of white boards, wall charts, and easel pads to ensure that all key pieces of information are visible at all times.
- We each must evolve our own style, often by watching and learning from what we like and do not like in other facilitators, from trying techniques that might or might not work, and by discovering what we feel comfortable doing. Facilitating decision conferences is far more of an art than a science.
- One of the most critical "rules" for the facilitator is to allow only one conversation at a time to allow everyone a fair opportunity to get their point across. That includes use of cell phones during the conference as well!
- One lesson that the author learned from Roy Gulick, is that to be successful, a decision conference facilitator has to be "45% decision analyst and 55% entertainer." Being an entertainer means being enthusiastic,

---

[2]Roy Gulick is a colleague and mentor of the author from days at Decisions and Designs, Inc., and is one of the most experienced decision conference facilitators in the world.

full of energy, exuding confidence that the session will succeed, engaging the "audience," using facial expressions and gestures, and using tricks of the trade when necessary.

C.7 SUMMARY

Decision conferencing was introduced to move beyond the traditional "doctor–patient" model of consultancy by providing group processes that enable client stakeholders to plan for and analyze their decisions collaboratively. Decision conferencing brings together key people who know the issues surrounding a decision for an open exchange of information via dialogue and discussion, to provide relevant data and judgments, and to make decisions. Decision conferences are intensive collaborative working sessions, typically 1–3 days in length, that bring together decision makers, stakeholders, subject matter experts, and a team of trained facilitators for a "structured conversation" with the goal of informed consensus. A "best practice" for decision conference facilitators is to use group consensus processes rather than Delphi techniques or nominal group processes.

Decision conferences can be conducted at the client's location, an off-site location, or at the facilitator's location. Technology also provides the opportunity for decision conferencing in a virtual environment. The use of explicit computer-based decision analysis models helps the participants to understand and focus on the most critical issues, to debate those issues logically and clearly, and to apply their priorities consistently.

Decision conferences are highly focused and productive; involve a broad base of technical expertise and management support throughout the organization leading to greater commitment and more successful implementation; are produced *with* the client instead of *for* the client; and provide a thorough audit trail of rationale for the group judgments and decisions. At the same time, it is sometimes difficult to get key personnel together for extended periods of time; labor commitments can be high; and there sometimes can be too much reliance on expert subjective judgments and not enough on hard verifiable data.

KEY TERMS

Consensus group process Open-group discussions aimed at producing a consensus view. They stress face-to-face exchanges of information, and direct interactions are encouraged. They are typically facilitated by a trained professional who focuses discussion and keeps any individual from dominating the group.
Decision conference An intensive working session, typically 1–3 days in length that brings together decision makers, stakeholders, and subject matter experts with a team of trained facilitators. The facilitator leads the group through

a structured conversation with the goal of informed consensus through an open exchange of information.

Delphi techniques A group processes in which each participant gives an individual opinion in an anonymous fashion, and is then shown all responses but not told who provided which response. Participants then revise opinions, and the assessment process goes through additional iterations.

Facilitation A process in which a person whose selection is acceptable to all members of the group, who is substantively neutral, and who has no substantive decision-making authority, diagnoses and intervenes to help a group improve how it identifies and solves problems and makes decisions, to increase the group's effectiveness.

Nominal group process Calls for all individuals to provide assessments to the group without discussion. The group then discusses *all* judgments for clarification and evaluation. Individuals may reconsider their assessments at this point, and any remaining differences are resolved mathematically through averaging.

REFERENCES

Adelman, L. (1982). Real-time computer support for decision analysis in a group setting: Another class of decision support systems. *ORSA/TIMS National Meeting*, pp. 18–21. Detroit, MI: ORSA/TIMS.

IDCF (2012). Decision conferencing. http://decisionconferencingforum.org/node/2, accessed July 2012.

Kramer, K. & King, J. (1983). *Computer Supported Conference Rooms*. Irvine: University of California Press.

Kuskey, K. (1983). *An Approach to Resource Allocation for Project Planning and Design*. McLean, VA: Decisions and Designs, Inc.

Kuskey, K. (2004). *Overview of Decision Facilitation*. Vienna.

O'Connor, M.F. (1984). Methodology for corporate crisi decision making. In S. Andriole (ed.), *Corporate Crisis Management*. New York: Petrocelli Books.

Phillips, L. (2007). Decision conferencing. In W. Edwards, R. Miles, & D. von Winterfeldt (eds.), *Advances in Decision Analysis*, pp. 375–399. Cambridge: Cambridge University Press.

Ring, R. (1980). A new way to make decisions. *Graduate Engineer*, November, 46–49.

Ulvila, J. (1984). Use of expert judgment. In T.A. Bresnick, J. Ulvila, D. Buede, R. Hullander, & R. Brown (eds.), *Problem Solving and Decision Making: Workshop Manual*. Falls Church, VA: Decision Science Consortium, Inc.

Index

Handbook of Decision Analysis, First Edition. Gregory S. Parnell, Terry A. Bresnick, Steven N. Tani, and Eric R. Johnson.

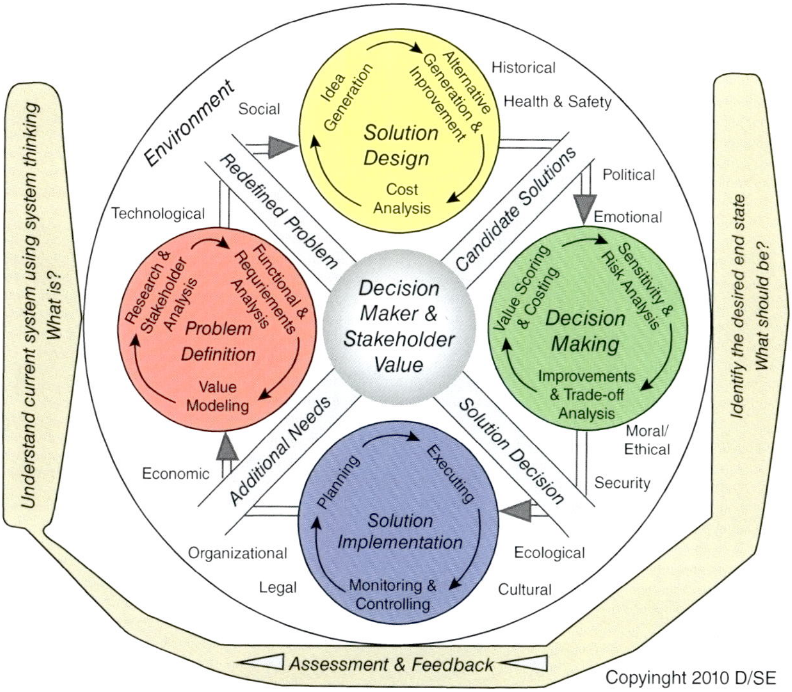

FIGURE 5.5 Systems decision process.

Handbook of Decision Analysis, First Edition. Gregory S. Parnell, Terry A. Bresnick, Steven N. Tani, and Eric R. Johnson.

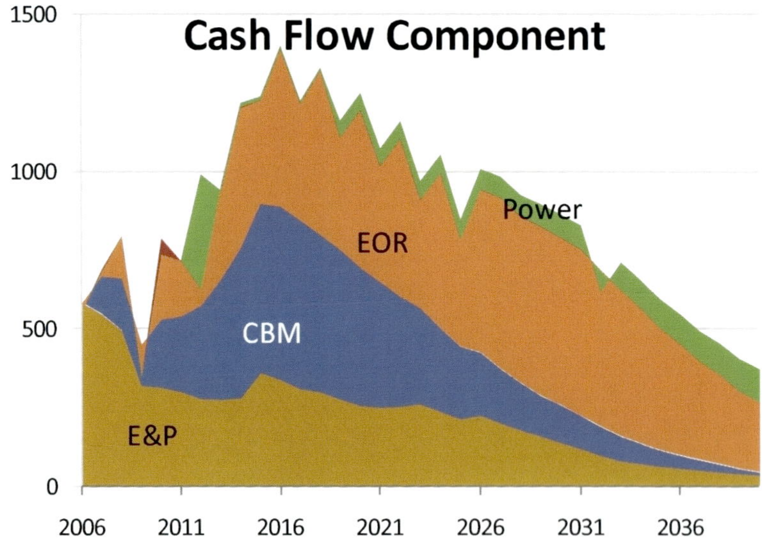

FIGURE 9.3 RNAS cash flow by business unit.

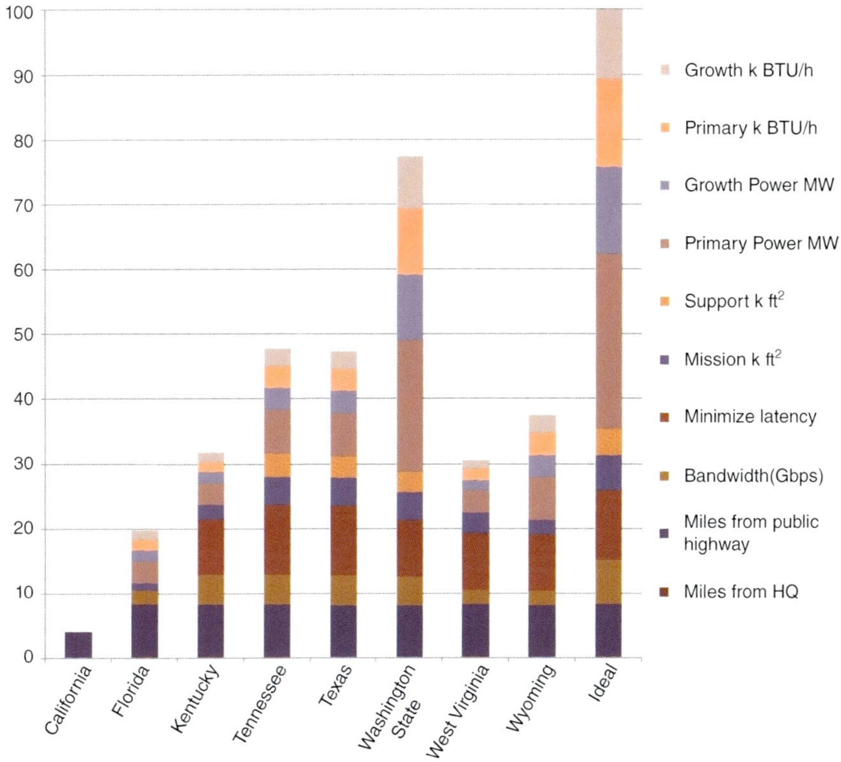

FIGURE 9.14 Data center value components chart.

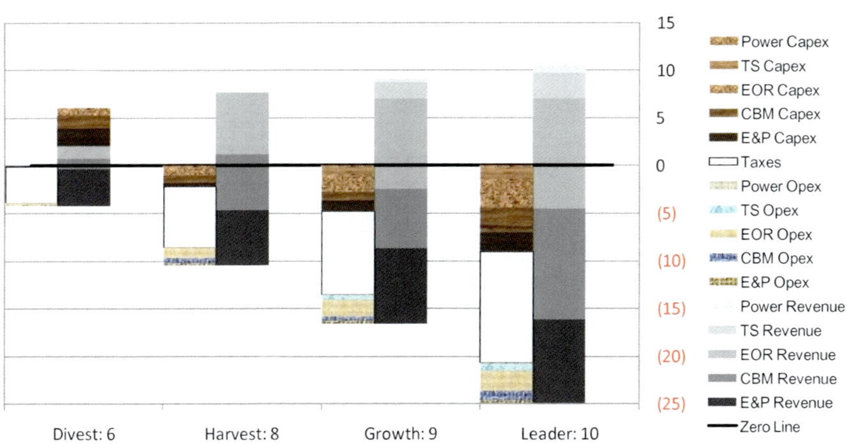

FIGURE 11.6 RNAS value components chart.

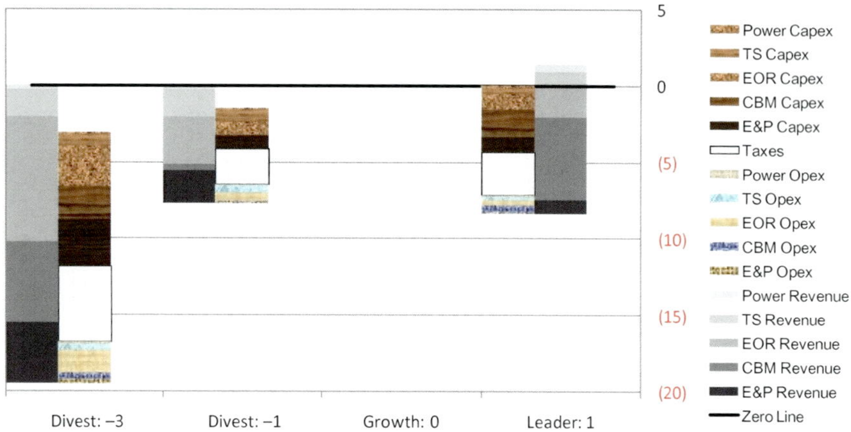

FIGURE 11.7 RNAS value components, as compared with the Growth strategy.

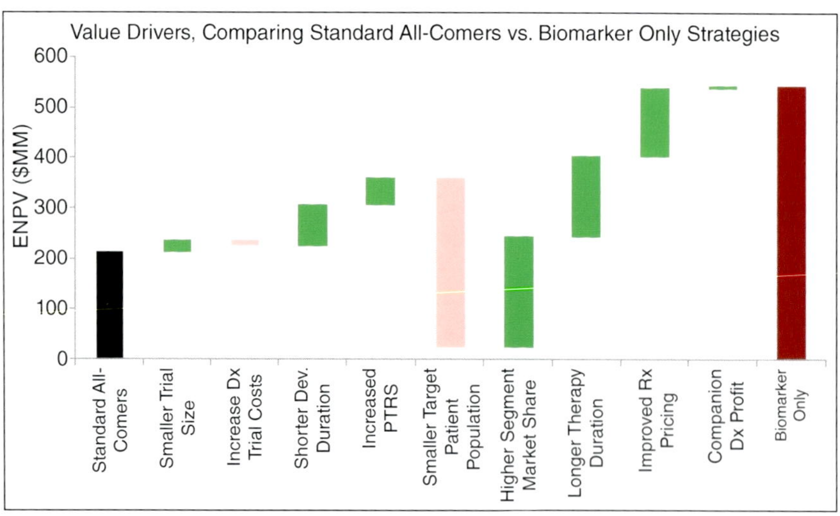

FIGURE 11.21 Geneptin waterfall chart.

Printed in the USA/Agawam, MA
August 7, 2020

759361.001